Textbook of
KINESIOLOGY

Textbook of
KINESIOLOGY

VD Bindal PhD LPT (USA)
Associate Professor and Head
Division of Physiotherapy
Lakshmibai National Institute of Physical Education
(Deemed to be University)
Gwalior, Madhya Pradesh, India

JAYPEE *The Health Sciences Publisher*

New Delhi | London | Panama

Jaypee Brothers Medical Publishers (P) Ltd.

Headquarters
Jaypee Brothers Medical Publishers (P) Ltd
4838/24, Ansari Road, Daryaganj
New Delhi 110 002, India
Phone: +91-11-43574357
Fax: +91-11-43574314
Email: jaypee@jaypeebrothers.com

Overseas Offices

J.P. Medical Ltd
83 Victoria Street, London
SW1H 0HW (UK)
Phone: +44 20 3170 8910
Fax: +44 (0)20 3008 6180
Email: info@jpmedpub.com

Jaypee Brothers Medical Publishers (P) Ltd
17/1-B Babar Road, Block-B, Shyamoli
Mohammadpur, Dhaka-1207
Bangladesh
Mobile: +08801912003485
Email: jaypeedhaka@gmail.com

Jaypee-Highlights Medical Publishers Inc
City of Knowledge, Bld. 235, 2nd Floor,
Clayton, Panama City, Panama
Phone: +1 507-301-0496
Fax: +1 507-301-0499
Email: cservice@jphmedical.com

Jaypee Brothers Medical Publishers (P) Ltd
Bhotahity, Kathmandu, Nepal
Phone: +977-9741283608
Email: kathmandu@jaypeebrothers.com

Website: www.jaypeebrothers.com
Website: www.jaypeedigital.com

© 2018, Jaypee Brothers Medical Publishers

The views and opinions expressed in this book are solely those of the original contributor(s)/author(s) and do not necessarily represent those of editor(s) of the book.

All rights reserved. No part of this publication may be reproduced, stored or transmitted in any form or by any means, electronic, mechanical, photocopying, recording or otherwise, without the prior permission in writing of the publishers.

All brand names and product names used in this book are trade names, service marks, trademarks or registered trademarks of their respective owners. The publisher is not associated with any product or vendor mentioned in this book.

Medical knowledge and practice change constantly. This book is designed to provide accurate, authoritative information about the subject matter in question. However, readers are advised to check the most current information available on procedures included and check information from the manufacturer of each product to be administered, to verify the recommended dose, formula, method and duration of administration, adverse effects and contraindications. It is the responsibility of the practitioner to take all appropriate safety precautions. Neither the publisher nor the author(s)/editor(s) assume any liability for any injury and/or damage to persons or property arising from or related to use of material in this book.

This book is sold on the understanding that the publisher is not engaged in providing professional medical services. If such advice or services are required, the services of a competent medical professional should be sought.

Every effort has been made where necessary to contact holders of copyright to obtain permission to reproduce copyright material. If any have been inadvertently overlooked, the publisher will be pleased to make the necessary arrangements at the first opportunity. The **CD/DVD-ROM** (if any) provided in the sealed envelope with this book is complimentary and free of cost. **Not meant for sale.**

Inquiries for bulk sales may be solicited at: jaypee@jaypeebrothers.com

Textbook of Kinesiology

First Edition: **2018**

ISBN 978-93-5270-452-1

Dedicated to
My Loving Grandchildren
Devashish, Aarna and Anaya

Preface

My interest in kinesiology prompted me to resign my 11 years long service at the State Government of Rajasthan and accept the assignment at the present institute where I have been working almost for 30 years now.

The human body has complex systems which is subject to both mechanical and biological laws and principles. How effectively and efficiently one performs depends upon both the mechanical and biological aspects of functions as they are directly related to physical performance.

The general approach to the book, *Textbook of Kinesiology* is to provide possible information needed mostly by the physical educators and physiotherapists though it has broad application across any discipline that incorporates movement science, such as the Sports Coaching, and the Athletics. The basic facts, concepts and principles of kinesiology and the science of human movement are primarily organized as to be contained in any standard textbook of physical education and physiotherapy.

The contents of the book are broken into 3 major Sections and further into 16 Chapters. The First Section, i.e. *Introduction and Fundamentals* contains 3 Chapters; Introduction to Kinesiology, Anatomical and Physiological Fundamentals of Human Motion, and Basic Biomechanical Concepts. The Second and a major Section of the book, i.e. *Kinesiology of Body Regions* is dealt with in 9 Chapters in great length covering various body regions, i.e. Shoulder Girdle, Shoulder, Elbow and Radioulnar Joints, Wrist and Hand, Pelvic Girdle, Hip, Knee, Ankle and Foot, and Spine. The Third Section, i.e. *Motor Skills: Principles and Application* includes 4 Chapters, one each on Posture, Locomotion, Application of Kinesiology to Selected Daily Life Activities and Sports Skills, and Prevention of Sports Injuries: The Mechanical and Kinesiological Viewpoint.

I do not claim this book entirely my original work, rather acknowledge with a great humility all the work and contribution that I have used of many great authors and scholars (some mentioned and many not) in preparing and organizing the subject material in this book. I am is particularly indebted to Shri Jitendar P Vij (Group Chairman) and Mr Ankit Vij (Group President) of M/s Jaypee Brothers Medical Publishers (P) Ltd, New Delhi, for their acceptance of this book for publication. I am very thankful to Mr Ajay Mathe, who typed the manuscript as directed, and also to Mr VS Parmar, Ashish Aronkar and many others, who helped directly or indirectly, in preparing and bringing about this book.

The book is comprehensive, readable and well illustrated and it is hoped that it will prove to be a useful textbook to the students and teachers of physical education and physiotherapy, to whom kinesiology as a subject is taught in various colleges and universities in India. The book also contains *Glossary*, and *Index* towards the end. The suggestions from the readers are welcome for further improvement.

VD Bindal

Contents

Section I Introduction and Fundamentals

Chapter 1: Introduction to Kinesiology   3
- Major Contributions: Past and Current 5
- Aims and Objectives of Kinesiology 9
- Practical Application of Kinesiology: In Physical Education and Physical Medicine 11
- Role and Importance of Kinesiology In Physical Education and Sports 12

Chapter 2: Anatomical and Physiological Fundamentals of Human Motion   14
- The Skeletal System 14
- Muscular System 27
- Neuromuscular Concepts 36
- Nervous Control of Voluntary Movements 45

Chapter 3: Basic Biomechanical Concepts   50
- Gravity 50
- Center of Gravity 50
- Line of Gravity 51
- Base of Support 51
- Axes and Planes 52
- Equilibrium and Stability 55
- Force 58
- Motion 62
- Levers 68

Section II Kinesiology of Body Regions

Chapter 4: Shoulder Girdle   79
- Movements 80
- Muscles of the Shoulder Girdle 81

Chapter 5: Shoulder   86
- Movements 87
- Muscles of the Shoulder Joint 88

Chapter 6: Elbow and Radioulnar Joints   94
- Elbow 94
- Radioulnar Joints 95
- Muscles of Elbow and Radioulnar Joints 95

Chapter 7: Wrist and Hand   100
- Ligaments of Wrist and Hand 101
- Movements of Wrist and Hand 102
- Muscles of Wrist and Hand 105

Chapter 8: Pelvic Girdle   112
- Movements of Pelvic Girdle 113
- Muscles of the Pelvis 115

Chapter 9: Hip   116
- Movements 116
- Muscles 117

Chapter 10: Knee   126
- Movements 127
- Muscles 128

Chapter 11: Ankle and Foot   131
- Bones and Joints of Foot 131
- Movements 133
- Muscles of the Ankle and Foot 137

Chapter 12: Spine   143
- Structure 144
- Movements 145
- Muscles 147

Section III Motor Skills: Principles and Applications

Chapter 13: Posture   157
- Introduction 157
- Types of Posture 158
- Static Versus Dynamic Posture 159
- Good Posture 161
- Poor or Bad Posture 162
- Static Positions 162
- Postural Defects 165

Chapter 14:	**Locomotion**	173	**Chapter 16:**	**Prevention of Sports Injuries: The Mechanical and Kinesiological Viewpoint** 208

Chapter 14: Locomotion — 173
- Walking 173
- Running 185
- Common Deviations of Normal Gait 189

Chapter 15: Application of Kinesiology to Selected Daily Life Activities and Sports Skills — 192
- Daily Life Activities 192
- Sports Skills 200

Chapter 16: Prevention of Sports Injuries: The Mechanical and Kinesiological Viewpoint — 208
- Dissipation of Force 208
- Landing 209
- Catching 210
- Summary of Mechanical and Kinesiological Principles Related to Prevention of Injuries 212

Bibliography 215
Glossary 217
Index 223

SECTION I

Introduction and Fundamentals

- Introduction to Kinesiology
- Anatomical and Physiological Fundamentals of Human Motion
- Basic Biomechanical Concepts

CHAPTER 1

Introduction to Kinesiology

DEFINITION

The term *"Kinesiology"* is a combination of two Greek verbs *"Kinein"*, meaning "to move", and *"logos"* meaning "to discourse" or "to the study". **As such, kinesiology may be defined as a science which investigates and analyzes human motion.**

The study of the human body as a machine for the performance of work has its foundations in three major areas of study namely, mechanics, anatomy, and physiology; more specifically, biomechanics, musculoskeletal anatomy, and neuromuscular physiology. The accumulated knowledge of these three fields forms the basis for the study of human movement.

Some authorities refer to kinesiology as a science in its own right; while there are others who claim that kinesiology is not a true science because the principles on which it is based are derived from basic sciences such as anatomy, physiology, and physics. In any event, the unique contribution of kinesiology is that it selects from many sciences those principles that are relevant to human motion and systematizes their application. As such, as a modern science, kinesiology not only must adapt the practices and standards, but must also make use of all those facts and principles gathered by other sciences which can contribute to a complete understanding of man's movements.

From the time we stretch and start to waken in the morning until we lie down again for sleep at night, practically everything we do is the result of changing muscular contractions. Even the thoughts of an inactive man produce changes in the muscular response. The period of rest and sleep is often thought of as one of the inactivity, but respiration must continue, heart action slows but never stops, and the body is rolled, turned, flexed, and extended during the sleeping period. These are all motor acts. When the individual opens his eyes or speaks, it is by motor acts. The muscular response stops to exist only upon the paralysis or death.

Movement is so generally observable in living organisms that it has attracted the attention of man ever since we have a record of his philosophic and scientific interests in himself and the world around him. Even the lower animals know that movement is a sign of life. They are always quick to detect any motion near them, especially if they are fearful of danger. Many will take a position of complete immobility when they are aware of danger. Or they will move very cautiously and slowly to prevent their movement from being clearly visible or remarkable.

Man seems to think in terms of movement. A study of the art of nearly all primitive people shows that it deals largely with active participants in war, sports, or routine occupations. We hardly find reclining, sitting or standing figures which are entirely passive in their attitude. The gods, in all the art forms, were usually taking part in man's activities. Man and animals, as depicted in carvings, sculptures, drawings and literature, are active creatures. From earliest times, motor skill and human movement had art, religious and militarily values.

It is obvious that how a machine can be used depends upon the way it is made; its structure, the material of which the parts are composed, and the way they are put together. It is also known that two machines may perform the similar function with different degrees of effort and efficiency. This too is the result of structural quality and arrangement. The human mechanism has many of the characteristics of other machines. Complete understanding of the human motion necessitates a knowledge of anatomy, both gross and microscopic.

Movement can take place only as normal body functioning proceeds. For example, muscle-contraction is not an isolated process, which occurs independently of the rest of the body. The nervous, circulatory, respiratory, digestive, and excretory systems all have their part to play in this phenomenon. As such, the science of physiology as a whole has basic contributions to make to kinesiology.

The field of mechanics is a logical source to kinesiology. One of the essential features of kinesiology is that it treats all motor acts as mechanical events. Man can be regarded as a machine, a device for transmitting energy. He takes in fuel, which can be converted to energy, and the energy can in turn be utilized to do work. Though the machine of man existed thousands of years before the man-made machines, understanding of mechanical principles was not derived from observation of human movement. Man observed the movement of external objects and means by which objects could be moved, and then formulated general laws and principles. These enabled him to construct simple and complex devices for doing work.

Fundamental to the understanding of man's movements is the realization that he is a living organism and that his structure and behavior have been shaped by his anthropologic ancestors. His backbone is credited to him by the fish family, who developed that structure when the surface of the earth was covered with water. When part of the earth's surface was raised above the water, the animal life, to survive on land, had to develop new structural forms for locomotion. The body needed to be lifted above the earth's surface and yet maintain some contact with it; fins were gradually shaped into legs and feet. Because air does not support the body mass, as does water, the backbone and the newly developed legs needed material to better resist gravitational pull, and thus the cartilaginous tissue changed to bone. To escape enemies and to secure food, rapid locomotion was needed, and primitive legs lengthened to the thighs and legs of present quadrupeds.

In these changes and in those that followed is seen the capacity of living organisms to adapt structure and behavior to the demands of the environment. The demand of environment, as quadruped walking changed to biped walking, resulted in structural modifications in pelvic, leg and toe bones, and in gluteal muscles, to meet out the functional requirements.

With the growth of plants and trees on the land, some animal forms escaped ground enemies by choosing an arboreal or tree-like life. Here a new means of locomotion developed, that of grasping branches and swinging the suspended body from limb to limb. This type of life modified body structure, the distal ends of the forelimbs were shaped into hands, and the proximal end of the forelimb developed a joint capable of great movement. Gravitational force pulled the torso or the trunk and lower limbs downward, providing a relationship of body segments that would enable descendents to stand on two rather than four feet.

When some of the tree-dwelling animals returned to ground life and found bipedal locomotion profitable, the human foot was developed. The heels developed and lengthened, providing a rear for the base of support. The forepart of the foot developed to provide the front of the base. According to many experts, the foot in human is one such peculiar structure that differentiates man from other animals. The bony structures between the heel and ball of foot are compactly fitted together, enabling them to bear the weight of the body and transmit it to the ball of the foot or the toes. This compactness enables man to rotate his weight about the metatarsophalangeal joints, an action that is impossible in animals such as the ape.

As the foot took over the weight-bearing function, the forelimbs were freed for manipulation of external objects. In such manipulative skills, man exceeds all animals, and according to many experts, the hand and the opposable thumb in man are his unique structural features. The primates that still live in trees have five digits at the distal ends of the limb, and the first of these digits can close against the other four digits as a branch of tree is clasped. *Man inherited his hand but developed his foot.*

Each species inherits a basic design that will be modified by its mode of maintaining life. As man's ancestors developed the ability to balance on two feet, they also modified their forms of locomotion, i.e. walking, running, leaping, and jumping. The free forelimbs developed the ability to throw and strike, to pull and push, and to lift. The practice of these skills further modified the structure, without a change in its basic design.

Apart from the changes and modifications in the bony structures, those parts of the body involved in movement, primarily the muscular and nervous systems, were also modified. Since structure and function develop simultaneously, they are interrelated to each other. Man does not choose his structure, that is inherited from species that preceded him. Similarly, he does not choose his basic movement patterns, as they too, are inherited. However, within limits, structure can be modified by environment, exercise, and nutrition. His movements also have limitations imposed by inheritance of structure of bones and joints, of muscles, and of nerve patterns. These can be modified to some extents, but not changed basically.

Motor skill must be studied with the realization that it is not the result of experiences of a single life span. From birth to maturity, each individual will have at his command motor responses that will to some degree meet the needs of each growth stage. Such responses are man's heritage, gifts from ancestors who inherited to their descendants bone, muscle, and nerve structures and patterns of motor behavior that they have found to be valuable. Each heritage is made with a wise condition; the successor must use and exercise the gift to bring to fulfillment its potential value.

Kinesiology is not an isolated science, sufficient in itself in methods, knowledge, or contribution. It is a composite of several of the sciences. It requires wide information and it offers opportunity for application of principles and laws.

Man designs the machine, he builds as best he can for the purpose he desires. We inherit our biological and physical make-up. But our human machinery is designed cleverly enough, for the demands we make upon it. However, if the demands continue beyond the scope of this human machine, nature makes every effort possible to make the necessary adaptations.

MAJOR CONTRIBUTIONS: PAST AND CURRENT

The origin and development of kinesiology are briefly traced chronologically and logically. The mention of the history of kinesiology may be found in treatises and medical histories. Kinesiology reaches back to the first scientific concepts of movement that man was able to comprehend, although some of the basic ideas had been known and used from olden times. Its roots are certainly deep in the beginnings of medical history.

Ancient Era

The *Greeks* were among the first to practice so-called scientific thinking, as opposed to thinking based on emotional and spiritual ideas. The Greek philosophers believed in the unity of body and mind. Their interpretation of man's activity was mechanistic, in accordance with their materialistic interpretation of the universe.

- *Hippocrates (460-370 BC)* advocated the concept that man should base observations on and draw conclusions from only what he perceived through his senses (particularly the senses of touch, sight, hearing, and smell) without recourse to the supernatural. He recognized the physiological effect of a common place activity such as walking. For example, he wrote that walking should be rapid in winter and slow in summer; unless it be under a burning heat; that fleshy people should walk faster, thin people slower.

  Hippocrates, and Aristotle a Century or so later, devised certain empirical theories of anatomical structure and human mechanics. Though there was much of truth in their ideas, but the beginnings of the scientific approach came much later.

- *Aristotle (384-322 BC)* is usually given the title *"father of kinesiology"*. About three centuries before Christ, Aristotle wrote, "the animal that moves makes its change of position by pressing against that which is beneath it. Hence, athletes jump further if they have the weights in their hands than if they have not, and runners run faster if they swing their arms, for in extension of the arms there is a kind of leaning upon the hands and wrists." *Hart* said, "from the point of view of mechanics, we may regard Aristotle's work as the starting point of a chain of thought which played an important part in the evolution of the subject.

  Aristotle was the first to analyze and describe the complex process of walking, in which rotatory motion is transformed into translatory motion. Aristotle's treatise, *Parts of Animals, Movements of Animals and Progression of Animals*, described for the first time the actions of the muscles and subjected them to geometrical analysis. The ideas expressed by Aristotle were the forerunners of the ideas of Newton, Borelli and others. His concepts of leverage, gravity, and laws of motion were remarkably accurate.

- Another Greek, *Archimedes* (287-212 BC), a renowned mathematician, determined hydrostatic principles governing floating bodies which are still accepted as valid in the kinesiology of swimming. The broad scope of Archimedes inquires included the laws of leverage and problems related to determining the center of gravity. Principles originally developed by Archimedes are still used in determinations of body composition.

- The *Romans*, who did not hesitate to recognizing and utilizing the best from the culture of the nations they conquered, brought to Rome concepts of Greek medicine. *Claudius Galen* (131-201 AD), a Roman physician brought the science of anatomy to Rome. His discoveries and forthright attitude made him the outstanding anatomist and physiologist until the fifteenth century. In his essay *"De Motu Musculorum"* he distinguished between motor and sensory nerves, and between agonist and antagonist muscles, described tonus, and introduced terms such as diarthrosis and synarthrosis which even today are of major importance in the terminology of arthrology. The idea that muscles were contractile seems to have originated with Galen. He taught that muscular contraction resulted from the passage of "animal spirits' from the brain through the nerves to the muscles.

Renaissance

Following Galen's muscle studies, kinesiology like other fields of science did not develop for over a thousand years due to the then prevailing neglect of physical exercise and bodily development. During the Renaissance, German, English, French, and Italian physiologists and physicists attacked the problem of analyzing animal and human movements. These studies form the real beginning of our modern understanding of the problem.

It was not until the time that the scientific awakening known as the Renaissance was perhaps initiated, and further advanced in the field of kinesiology by one of the world's greatest scientists in the history, *Leonardo da Vinci* (1452-1519). He was an artist, engineer and scientist who is given credit for developing modern science of anatomy. He studied the structure of man, especially noting the relation of the center of gravity to balance and motion during different movements. He made these observations while developing a treatise on painting.

Da Vinci's ability to draw the action of muscles when the human body was performing a dynamic act was of great value to medical students and to the science of kinesiology. He described the mechanics of the body in standing, walking up and down hill, in rising from a sitting position and in jumping. Da Vinci was probably the first to record scientific data on human gait.

- Since scientific contributions of Da Vinci were hidden from the world almost for 200-300 years after his death and which have only recently been discovered and published, it was left to others to add to man's knowledge in this area. *Andreas Vesalius* (1514-1564), considered to be the developer of the modern concept of anatomy, was one of them. His drawings, which portray muscles in action in a living, moving human being, are in keeping with the spirit of the Renaissance.
- *Galileo Galilei* (1564-1642) was a noted mathematician and astronomer. Although he is known as one of the experts of natural law and made accessible to man two instruments, the telescope and microscope, his contributions to the field of kinesiology are enormous. He is credited with "laying firm foundation of mechanics." Galileo's alleged experiments on the rate of acceleration of falling bodies from the leaning Tower of Pisa in 1590 or 1591 laid the basis for our present concept of the rate of falling objects in sports and athletics. He demonstrated that the acceleration of a falling body is not proportional to its weight and that the relationship of space, time and velocity is the most important factor in the study of motion. Galileo also proved that the trajectory of a projectile through a resistance-free medium is a parabola. His work gave impetus to the study of mechanical events in mathematical terms, which in turn provided a basis for the emergence of kinesiology as a science.
- *Francis Glisson* (1597-1677) demonstrated through phelthysmographic experiments that muscles do contract, and that viable tissue had the capacity to react to certain stimuli.
- *Alfonso Borelli* (1608-1679), an Italian physicist and one of the pupils of Galileo, combined the sciences of mathematics, physics, and anatomy in first treatise on kinesiology, "*De Motu Animalium*", published in 1630 or 1631. Borelli applied Galileo's mathematical principles to movement and sought to demonstrate that animals are machines. One expert Steindler regards Borelli as the *father of modern biomechanics* of the locomotor system. Borelli recognized that bones serve as levers and are moved by muscles in accordance with mathematical laws. He also believed that the movements of animals are affected by other forces, such as air and water resistance, and poor or good mechanical position. *Because of these beliefs, Hirt called Borelli the father of modern kinesiology.* Borelli has even been given credit for discovering the reciprocal action of muscles, a concept which Sherrington is thought to have conceived much later.
- *Isaac Newton* (1642-1727) laid the foundation of modern dynamics. Particularly important to the future of kinesiology was his formulation of the three laws of motion which express the relationships between forces and their effects and are in use today. Newton is also credited with the first correct general statement of the parallelogram of force, based on his observation that a moving body affected by two independent forces acting simultaneously moved along a diagonal equal to the vector sum of the forces acting independently. And thus basis for the mechanical analysis of movement was established. Since two or more muscles may pull on a common point of insertion, each at a different angle and with different force, the resolution of vectors of this type is a matter of considerable importance in the solution of academic problems in kinesiology.
- *James Keill* (1674-1719), in his studies of muscular contraction, calculated the number of fibers in certain muscles, and assumed that on contraction, each muscle fiber became spherical and shortened. He also established the amount of tension developed by each fiber to lift a given weight.
- *Albrecht Haller* (1708-1777), a great Swiss physiologist of the eighteenth century brought into clear focus the concept of independent irritability and excitability of muscle tissue. The idea that contractility is an innate property of the muscle was first established by Haller.
- *Luigi Galvani* (1737-1798), a professor of anatomy at the University of Bologna, Italy, made the discovery

that was the forerunner of the concept of irritability of muscles. During experiments on muscle and nerve preparations, he noted the contraction of muscle when the leg of a frog contacted metal and devised an arc of two metals with which muscle contractions could be induced.
- In Lecture series on muscle motion, *John Hunter* (1728-1793), a great anatomist, described the structure and muscular function in considerable detail, including the origin, insertion, and shape of muscles, the mechanical arrangement of their fibers, the two joint problem, contraction and relaxation, strength, hypertrophy, and many other aspects of the subject. The lectures of Hunter are regarded as summarizing all that was known about kinesiology at the end of the eighteenth century.

Nineteenth Century Onwards

The nineteenth century saw still greater contributions. The basic facts of neuromuscular functioning were revealed by the work of such men as the Weber brothers, Sherrington, and Helmholtz. Additional experimental evidence through following decades re-affirms many theories of nervous stimulation and inhibition, and muscular reaction. Scientists such as Braune, Fischer, Duchenne, Marey, and many others studied the problems of muscle mechanics, of body balance, and of locomotion. It was in these years that the science of kinesiology was really founded.
- *Guillaume Benjamin Amand Duchenne* (1806-1875) was a great investigator, who devoted much of his time and effort to discovering the function of isolated muscles by stimulating them electrically. He initiated to classify the functions of individual muscles in relation to body movements. Duchenne's book *"Physiologie des movements"* published in 1865 has been claimed to be one of the greatest books of all times.
- *Adolf Eugen Fick* (1829-1901) made important contributions to our knowledge of the mechanics of muscular movement and evolved the terms, *isotonic* and *isometric*. The present theory of resistive exercises is also based on his contributions.
- The *Weber brothers, Ernst Heinrich* (1795-1878), *Wilhelm Eduard* (1804-1891), and *Eduard Friedrick Wilhelm* (1806-1871) investigated the influence of gravity on limb movements in walking and running. And that the gravity is often the force that propels a walker forward, causing him to fall unless balance is reestablished. They were also among the first scholars to study the path of the center of gravity during movement. They also believed that the body was maintained in the erect position primarily by tension of ligaments, with little or no muscular exertion. The publication of *"Die Mechanik der menschlichen Gerverkzeuge"* published by the Webers in 1836 still stands as the classical work, which firmly established the mechanism of muscular action on a scientific basis.
- *Jues Amar* is perhaps one of the greatest contributors to the field of kinesiology in the area of efficiency of work and body mechanics. His book, *"The Human Motor"* published in 1914, and translated into English in 1920, was an attempt to bring in one volume all the known physiologic and kinesiological principles involved in industrial work and in the performance of certain sports movements.
- *John Hughlings Jackson* (1834-1911), the father of modern neurology, made definite contributions to knowledge pertaining to the control of muscular movement by the brain. Jackson is given credit for the famous expression "nervous centers know nothing of muscles; they only know of movements."
- *Charles Edward Beevor* (1854-1908), after careful study of the muscular actions involved in the movements of certain joints, proposed that the muscles be classified as prime movers, synergic muscles, fixators, or antagonists. He was of the opinion that the antagonistic muscles always relaxed in strong resistive movements. In this respect, Beevor was influenced by the work of charles Sherrington (1857-1952), who advanced the theory of the reciprocal innervation and antagonistic muscles in a number of papers published near the end of the Nineteenth Century.
- *Karl Culmann* (1821-1881), a German engineer, developed a hypothesis that led to the trajectory theory of the architecture of bones. This led *Julius Wolff* (1836-1902) to develop his famous *"Wolff's Law"*, "Bones in their external and internal architecture conform with the intensity and direction of the stresses to which they are habitually subjected". Wolff believed that the formation of bone results both from the force of muscular tensions and from the weight of the body coupled with gravitational pull, which is considered a major contribution on skeletal development.
- *Bassett* has proposed a restatement of Wolff's law in modern terms: "The form of the bone being given, the bone elements place or displace themselves in the direction of the functional pressures and increase or decrease their mass to reflect the amount of functional pressure.
- *John C Koch*, in his paper, "Laws of Bone Architecture" concluded that the compact spongy materials of bone

are so composed as to produce maximum strength with a minimum of material and that in form and structure bones are designed to resist in the most economical manner the maximum compressive stresses normally produced by the body weight. Koch also commented that alterations in posture increase the stress in certain regions and decrease it in others, and that if postural alterations are maintained, the inner structure of the affected bones is altered.
- *F Pauwels* attempted to demonstrate that muscles and ligaments act as traction braces to reduce the magnitude of stress in the bones. Though Pauwels work was criticized by *F Gaynor Evans*, nevertheless, the theory of functional adaptation to static stress remains a major hypothesis in the study of skeletal development. *JH Scott* has reviewed the material in the field in an effort to construct a working hypothesis of the developmental and functional relationships which exist between the skeletal system and the neuromuscular system.
- *Henry Pickering Bowditch* (1814-1911), demonstrated the treppe phenomenon (1871), the "all or none" principle of muscle contraction (1871), and the indefatigability of the nerves (1890).
- *Ivan Mikhailovich Sechenov* (1829-1905), a famous Russian physiologist declared in 1863, "all the endless diversity of the external manifestations of the activity of the brain can be finally regarded as one phenomenon- that of muscular movement."
- *Etienne Jules Marey* (1830-1904), a French physiologist was so interested in human movement that he developed photographic means for use in biological research. *Marey*, and *Robinson* both were convinced that human movement was the important function of man and affected all his other activities.
- *Eadweard Muybridge* (1831-1904), also through his photographic skill, brought a new tool to kinesiological investigation. *Animal Locomotion* (1887), an eleven volume work was one of the publications of Muybridge, and *The Human Figure in Motion* contains much of his original work. Using 24 fixed cameras and two portable batteries of 12 cameras each, he was able to take pictures of animals and people in action.
- *Angelo Mosso* (1848-1910), a scientist of the Nineteenth century made an important contribution to the study of kinesiology, the invention of the ergograph in 1884. This instrument, now available in several specialized versions, has become a nearly indispensable tool for the study of muscular function in the human body.
- The study of developmental mechanics was introduced by *Wilhelm Roux* (1850-1924) who stated that muscular hypertrophy develops only after a muscle is forced to work intensively, which was later demonstrated experimentally by *Werner W Siebert*.
- *B Morpurgo* showed that increased strength and hypertrophy are a result of an increase in the diameter of the individual fibers of a muscle, not a result of an increase in the number of muscle-fibers.
- The photographic techniques of Marey and Muybridge opened the way for two German scientists and anatomists, *Christian Wilhelm Braune* (1831-1892) and *Otto-Fischer* (1861-1917), to study the human gait by means of photographic devices. They also developed an experimental method to determine the center of gravity of the human body, and the report of which they published in 1889 became very famous. Today's concepts on posture appear to have had their origin in the experiments of Braune and Fischer.
- *Rudolf A Fick* (1866-1939), a German who followed on the work of Braune and Fischer, eventually became one of the outstanding authorities in the field of mechanics of joint and muscular movement. Fick's contention that for people of different races and cultures, there is no one posture that is normal for all still holds true.
- The late Nineteenth and early Twentieth Centuries were most productive of physiological studies closely related to kinesiology. *L Ranvier*, about 1880, discovered the difference in the speeds of contraction of red and white muscle, which according to *Granit*, brought functional aspects into the focus of subsequent research.
- *Wedenski* demonstrated during 1880 the existence of action currents in human muscles, although practical use of this discovery had to await the invention of a more sensitive instrument, until *W Einthoven* developed the string galvanometer in 1906. The physiological aspects of *electromyography* were first discussed in a paper by *H Piper*, of Germany, during 1910-1912, however, interest in the subject did not become widespread in the English speaking countries until the publication of a report by *ED Adrian* in 1925. By utilizing electromyographic techniques, Adrian demonstrated for the first time that it was possible to determine the amount of activity in the human muscles at any stage of a movement. The development of the electromyograph represents one of the greatest advances in kinesiology.
- In the study of the physiologic aspects of striated muscular activity, three names are very important. The brilliant studies of *Archibald V Hill* in the oxygen consumption of muscle, of *Hugh E Huxley* in the ultrastructure of striated muscle, and of *Andrew F Huxley* in the physiology of striated muscle distinguish

them as the world's leading authorities in their respective fields.
- *Arthur Steindler* (1878-1959)'s publications "*Mechanics of Normal and Pathological Locomotion in Man*", in 1935, and later a larger and more complete Volume, "*Kinesiology of the Human Body under Normal and Pathological Conditions*" in 1955 have become classics in the field and an important contribution to our understanding of body mechanics. *Goldthwaite* also wrote on posture and body mechanics in connection with health and disease. The ideas of Steindler and Goldthwaite were later transmitted from a static to a dynamic concept. *McCloy, Fenn Cureton, Elftman, Karpovich* and others were very much interested in the mechanics of human movement and moved the field of kinesiology into a new era.
- The interest in the subject of posture has declined among kinesiologists in the USA during few decades probably, partly due to the general acceptance of a saying that "the physiological benefits obtained from correction of common postural defects are mostly imaginary", and partly due to the growing realization that individual differences almost prevent valid generalizations. Perhaps much of the effort in earlier times was devoted to the study of static posture, which is now being directed to research concerning dynamic locomotion. *Wallace Fenn* (1893-1971), *Plato Schwartz, Verne Inman, Herbert Elftman, Dudley Morton,* and *Steindler* are among the few scientists who have made important contributions to knowledge concerning this aspect of kinesiology.
- The use of *cinematography* for kinesiological studies of athletes and industrial workers has become quite common. A relatively recent and important development in the study of human motion is the use of *cineradiographic techniques*. Advances in techniques, in future may make it possible to record the complete sequence of musculoskeletal movements rather than only a fraction of them. A fascinating new parameter was opened up with the invention of the *Electronic Stroboscope* by *Harold Edgerton*. This instrument which is capable of exposures as short as one-millionth of a second, can record in a series of instantaneous photographs, an entire sequence of movement. This apparatus seems particularly promising for analysis of the various sequences of skilled movements.

Now, psychologists, psychoanalysts, psychiatrists, and other social scientists have become interested in investigating the psychosomatic aspects of kinesiology. The studies of JH Van Den Berg, Edwin Straus, and Temple Fay are some of the representative analyses which have significantly contributed to our knowledge concerning the "Why" of human movement.

The modifications which a man makes in his environment cause a change in his structure. Alterations of structure affect the relationship between the various components and result in changes in function. Thus, man to some extent is his own architect.
- The kinesiologists are no longer satisfied to deal merely with the mechanical analysis of human movement. The Society for Behavioral Kinesiology defines Behavioral Kinesiology as "the science of the structures and processes of human movement and their modification by inherent factors, by environmental events, and by therapeutic intervention".

The kinesiology students of the future may be required to distinguish five subdivisions within the discipline; (1) *Structural and Functional Kinesiology*, dealing with the inter-relations between the structure and function of the body; (2) *Exercise Physiology*, or the correlation between kinesiology and basic sciences such as physiology and biochemistry; (3) *Biomechanics*, the investigation of human movement by means of the concepts of classical physics and their derivatives in the practical engineering; (4) *Developmental Kinesiology*, the relation of kinesiology to growth, physical development, nutrition, aging etc; and (5) *Psychological Kinesiology*, the study of the mutualities of movement with such topics as body image, self image, esthetic expression, personality, cultural communication, motivation etc.

Currently new devices for studying human movement are being used by many investigators which include cinematography, electronic stroboscope, force platforms, electromyography, electrogoniometer, etc. The researchers in kinesiology are sharing technical skills and equipment with anatomists, engineers, mathematicians, physicists, physiologists, and psychologists, etc. in furthering and advancing the science.

AIMS AND OBJECTIVES OF KINESIOLOGY

Some of the important aims and objectives of kinesiology are described below:
- ***To understand the human structure, and analyze the human movements, and their underlying principles:*** This is probably one of the most important aims and objectives of kinesiology. Kinesiology is the basic science in preparation of professionals of human motion, whether they are in physical education, physical therapy, athletic training or other related profession. It provides us the knowledge about various parts of the locomotor system. In kinesiology we get

to learn what muscles, bones and joints are involved in a particular movement, and to what extent; what principles of mechanics are involved in the movements or the activities; what is the effect of gravity and other forces on the muscular system; and how the bones serve as the anatomic levers in the human body and how the muscles provide the necessary force to move the body levers. Kinesiology thus helps us to learn and analyze all these aspects and the movements of the human body and discover their underlying principles to improve performance.

- ***It helps to organize, integrate, and make application of facts and principles of basic and contributing sciences such as anatomy, physiology and mechanics in the analysis of the human motion:*** Anatomy deals with the description of internal structure of muscles, bones and other tissues, however it is not analytical. Physiology tells us about the properties of the muscles and that all muscles possess a certain degree of tonus, but does not enable us to know the interrelationship of these and other facts to the problem of posture, to the alteration of postural habits, and to the effectiveness of all motor performances. Similarly, with the mechanics or the physics, we learn the law of gravity, the law of inertia and different types of levers etc., however they are mostly applied only to the nonliving objects. As such, the kinesiology is not an isolated science, rather it attempts to integrate all the relevant information from all the contributing fields for its direct application to the problems of the teachers of physical education, coaches and the therapists.
- ***It contributes to successful participation in various physical activities:*** The knowledge of analysis of human motion helps how motor skills and techniques can be improved to ensure successful participation in various physical activities.

With the knowledge of kinesiology, a teacher learns the nature and effects of each physical activity so as to select intelligently the activity which will contribute to achieve the targeted aims for an individual. If the physical activities and skills are to be taught and poor performances corrected, then the teacher must be able to break the skill and activity down into the parts, analyze them, and finally the co-ordination of these parts will help in proper learning of the skills and ultimately the performance and participation is ensured. In order to understand the physiological, developmental, or therapeutic effects of an activity, it needs to be analyzed and compared with other activities, and the kinesiology thus helps in all these aspects.

- ***To improve the human structure and fitness:*** This is also one of the important objectives and functions of the kinesiology to aid the improvement of human structure through the intelligent selection of activities and the efficient use of the body. The human structure improves with use provided it is used in accordance with the principles of kinesiology and efficient human motion.

  Kinesiology also helps to improve the general physical condition and fitness of the individuals through systematic exercise program.

- ***To help understand the problems of movement efficiency and economy:*** Kinesiology contributes to help analyze the physiological cost, energy budgeting and muscular timing of the physical activity and movements. The structure and mechanics of human performance are not ignored in the economic world as well. Many industries employ an efficiency expert who is usually both a psychologist and an engineer, and whose responsibility includes determination of ways and means to secure the most efficient and economical performance of work by the employees. This includes working position, speed, load, and flow of movement as related to both production and fatigue.

  Kinesiology contributes in the design of furniture and automobiles in consideration of human anatomy and human comfort. However, unfortunately, the comfort does not always ensure the best effect on the body from the use of the article. For example, a big, easy and thoroughly comfortable chair, desirable for a period of rest and relaxation may not be suitable for constant use of anyone particularly a growing child due to the considerable flexion of the spine in these chairs.

  Thus, the kinesiology helps the manufacturers design the chairs which give support at the proper point in the back, improve and adjust the height of tables, sinks and wash basins, kitchen platform (for stove), telephones, and other appliances. The present day office furniture, cars, and the work equipment are also designed to satisfy anatomical and mechanical requirements of the human structure, which are also adjusted for individual differences. The kinesiology contributes in all these areas to make proper adjustment for efficiency, minimizing the fatigue and maintenance of good working postures.

- ***To help recognize, and correct the irregular, awkward movement:*** The knowledge of kinesiology assists in recognizing and analyzing the quality of awkward and skillful movements and correct the

irregular movements in accordance with the principles of kinesiology.
- **To analyze the posture and body mechanics:** The knowledge of kinesiology is helpful in analyzing the posture and body mechanics. The faults of posture can be identified earlier and accordingly scientifically designed exercises and activity can be imparted for their correction.
- **To help understand the nature of common musculoskeletal athletic injuries, and thus help in their prevention and rehabilitation:** With the knowledge of kinesiology, one can understand the nature and mechanism of most of the common musculoskeletal injuries. The appropriate preventive conditioning, and flexibility and muscle-strengthening exercises help in preventing the athletic injuries to some extent. The application of kinesiological principles to the acts of landing, falling, catching, etc. also, to some extent prevent the occurrence of injuries in the sports fields. Similarly, know how of the muscles will help in designing suitable activities and exercises for re-educating the weak muscles during the treatment and rehabilitation of the injuries.
- *It provides the educational experience to the students of physical education, physiotherapy and other related professions:* The study of kinesiology is also an essential part of the educational experience of students of physical education, physiotherapy, athletic training, occupational therapy and other related professions. The students of orthopedics, recreation therapy etc. also study kinesiology as part of their professional training.
- **Kinesiology is useful in the daily life activities:** The knowledge and principles of kinesiology help apply mechanically and economically efficient methods of using the body in various activities of daily life such as sitting, rising, stair climbing, lifting, carrying loads, pushing, pulling, etc. Not only the body can conserve energy but also the acts can be performed safely without strain, fatigue or injury.

PRACTICAL APPLICATION OF KINESIOLOGY: IN PHYSICAL EDUCATION AND PHYSICAL MEDICINE

Kinesiology finds its greatest practical applications in the professions of physical education and sports; and physical medicine. The experts of both of these professions are confronted to teaching the individuals to make the most effective use of their bodily machines. Whereas in the physical education and sports, with the study of the mechanical principles, movements and techniques of the body and of the implements, balls and other equipment, kinesiology helps to prepare physical educators and coaches to teach effective performance in both fundamental and specialized motor skills of the sports. On the other hand, in physical medicine which mainly comprise the physiotherapists and occupational therapists, the study of the kinesiology by these therapists helps them to evaluate and apply the effect of therapeutic exercises and their other techniques upon the human body, with the sole purpose of restoration of impaired function and application of methods of compensating the lost function of the patients. The therapists work with individuals having injuries, diseases or congenital defects affecting the motor mechanism. They must know the extent of the dysfunction, the reaction to expect from the muscles involved, what forces to oppose, how to provide substitute motions, and where and how to fit artificial supports. *As such, in physiotherapy and occupational therapy, kinesiology aims to apply mechanical principles and movements for muscle-reeducation, postural corrections, gait abnormalities, for the use of tools and household implements, and to the modifications of vocational and home making activities caused due to the limitations in neuromuscular capacity and skeletal structure.*

The goal of "effective performance" for the therapists does not refer so much to the *"skillful performance"* in sports activities as it does to the physical educators and coaches, rather it mainly focuses to the *"adequate performance"* in the activities associated with daily living.

Though the physical educators and coaches apply the knowledge of kinesiology mainly to the movements of the normal body; and the therapists are mainly concerned with the movements of a body which has suffered an impairment in function; the difference lies in the emphasis and methods used, rather than in the purpose. Both the physical educators and the therapists, however have one common application in studying kinesiology; they are both concerned with posture and body mechanics of daily life skills, and analyzing the anatomical and mechanical bases for training, and then to the intelligent selection of exercises, activities and other mechanically efficient methods, for using the body in daily life skills and sports as according to the individual need.

The knowledge of kinesiology has a three fold purpose both for the professionals of physical education and physical medicine for the analysis and modification of human movement. The kinesiology should enable them to help their students or clients perform with optimum "safety", "effectiveness", and "efficiency". "Safety" should be a great concern for all the movement professionals to design or select the movements or activities in such a way as to avoid doing any harm to the body. Both the educators and the therapists should also set goals for "effective" performance, which is judged by success or failure in meeting the set goals. While producing an effective performance, the movement specialists should also focus to achieve their movement goals with the least amount of effort, as "efficiently" as possible.

The analysis of motion alone as an aim of kinesiology should not be an end in itself, rather it should be a means of learning new movement patterns and improving the safety, effectiveness and efficiency of old ones. Kinesiology serves only half its purpose when it provides information of value for learning or teaching motor skills. It must also serve to lay the foundation for perfecting, repairing, and keeping in good condition the human body, which is an incomparable machine.

ROLE AND IMPORTANCE OF KINESIOLOGY IN PHYSICAL EDUCATION AND SPORTS

Kinesiology is the basic science of the physical educator who deals with the motor performance as a means toward the development of the total individual. The development of total individual through physical activity, and skills is a unique contribution of physical education which perhaps no other branch of education aims to attain so much. Therefore, the physical education teacher must have thorough knowledge of, and ability to analyze the motor performance which kinesiology provides to him. Then only he will be able to guide toward the most effective learning, and provide the greatest benefit to the human body.

- ***Kinesiology serves the dual purpose for physical educator; it perfects the performance in motor skills as well as perfects the performer:*** Kinesiology helps to prepare the physical educator to teach effective performance in both fundamental and specialized motor skills. Perfecting the performance refers to mastery and perfection in the technique, and to define the standards of a skill. On the other hand, perfecting the performer means that an individual sports person is made perfect in the act. The intelligent selection of the methods, skill, and activity will help perfect both the performer and performance.
- ***Kinesiology helps the physical educator and coach to analyze the activity for better and easier teaching:*** The knowledge and application of kinesiological principles can contribute immensely in teaching various skills and techniques during sports coaching. It is however very important for the physical educator and coach to determine what kinesiological knowledge to select and how to apply it in a given teaching situation.

The coach should be able to explain and demonstrate the desired performance to the learner, and also analyze the learner's performance so that he can focus on the factors responsible for errors and successes and thus provide a basis for subsequent and more successful attempts by the learner.

- ***It is helpful in evaluating the effect and usefulness of activities:*** The physical educators or coaches, while they deal with physical development or motor skills, the knowledge and an understanding of kinesiological principles help them assess and evaluate the extent of effect produced by the exercise and movement to achieve the purpose for which these were prescribed.
- ***Kinesiology assists the coach and physical educator to assess the kinesiological requirements of the activity:*** For successful performance, each separate motor skill demands its own combination of kinesiological abilities and characteristics. For example, considerable body mass is a necessity for inside line play in the football, but it is hindrance in gymnastics. Though both these activities require great strength, but it should be predominantly in the legs for football and, in the arms for the gymnastics.

The knowledge that the motor skills can be learnt at highly effective levels by the performers and not merely through the academic knowledge, assists the coaches and physical educators in imparting the training and getting the desired performance.

- ***Kinesiology helps the physical educator and coach to assess the activity aptitude of the performer:*** Since each performer be it a student or sports person has his own abilities and potentialities, the kinesiology helps the physical educator and coach to match the performer to the activity, and the activity to the performer. For example, a short, stocky boy would be much more appropriate to be a successful gymnast than he would prove to be a successful high jumper. Similarly, a basketball coach will like to select a team of predominantly tall, heavy, slow-moving men as his defensive system.
- ***Kinesiology is also greatly helpful in the design and selection of sports clothings, equipment and other facilities:*** The designers of sports clothing and equipment are now more alert and keen to apply the knowledge of kinesiology so as to ensure the free action and movement of the sports persons during the execution of sports activity. The clothing which binds tightly through the arms and shoulders

is more fatiguing and reduces speed of action. The clothings should be designed to provide freedom of movement, avoid strain, and avoid weight on shoulders and back. The quality of the material for the clothing, and sports equipment, for example for the bats and racquets, their size, shape, and design, etc. all employ the kinesiological considerations for the efficient and successful execution of sports skills and techniques by the sports persons. All these ultimately help in enhancing the sports performance safely and effectively.

- ***Kinesiology aids in the prevention, first aid and rehabilitation of musculoskeletal athletic injuries:*** Since the athletic injuries are almost and invariably associated with intense physical and sports activities, and a physical educator or a coach generally would not be expected, by his limited knowledge in this area, to make a final diagnosis of an injury. He, however, needs to have an understanding of the nature of trauma. Kinesiological knowledge helps him anticipate, prepare and be alert for the type of an injury which may occur in a given situation.

By ensuring proper conditioning and flexibility exercises, sports implements, and the knowledge of nature of operating forces, avoidance of fatigue all can assist a physical education teacher or coach prevent the injuries.

Similarly, with his kinesiological knowledge he can also assist in the first aid and therapeutic exercise during the later phase of rehabilitation as guided or directed by the physician.

- ***Kinesiology also plays an important role in following aspects related to physical education and sports:***
  - Enables physical educator and coach to provide effective scientific training to the players and get the optimal performance of the sports skills and techniques.
  - Selection of exercises and preparation of activity programme on the basis of individual needs, age, sex, etc.
  - Helps to avoid unwanted movement, errors and faults of sports skills and techniques.
  - Identification of postural faults and correction through suitably designed therapeutic exercises and activity.

CHAPTER 2

Anatomical and Physiological Fundamentals of Human Motion

In order to study kinesiology effectively, the knowledge and understanding of those anatomical and physiological aspects which are most relevant to kinesiology is very essential. Though the bones, joints, muscles, connective tissue, blood vessels and nerves are all vital elements of human motion, however to analyze the human movements, the knowledge of musculoskeletal system is probably of great importance. It is very essential to understand how the bones, joints, and muscles serve as elements in anatomical levers, which act according to the laws of mechanics.

The skeletal and muscular framework is an arrangement of bones and muscles. Adjacent bones are attached to one another by joints which provide for the motion of the articulating bones. And the muscles that cover or extend the joints provide the necessary force for moving the bones to which they are attached. In other words, speaking mechanically, the total bone-joint-muscle structure is a complex combination of human levers that makes possible a great number of coordinated movements, ranging from the small hand and finger motions used in assembling a television set to the total body movements of a swimmer. In the human body, a bone that engages in a turning or angular type of movement forms the lever, and the muscle, attached to the bone when it contracts supplies the force to move the bony lever. This force is always a "pulling" force as the muscles being flexible can only pull, as they are unable to push.

In this chapter, therefore the essential anatomical and physiological fundamentals most relevant to kinesiology, i.e. the study of the skeletal system, (the bones and joints), the muscular system, certain neuromuscular concepts, and the nervous control of voluntary movements are presented.

THE SKELETAL SYSTEM

The human skeleton, or the skeletal system, which is made up of numerous bones, is the rigid framework of the human body. Bones vary in size according to sex and race as well as among individuals. It is the bones which primarily determine the characteristics of body build; the tall or short stature; the large or small thorax; big or small ankles, knees and wrists, slender or broad feet, spreading large or slender hands and so on. These factors are important in the movements of the person as well as in the way he appears to the observer.

Functions of the Skeleton (Bones)

The bones and the skeletal system serve a number of purposes:
- They give support and shape to the body.
- The bones enable the movements by providing a rigid structure for muscle attachment and leverage.
- They protect the vital organs such as the brain, spinal cord, heart and the lungs.
- The skeletal system is also involved in the process of blood formation. It mostly occurs in the red bone marrow of femur, humerus, ilium, vertebral bodies, sternum and ribs.
- Calcium and phosphorus minerals are also stored throughout all the bones of the skeletal system.

The bones therefore should be rigid and strong to withstand the force of the muscles which pull upon them, and also to bear the stress of the load which the levers bear. All these demands and challenges are infrequently met out as the bones acquire strength through the daily stress of motor action.

Types of Skeletons

According to the anatomy textbooks, there are 206 bones in the human skeleton. However, only 177 bones engage in voluntary movements. The bones that do not participate in the movements are hyoid, the coccyx, 6 ossicles, and 21 skull bones).

The human skeleton consists of two major parts:
1. *Axial skeleton.*
2. *Appendicular skeleton.*

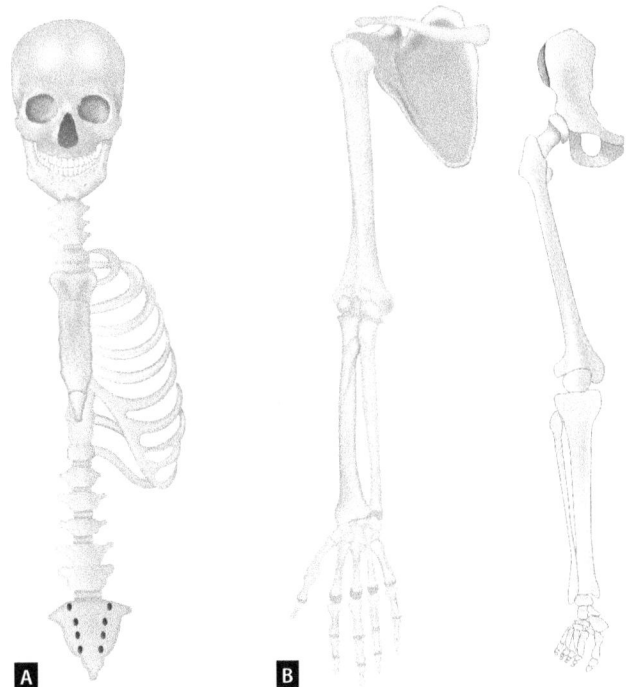

Figs. 2.1A and B: (A) Axial skeleton; (B) Appendicular skeleton

The axial skeleton (**Fig. 2.1A**) forms the upright part of the body and consists of approximately 80 bones of the skull, spinal column, sternum and ribs. The appendicular skeleton (**Fig. 2.1B**) attaches to the axial skeleton and contains 126 bones of both the extremities. The bones of upper extremity include the scapula, clavicle, humerus, radius, ulna, carpal bones, metacarpals and phalanges. The bones in lower extremity include three fused bones of the pelvis (ilium, ischium, and pubis), the femur, tibia, fibula, seven tarsal bones, metatarsals and phalanges.

- Since the pelvis is actually a link between the axial skeleton and the lower extremity (a part of the appendicular skeleton), and is equally important functionally to both of these, the pelvis may be classified with either the axial or the appendicular skeleton.
- Certain individuals may have additional sesamoid bones, such as in the flexor tendons of the great toe and of the thumb.

Composition and Gross Structure of Bone

Bones in a sense can be considered organs as they are made up of several different types of tissues (i.e., fibrous, cartilaginous, osseous, nervous, and vascular), and they function as integral parts of the skeletal system.

Bone is made up of about one-third *organic* (living) material and, two-thirds *inorganic* (nonliving) material. About 25–30% of bone is water. Approximately 60 to 70% is composed of minerals (calcium phosphate, and calcium carbonate), the inorganic or nonliving material which gives the bones hardness, and strength—an ability to resist compression. Generally speaking, a bone can withstand about six times the stresses to which it is subject in ordinary activities.

The *organic* or living material of the bone comprises of the cells, the fibrous matrix or collagen, and the ground substance. The organic material constitutes a small fraction of the total weight of the bone. All this organic matter is impregnated with the inorganic bone salts. The collagen provides the bone some flexibility and strength to resist tension. After maturity, the proportions of fluid and of organic material gradually decrease with age. For this reason, the bones of aged people become brittle and healing becomes difficult.

There are two forms of bony tissues:
1. *The compact, dense or cortical bone*, and
2. *The cancellous, or spongy bone.*

The **compact bone** makes up a hard, dense outer shell. It always completely covers bone and tends to be thick along the shaft and thin at the ends of long bones. About 5–30% of the compact, cortical bone is porous having fewer spaces. The hard, compact bone withstands greater stress, but less strain.

The ***cancellous bone*** is about 30–90% porous and spongy called "trabeculae" in the interior of a bone. The cancellous bone has larger cavities and spaces. The trabeculae are arranged in a pattern that resists local stresses and strains. These trabeculae tend to be filled with marrow and make the bone lighter. Cancellous bone makes up most of the articular ends of bones.

- The relative quantity of compact and cancellous tissue varies in different bones and in different parts of the same bone, depending on the need for strength or lightness. The shafts of the long bones are made up almost entirely of compact tissue, except that they are hollowed out to form a central canal, the medullary canal.

  Both compact and cancellous bone respond to the pressure and tension forces acting on the bone. As described above, the compact bone may be considered as contributing to strength and protection, whereas the cancellous bone is designed to provide strength with economy in weight.
- The ***epiphysis*** is the end of a long bone at each end of the diaphysis, which tends to be enlarged and wider than the shaft and shaped to make a joint with adjacent bone. The epiphysis is covered by hyaline cartilage to facilitate smooth movement, and also reduces friction. In adult bone, the epiphysis is osseous, but in growing bone the epiphysis is cartilaginous material called as *epiphyseal plate*. Longitudinal growth occurs here through the manufacturing of new bone.

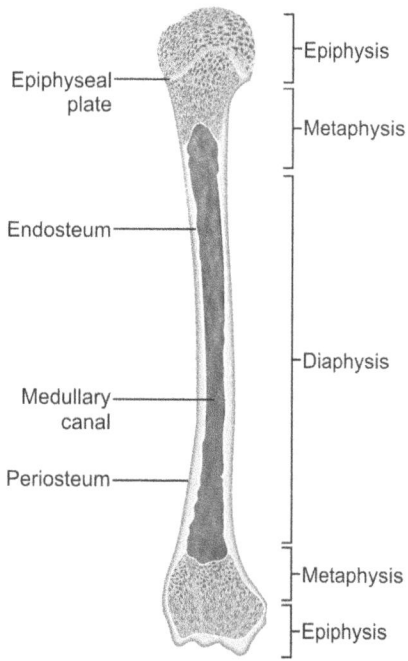

Fig. 2.2: Gross structure of a long bone (longitudinal cross section)

- The ***diaphysis*** is the main, long cylindrical shaft of bone. It is made up mostly of compact bone, which gives it great strength. Its center, the ***medullary canal*** is hollow which decreases the weight of the bone. This medullary canal contains marrow and provides passage for nutrient arteries. The ***endosteum*** is a membrane that lines the medullary canal. It contains *osteoclasts* which are mainly responsible for bone resorption.
- In long bones, the wider part at each end of the diaphysis is called the ***metaphysis***. It is mostly made up of cancellous bone and functions to support the epiphysis.
- ***Periosteum*** is the thin fibrous membrane covering all of the bone except the articular surfaces that are covered with hyaline cartilage. The periosteum contains nerve and blood vessels that are important in providing nourishment, promoting growth in diameter of immature bone, and repairing the bone. It also serves as an attachment point for tendons and ligaments.

The gross structure of a long bone (longitudinal cross section) is shown in **Figure 2.2.**

Bone mass generally increases with increased stress to it. The size and shape of the bone is influenced by the direction and magnitude of the forces habitually applied to it.

Types of Bones

Despite the great variety of shapes and sizes of bones, there are mainly four types of the bones, i.e. long bones, short bones, flat bones, and irregular bones **(Figs. 2.3A to D)**. A fifth type, the small sesamoid bones are also found in some individuals.

1. ***Long bones:*** These bones are named because their length is greater than their width. These bones are the largest bones in the body and form the framework of the appendicular skeleton.

    The long bones are characterized by a long, roughly cylindrical shaft, with relatively broad or bulbous ends. The *shaft* or the body of these bones has thick walls and contains a central cavity called as *medullary cavity* or canal. The examples of long bones in the upper extremity are the clavicle, humerus, ulna, radius, metacarpals and phalanges. In the lower extremity, the femur, tibia, fibula, metatarsals and phalanges are all the examples of long bones.

    The long bones are adapted in size and weight for specific biomechanical functions. The tibia and femur are large and massive to support the weight of the body, whereas the long bones of the upper extremity, i.e. the humerus, radius and ulna are smaller and lighter to promote ease of movement.

2. ***Short bones:*** These bones are relatively small and solid bones. The short bones tend to have more or less equal dimensions of height, length, and width, giving them shape like a cube. They have a great deal of articular surface, and usually articulate with more than one bone. They are mostly cancellous bone with a thin outer layer of compact tissue. These bones provide limited gliding motions and serve as shock absorbers. Examples of short bones include the carpals (wrist bones) and the tarsals (ankle bones).

3. ***Flat bones:*** These bones have a broad surface but are not very thick. They are thin but seldom flat. They tend to have a curved surface rather than a flat one. They are made up of two layers of compact bone with cancellous bone and marrow in between. These bones protect underlying organs and soft tissues and also provide large areas for muscle and ligament attachments. The examples of flat bones are sternum, scapula, ilium, and ribs.

4. ***Irregular bones:*** As the name implies, these bones are of highly different and mixed shapes that fulfill the special functions in the human body. For example, the vertebrae provide a bony, protective tunnel for the spinal cord, offer several processes for muscle and ligament attachments, and support the weight of the superior body parts while enabling the movement of the trunk in all three planes. The examples are bones such as of the spinal column (all 24 vertebrae), the sacrum, the coccyx and maxilla that do not fit into the other types. These bones are composed of cancellous bone and marrow encased in a thin layer of compact bone.

5. ***Sesamoid bones:*** These small bones resemble the shape of sesame seeds which are embedded within the tendons of musculotendinous unit, and located near the ends of long bones in the extremities. These accessory bones develop within the tendon and protect it from excessive

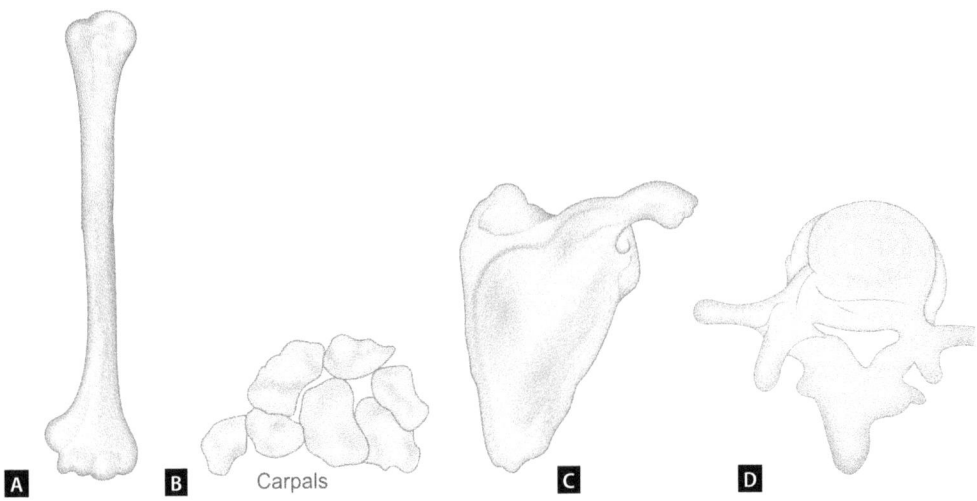

Figs. 2.3A to D: Types of bones: (A) Long; (B) Short; (C) Flat; (D) Irregular

wear. The sesamoid bone like patella also changes the angle of attachment of a tendon and increases the mechanical advantage.

In the lower extremity, sesamoid bones are found in the flexor tendons that pass posteriorly into the foot on either side of the ankle, i.e. flexor hallucis longus. In the upper extremity, these bones are found in the flexor tendons of the thumb near the metacarpophalangeal (MP) and interphalangeal (IP) joints.

The human skeleton provides several examples of the inter-relationships of structure and function. The lower extremity is for better adapted to weight bearing than the upper extremity. The pelvic girdle is a complete ring and relatively solid in the pubic and the two sacroiliac articulations. The condyles of tibia are broad and flat for the superimposed segment. The foot projects both forward and back from the ankle to provide a better base of support. Conversely, on the other hand, the incomplete and mobile pectoral girdle gives great freedom of motion, and the radius swinging freely around the ulna carries the hand through a wide range for placement and use of hand or fingers.

Joints

All bones have articulating surfaces so that various body movements may be accomplished. These surfaces form joints or articulations, some of which are freely movable, others slightly movable, or immovable. All kinds of joints are important for smooth co-ordinated movements of the body. **A joint is a union or articulation of two or more bones**. Although joints have several functions, perhaps the most important is to allow motion. Joints also help to bear the weight of the body and to provide stability. This stability may be mostly due to the shape of the bones making up the joint, or the soft tissue features. Joints also contain synovial fluid, which lubricates the joint and nourishes the cartilage.

The structures of the bones forming a joint determine the kind and degree of movements possible at a particular joint. In other words, it can also be said that the type of movements possible at a joint and degree of freedom depend upon the bony structure of a joint, and the positioning of the ligaments and muscles surrounding the joint.

- The freedom of movements, and their excursion (range) vary slightly in different individuals due to limitations imposed by ligaments, and muscles, and very slight variations in joint structures.
- The joints of the body are the sites of motion just as the hinge on the door, or the axle of the wheel is the point of action for these two objects. If a hinge of a different type is put on a door, it will result in a different use of the door. The revolving door, the swinging door, and the ordinary door with the one way hinge all behave differently because of the differences in their attachments to the adjoining door frame.
- The modern engineer is often called upon to build two types of structures. One type of structure must withstand the stress and strain of superimposed parts but there is little or no need for movement of component parts. Here, the emphasis is for strength and permanence of the connecting parts and a firm base for the superimposed parts. The second type of structure is a moving one with requirements for controlled freedom, variety in action, efficiency in operation, and interrelating action in various parts. Such a structure not only calls for a firm base but also for centers of prescribed motion that are protected against the forces of action.

The human frame must achieve this dual demand. The joints are fitted together or tied together as firmly as possible but are also free for certain dynamic functions. However, in certain most critical points of stability, or for protection of internal functions, movement is sacrificed and we find the rigidity and fixation characteristics of the static structure of the joints.

Classification of Joints

The structure and function of joints are so interrelated that it becomes difficult to describe them separately. A joint may allow a great deal of motion, as in the shoulder, or very little motion, as in the sternoclavicular joint. **A joint that allows a great deal of motion will provide very little stability. Conversely, a joint that is quite stable tends to have little motion.** The classification and terminology of joints are presented differently by various authors. However, here the joints are described and classified on the basis of their structure, and also on the function (amount of motion) allowed by these joints.

Structural Classification of Joints

On the basis of different patterns of joints structure, the joints are classified as:
- Fibrous
- Cartilaginous
- Synovial joints.

Fibrous Joints: These types of joints are joined together by connective tissue fibers, or have a thin layer of fibrous periosteum between the two bones **(Figs. 2.4A and B)**. *These are generally immovable joints.* Examples are sutures of the skull, gomphosis (joint formed between a tooth and its socket).

Cartilaginous Joints: These types of joints are joined together either by hyaline cartilage or fibrocartilage. *These joints allow very slight degree of movements* such as bending, twisting and some compression. These joints provide a great deal of stability. These joints are also called as *"amphiarthrodial joints".*

The vertebral joints **(Fig. 2.5)** are examples of joints in which fibrocartilage (disc) are directly connecting the bones; and the first sternocostal joint an example of hyaline cartilage connecting the sternum and the first rib.

Synovial Joints: A synovial joint has no direct union between the bony ends, rather there is a cavity filled with synovial fluid contained within a sleevelike capsule of the joint. The outer layer of the capsule is made up of a strong fibrous tissue that holds the joint together. The inner layer of the capsule is lined with a synovial membrane that secrets the synovial fluid. The articular surface is very smooth and covered with cartilage called "hyaline" or "articulating cartilage". A typical example of synovial joint is shown in **Fig. 2.6**.

The synovial joints are freely movable joints and also called as "diarthrodial joints".

These joints are however not as stable as the other types of joints but do allow a great deal of motion.

On the basis of number of axes, and the shape of the joint, these joints can be further classified:

(A) On the basis of number of axes, the joints are classified as below:
- Nonaxial
- Uniaxial
- Biaxial
- Triaxial joint.

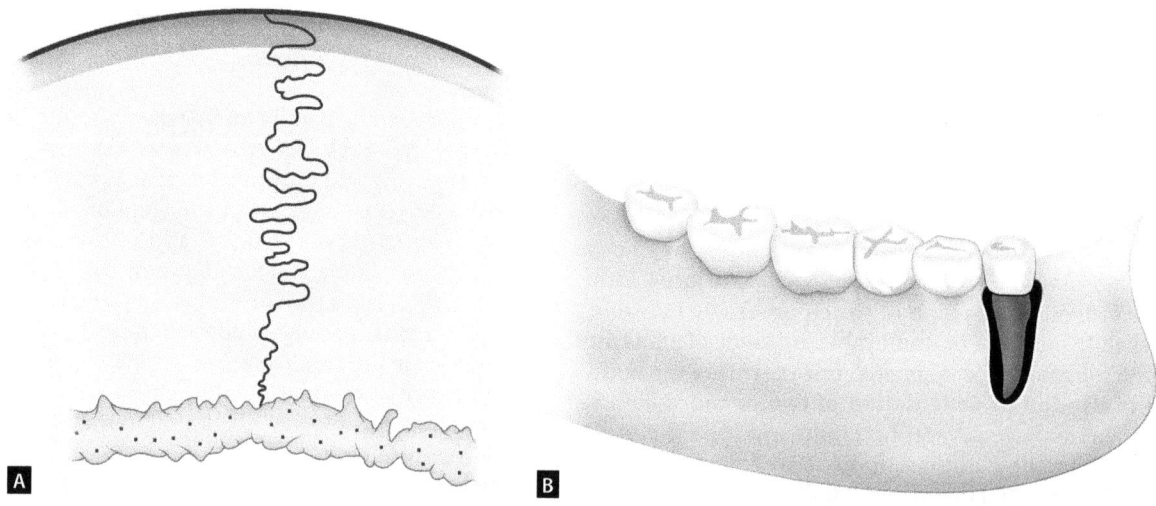

Figs. 2.4A and B: Fibrous joints: (A) Suture of skull; (B) gomphosis (joint between a tooth and its socket)

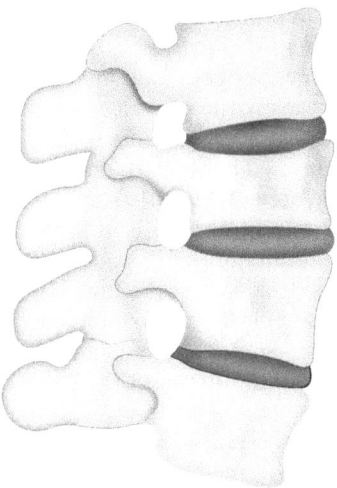

Fig. 2.5: Cartilaginous joint (vertebral Joints)

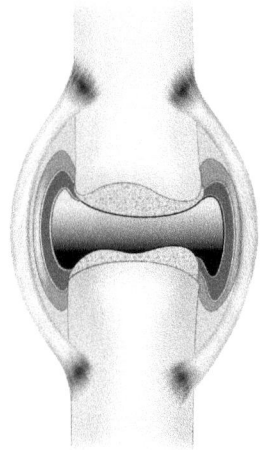

Fig. 2.6: A typical example of synovial joint

- **Nonaxial joint:** The joint surfaces are relatively flat and glide over one another instead of one moving around the other. The movement tends to be linear instead of angular. The motion that occurs between the carpal bones is an example of this type of joint. The nonaxial motion occurs secondary to other motion, for example, the motion of the carpals during the flexion/extension of the wrist.
- **Uniaxial joint:** This type of joint permits angular motion occurring in one plane around one axis, much like a hinge joint. The elbow or the humeroulnar joint is a good example of uniaxial joint in which the convex shape of the humerus fits into the concave shaped ulna. The only movements possible are flexion and extension, which occur in the sagittal plane around the frontal axis. The other examples are the knee joint, and interphalangeal joints of the hand and foot. The proximal radioulnar joint at the elbow is also a type of uniaxial joint permitting pronation and supination movements of the forearm, occurring in the transverse plane around the longitudinal axis.

    The rotation of the first cervical vertebra (atlas) over the second cervical vertebra (axis) which occurs during the rotation of the head is an another example of motion occurring at the uniaxial joint.
- **Biaxial joint:** Such joints allow movements in two different directions. The example is the wrist joint which permits flexion/extension movement around the frontal axis, and the radial and ulnar deviation occurring around the sagittal axis. The other examples of biaxial joints are metacarpophalangeal (MP) joints, and the carpometacarpal (CMC) joint of the thumb.
- **Triaxial joint:** This type of joint which is sometimes also referred to as a "*multiaxial joint*" permits motion actively in all three axes. The hip and shoulder joints are the best examples of this type of joints, which provide flexion/extension movements around the frontal axis, abduction/adduction movements around the sagittal axis, and the movement of rotation around the vertical axis. As such, it is clear that triaxial or multiaxial joints allow more motion than any other type of joints.

(B) On the basis of shape of joint, the joints are classified as under:
- **Plane, irregular (arthrodial):** This type of joint is characterized by two flat or irregular bony surfaces making the joint **(Fig. 2.7)**. This type of joint only permits limited gliding movement. Therefore, it is also referred to as a nonaxial joint. The examples are the joints of carpal bones of the wrist, and tarsometatarsal joints of the foot.
- **Hinge (ginglymus):** In this type of joint of which elbow joint is a good example, one surface is spool like, and the other is concave that fits over the spool-like process **(Fig. 2.8)**. A hinge joint permits wide range of movement only in one plane (sagittal plane) about a single joint axis (frontal axis), hence it is a uniaxial joint. The movements that occur are flexion and extension. The other examples of this type of joint are knee and ankle joints.
- **Pivot (trochoidal, screw):** This type of joint is characterized by a peg like pivot, as in the joint between the first cervical (atlas) and second cervical vertebra (axis), or by two long bones fitting against each other near each end in such a way that one bone can roll around the other one. The radioulnar joint **(Fig. 2.9)** of the forearm is an example of this type of joint in which the head of radius rotates around the stationary ulna in pronation and supination movements. As such, a pivot joint is also a uniaxil joint, permitting only rotation movement in the transverse plane around a longitudinal axis.

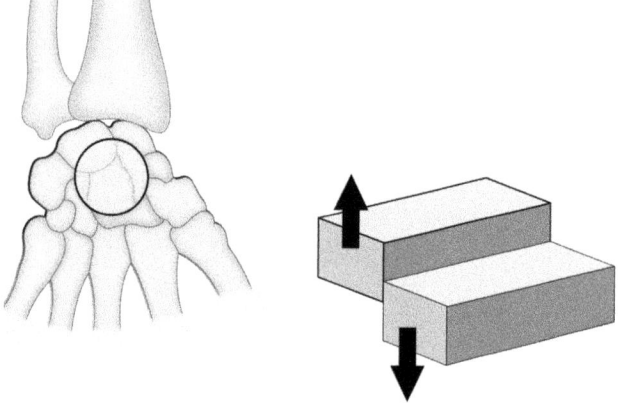

Fig. 2.7: Plane joint (intercarpal joints of wrist)

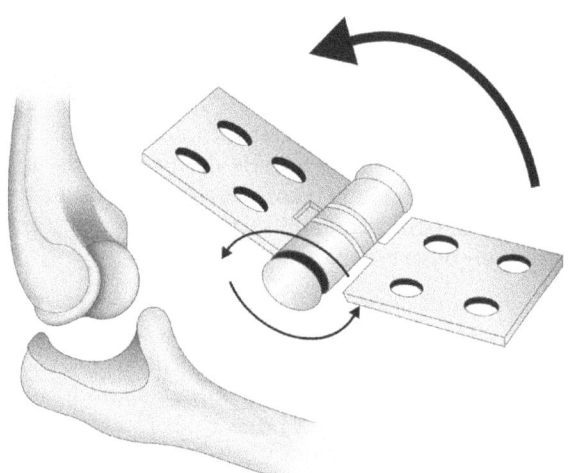

Fig. 2.8: Hinge joint (elbow joint)

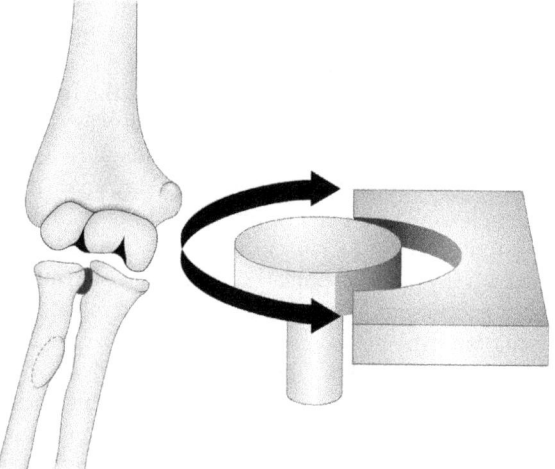

Fig. 2.9: Pivot joint (superior radioulnar joint)

- **Condyloid (ovoid, ellipsoidal):** In this type of joint, an oval or egg shaped convex surface fits into a reciprocally shaped concave surface. Such joint permits movements around two axes, therefore it is a biaxial joint. The movements that occur in a condyloid joint such as the wrist joint (formed between the radius and the proximal row of carpal bones) **(Fig. 2.10)** are the flexion/extension occurring around frontal axis in the sagittal plane, and abduction/adduction or the lateral flexion occurring around the sagittal axis in the frontal plane. When these movements are performed sequentially, they constitute circumduction. Some experts also refer condyloid joint a biaxial type of "ball and socket" joint that does not permit rotation movement.

  The other examples of this type of joint are the second to fifth metacarpophalangeal joints (MP joints) of the hand.

- **Saddle (sellar):** This type of joint is thought of as a modification of a condyloid joint in which the articulating bones provide for reciprocal reception to each other. The carpometacarpal joint of the thumb **(Fig. 2.11)** is only an example of this type of joint in which the articular surface of each bone is concave in one direction and convex in the other. Both the articulating bones fit together like a horse back rider fits in a saddle, which is why this joint is called a "saddle joint". As such, this is also a biaxial joint, permitting flexion/extension, abduction/adduction, and circumduction. The saddle joint differs from the condyloidal joint in the sense that a greater freedom of movement occurs in the saddle joint than a condyloid joint.

- **Ball-and-socket (spheroidal):** In this type of joint, the spherical or round shaped head of one bone fits into the cuplike cavity of the other bone. It is a triaxial or multiaxial type of joint of which the shoulder (glenohumeral) and hip joints **(Fig. 2.12)** are the examples. This type of joint permits movements about all three axes, and the movements occurring are flexion/extension, abduction/adduction, and rotation. The circumduction which is a sequential combination of all four movements flexion, extension, abduction and adduction, as well as the horizontal flexion/extension also occurs in this type of joint.

Functional Classification of Joints

On the basis of the amount of movement possible, the joints are classified into following three types:
1. Synarthrodial, or immovable joints
2. Amphiarthrodial, or slightly movable joints
3. Diarthrodial, or freely movable joints.

The first two types of joints, i.e. synarthrodial and amphiarthrodial do not have a true joint cavity. The third type, i.e. diarthrodial joints which are also called as the "synovial joints" have a joint cavity and permit free range of motion.

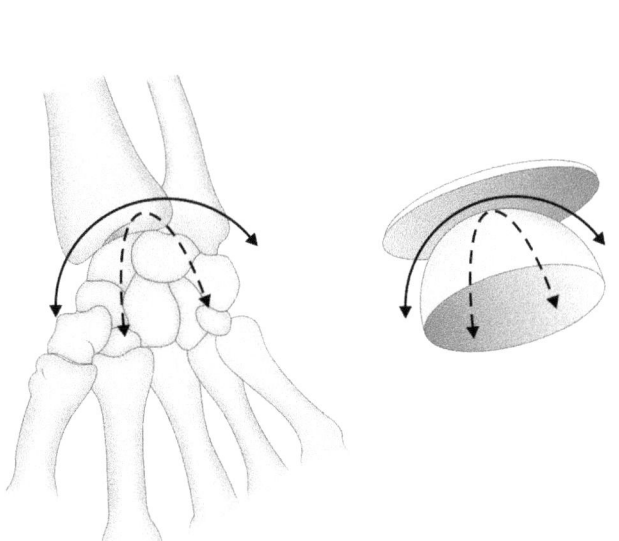

Fig. 2.10: Condyloid joint (wrist joint)

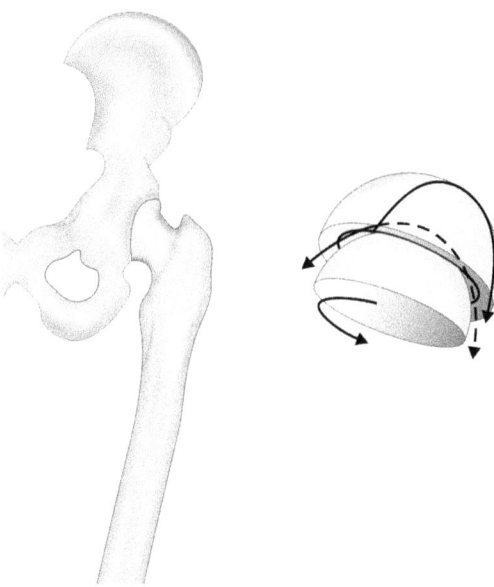

Fig. 2.12: Ball-and-socket joint (hip joint)

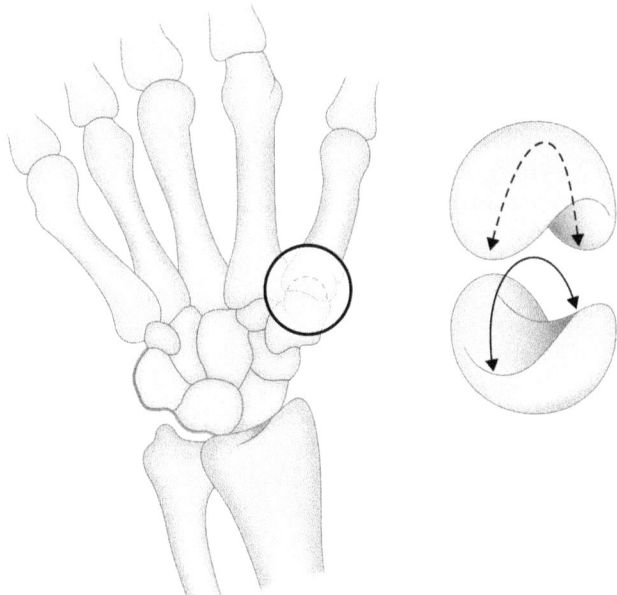

Fig. 2.11: Saddle joint (carpometacarpal joint of thumb)

Synarthrodial, or Immovable Joints: These joints are also called as the **"fibrous joints".** In these joints, the two bones grow together, with only a thin layer of fibrous periosteum between them. As the name indicates, there is no movement or motion possible at these joints. Examples are the sutures of the skull, sockets of teeth.

Amphiarthrodial, or Slightly Movable Joints: As according to the name, some movement is possible at these type of joints. These joints are further divided into two types:
1. Syndesmosis (or ligamentous).
2. Synchondrosis (symphysis or cartilaginous).
- In syndesmosis or ligamentous type of joints, slight movement is permitted by the elasticity of a ligament which holds or joins the two bones. The examples are, coracoacromial joint, inferior tibiofibular joint.
- In synchondrosis or cartilaginous type of joints, the bones are coated with hyaline cartilage, separated by a fibrocartilage and joined by ligaments. Movement or motion is allowed only by deformation of the disc. Examples are pubic symphysis, between bodies of vertebrae, joints of ribs with sternum, etc.

Diarthrodial, or Freely Movable Joints: These types of joints are also known as **synovial joints**. As described above under the heading of "synovial joints", a sleeve like covering of ligamentous tissue (called joint capsule) surrounds the bony ends that form the joint. This joint capsule is lined with a thin vascular synovial membrane that secrets synovial fluid to lubricate the area inside the joint capsule (called as the joint-cavity).

In certain areas, the joint capsule is further thickened to form tough, nonelastic ligament to provide additional support preventing abnormal movements or joint-opening. The articular surfaces on the bony ends inside the joint cavity are covered with hyaline cartilage, which functions as a shock-absorber for the joint. Structurally, these type of joints can be further subclassified, as described above under

Table 2.1: Summary of classification of diarthrodial/synovial joints

Number of axes	Type of joint (shape)	Joint movement	Example
Nonaxial	Irregular (plane)	Gliding	Intercarpals
Uniaxial	Hinge	Flexion/extension	Elbow, knee
	Pivot	Rotation	Atlas/axis, radius/ulna
Biaxial	Condyloid	Flexion/extension, abduction/adduction	Wrist, MP joints
	Saddle	Flexion/extension, abduction/adduction, rotation (accessory)	Thumb, CMC
Triaxial (multiaxial)	Ball and socket	Flexion/extension, abduction/adduction, rotation	Shoulder, hip

"synovial joints". The summary of classification of diarthrodial/synovial joints is given in **(Table 2.1)**.

Note: Since there is a strong relationship between the structure and function of the joints, there is a significant overlapping between both the systems of classification described above. Various authors also classify the joints differently. As such, all the joints of the body do not clearly fit into these systems of classification.

Joint Stability

Though the primary function of the joints is to provide the bones of being moved. However, with the provisions of movements, there is a threat of instability to the joints. As such, the joints do have a secondary function of providing stability to the joint, without interfering with the desired motions or movements.

All the joints of the body do not have the same degree of stability. Joints like hip or elbow, are fairly stable. On the other hand, the joints such as the shoulder or knee are less stable and therefore more easily injured. It is said "for everything that is given, something is taken". *As such, in the shoulder, movement is gained at the expense of stability, while in the hip, movement is sacrificed for stability.*

There are numerous factors that contribute to the stability and integrity, particularly of synovial joints. They include:
- **The shape of bony structure** (hip joint is more stable than the shoulder joint though both are ball-and-socket type of joints.
- **Ligamentous arrangement** (these bind adjacent bones tightly together and help to maintain the right relationship of the bones forming the joint. They also cheek the movement when the joint reaches its normal limits for which the joint is not constructed).
- **Muscular arrangement** (the muscles and muscle tendons extending the joint play a great role in stabilizing the joints, particularly to the shoulder and knee joints whose bony structure contributes very less to the stability).
- **Fascia** (holds the bones together and sometimes functions similar to ligaments within limits. The iliotibial band of the fascia lata is an example of fascia that helps to stabilize the knee joint).
- **The atmospheric pressure** (resists dislocation of a joint as the suction force in the joint cavity is a powerful inhibitor of traction or dislocation).

Mobility of Joints (Range of Motion)

All the joints in the same individual do not have the same amount of movement, and the same joint in different individuals do not have the same amount of movement or range of motion (ROM). The range of motion (ROM) is dependent upon several factors; shape of joint surfaces, limiting effect of the ligaments, controlling action of the muscles, body build, gender, amount of fat present, occupation, age, personal exercise habits, current state of physical fitness etc.
- The strength of muscles is undoubtedly the single most important factor that helps maintain the active mobility or range of motion in the joints. Exercising the muscles on all sides of a joint can contribute to both strength and mobility. Well developed or bulky muscles and excessive fatty tissue will also restrict motion, as is evident when bulky arm muscles limit elbow flexion, or well developed calf muscles limit the knee flexion. The mobility or range of motion is also relatively more in girls than boys, in slender than the obese individuals, in youngers than the older people and so on. Similarly, those engaging actively in certain sports like gymnastics have more flexibility than those involved in weight lifting, wrestling or judo. It is however, important to remember that the mobility or flexibility should not exceed the ability of the muscles to maintain the integrity of the joints.
- The specific amount of movement possible in a joint may be measured by using an instrument known as *"goniometer"* to compare the change in joint angles. It is to be remembered that the normal range of motion (ROM) for a particular joint varies to some degree from person to person. While measuring the range of motion in extremities, the opposite extremity of the individual is perhaps the best norm.

Terminology of Fundamental Joint Movements

The students of kinesiology should be familiar with the terminology to describe the joint movements and muscle actions. When using movement terminology, it is important to understand that the terms are used to describe the actual change in position of the bones relative to each other. That is, the angles between the bones change, whereas the movement occurs between the articular surfaces of the joint.

The movement terms are used to describe movement occurring throughout the full range of motion or through a very small range. Using the knee flexion for example, we may flex the knee through the full range by beginning in full knee extension and flexing it fully, so that the heel comes in contact with the buttocks; or we may begin with the knee in 90° of flexion and then flex it 30° more, this movement results in a knee flexion angle of 120°, even though the knee only flexed 30°.

Some movement terms may be used to describe motion at several joints throughout the body, whereas other terms are relatively specific to a joint or group of joints.

Bones forming the joints are capable of different kinds of movements. For defining the joint movements, it is assumed that the body is in the anatomical position. **The main fundamental movements that occur in the human body are as follows.**

Flexion

Flexion takes place at any joint when a body segment is moved in an anteroposterior or sagittal plane. Flexion usually involves bending movement that results in the decrease of the angle in a joint by bringing bones together. An example of flexion at the elbow joint is when the hand is drawn toward the shoulder **(Fig. 2.13A)**.

Extension

This is the straightening movement from the flexed position, that results in an increase of the angle in a joint by moving bones apart, usually in the sagittal plane. Using the elbow, an example is when the hand moves away from the shoulder **(Fig. 2.13B)**.

Continuation of extension beyond the starting position is referred to as *"hyperextension"* such as takes place at the shoulder, knee, and at elbow to some extent. Similarly, *hyperflexion* is the movement which refers strictly to the movement of the upper arm at the shoulder joint beyond the vertical. Therefore, the prefix "hyper" does not indicate a different motion, but only the excessive continuation of the movement involved.

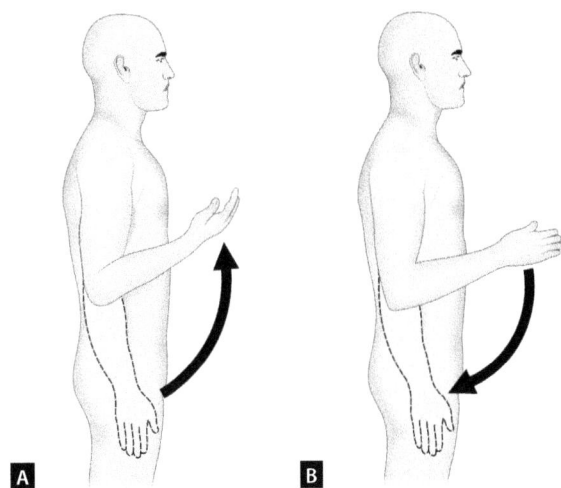

Figs. 2.13A and B: (A) Flexion of elbow; (B) Extension of elbow

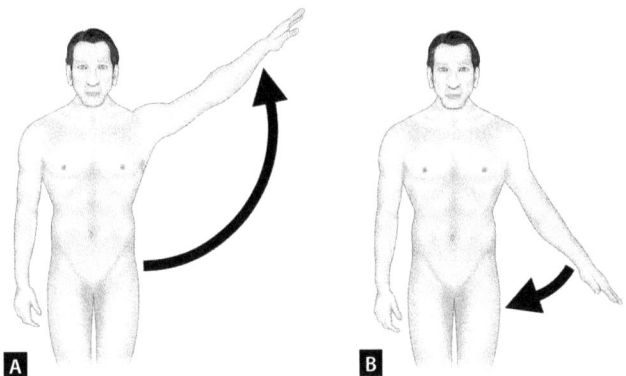

Figs. 2.14A and B: (A) Shoulder abduction; (B) Shoulder adduction

Abduction

This term refers to a lateral movement of a body segment, away from the midline of the body in the frontal plane about a sagittal axis. The example is the raising of the arm **(Fig. 2.14A)** or leg to the side horizontally.

In case of abduction of the arm at the shoulder beyond the 90° angle, it is still called the abduction though the part seems to be coming back toward the midline of the body. Similarly, the *"hyperabduction"* is the movement referred to the movement of the arm when it is abducted beyond the vertical.

Adduction

This is the movement in which the body segment moves or returns medially toward the midline of the body from

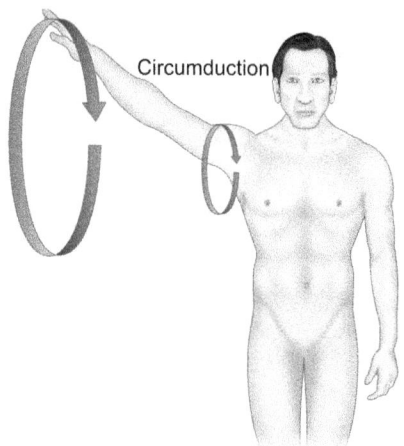

Fig. 2.15: Circumduction of shoulder

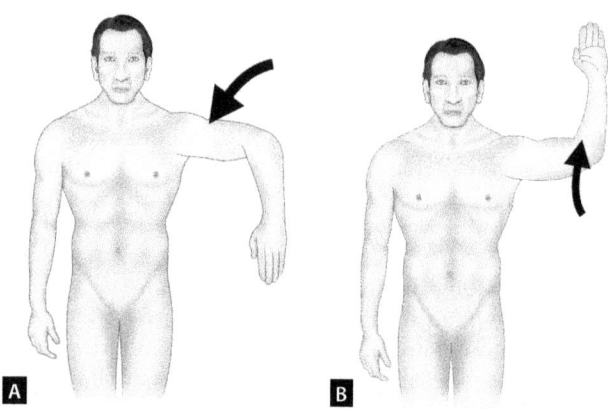

Figs. 2.16A and B: (A) Medial or internal rotation of shoulder; (B) Lateral or external rotation of shoulder

the abducted position, in the frontal plane. An example is lowering the arm to the side **(Fig. 2.14B)** or the thigh back to the anatomical position.

Abduction/adduction movements do not occur at the elbow or knee joints. At the wrist, abduction refers to the radial flexion, and the adduction to the ulnar flexion. Similarly, the terms, abduction and adduction have different meaning in case of the fingers and the toes. In the hand, the abduction refers to the movement of the fingers away from the midline of the hand, that is away from the middle finger. Similarly, the adduction movement would be called when other fingers move toward the middle finger. In case of the foot, the midline is said to pass through the second toe. As such, the movement of other toes toward, and away from the second toe is termed as adduction and abduction respectively.

Circumduction

This is a circular movement of a limb combining flexion, extension, abduction, and adduction, performed either in clockwise or anticlockwise direction. In this movement, the body segment describes a cone with its base at the distal end of the part, and the apex at the joint. An example occurs when the shoulder joint **(Fig. 2.15)** or the hip joint moves in a circular fashion around a fixed point.

Rotation

This is a type of movement which takes place around the long central axis of the bone. "*Medial or Internal Rotation*" refers to a rotary movement around the longitudinal axis of a bone toward the midline of the body **(Fig. 2.16A)**. It occurs in the transverse plane. Medial, internal, or inward rotation of the upper or lower limb occurs when the anterior surface of the limb or segment is turned medially or inward. "*Lateral or External Rotation*" refers to a rotary movement around the longitudinal axis of a bone away from the midline of the body, occurring in the transverse plane **(Fig. 2.16B)**. Lateral, external or outward rotation of the upper or lower limb occurs when the anterior surface of the limb or segment in turned laterally or externally.

Terms Describing Movements of Forearm (Radioulnar Joints)

Supination

This movement refers to external rotation of the radius along its long axis in the transverse plane so that it lies parallel to the ulna, and resulting in the palm-up position of the forearm, and the thumb returning to a lateral position from the medial one **(Fig. 2.17A)**.

Pronation

This movement refers to internal rotation of radius along its long axis in the transverse plane, so that it lies diagonally across the ulna, and resulting in the palm-down position of the forearm, and the thumb moving medially from the lateral side **(Fig. 2.17B)**.

Terms Describing Movements of Ankle

Dorsiflexion

This refers to a flexion movement of the ankle in which the dorsum of the foot moves towards the anterior aspect of the leg in the sagittal plane **(Fig. 2.18A)**.

Plantarflexion

This is the extension movement of the ankle in which the dorsal aspect of the foot and toes is lowered, and moved away

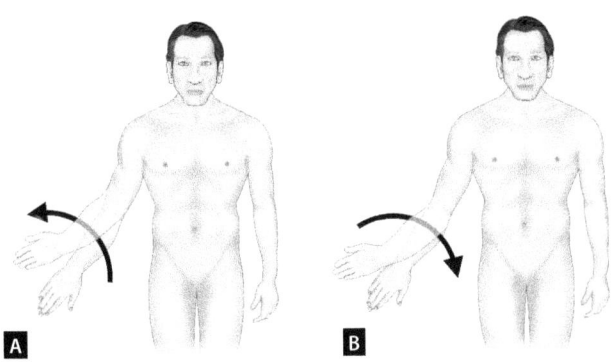

Figs. 2.17A and B: (A) Forearm supination; (B) Forearm pronation

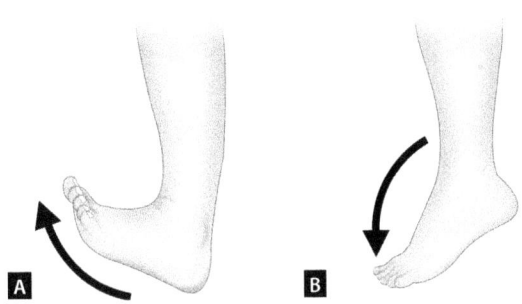

Figs. 2.18A and B: (A) Dorsiflexion; (B) Plantarflexion

from the body or the anterior aspect of the leg, in the sagittal plane **(Fig. 2.18B)**.

Terms Describing Movements of Foot

Inversion
This movement occurs when the sole of the foot turns medially or inward in the frontal plane. An example is standing with the weight on the outer border of the foot **(Fig. 2.19A)**.

Eversion
This refers to a movement in which the sole of the foot is turned laterally or outward in the frontal plane. An example is standing with the weight on the inner border of the foot **(Fig. 2.19B)**.

Terms Describing Movements of Shoulder Girdle (Scapula)

Elevation
This refers to a superior or upward movement of the shoulder girdle. An example is shrugging the shoulder.

Depression
This refers to an inferior or downward movement of the shoulder girdle. An example is returning to the normal position from a shoulder shrug.

Protraction (Abduction)
This refers to a forward movement of the shoulder girdle in the horizontal plane away from the spine. This is actually the abduction of the scapula, away from the spine **(Fig. 2.20A)**.

Retraction (Adduction)
This is backward movement of the shoulder girdle (from the forward position), in the horizontal plane toward the spine.

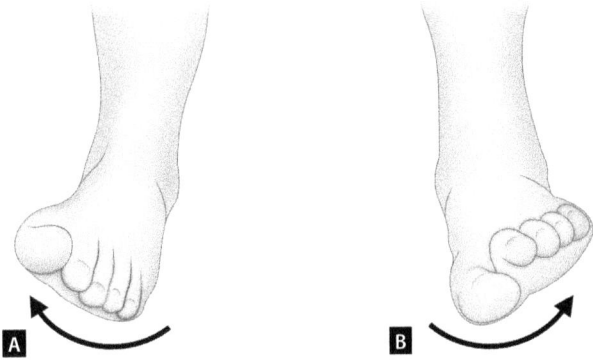

Figs. 2.19A and B: (A) Inversion; (B) Eversion

This is actually an adduction movement of the scapula, toward the spine **(Fig. 2.20B)**.

Upward Rotation
This refers to a rotary movement of the scapula in the frontal plane with the inferior angle of the scapula moving laterally and upward.

Downward Rotation
This is a downward, rotary movement of the scapula from the upward rotatory position, in the frontal plane, and the inferior angle of the scapula returns or moves medially and downward.

Terms Describing Movements of Wrist and Hand

Palmar Flexion
This is actually the flexion movement of the wrist in the sagittal plane with the anterior aspect of the hand moving toward the anterior side of the forearm **(Fig. 2.21A)**.

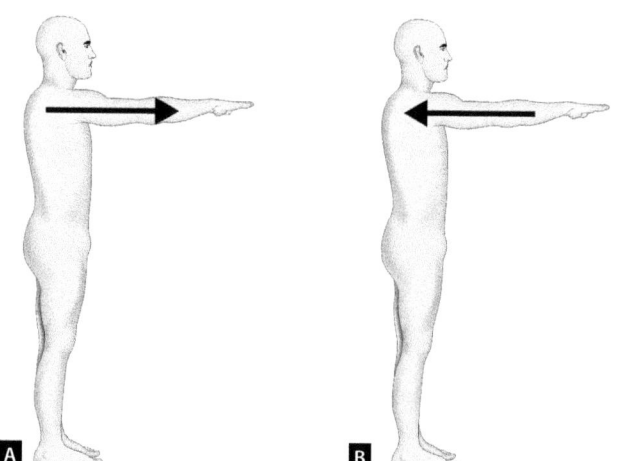

Figs. 2.20A and B: (A) Protraction of scapula; (B) Retraction of scapula

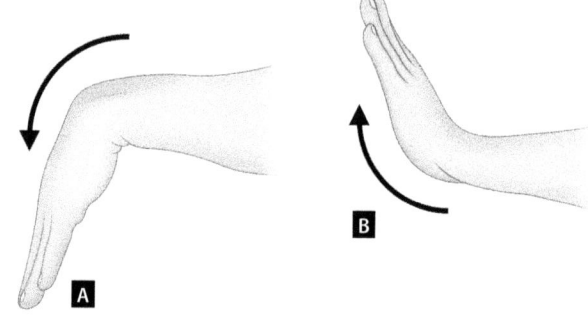

Figs. 2.21A and B: (A) Palmar flexion of wrist; (B) Dorsal flexion of wrist

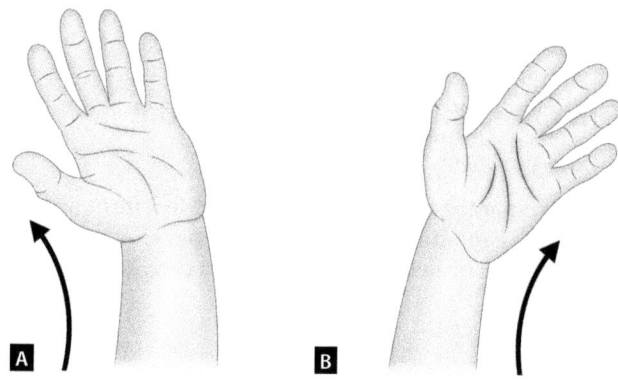

Figs. 2.22A and B: (A) Radial flexion of wrist; (B) Ulnar flexion of wrist

Dorsal Flexion

This is the extension movement of the wrist in the sagittal plane with the dorsal or posterior aspect of the hand moving toward the posterior side of the forearm (**Fig. 2.21B**).

Radial Flexion (Radial Deviation)

This is the abduction movement at the wrist in the frontal plane with the thumb side of the hand moving laterally toward the forearm. This is also referred to as the *lateral flexion* of the wrist (**Fig. 2.22A**).

Ulnar Flexion (Ulnar Deviation)

This is the adduction movement at the wrist in the frontal plane with the little finger side of the hand moving toward the forearm. This is also referred to as the *medial flexion* of the wrist (**Fig. 2.22B**).

Opposition of the Thumb

This refers to a diagonal movement of the thumb across the palmar surface of the hand to make contact with the fingers.

Reposition of the Thumb

This is a movement in which the thumb returns diagonally to the anatomical position from opposition, with the hand and/or fingers.

Though various authors differ for the values of normal range of motion for different movements and joints of the body, and the standardized normal values according to age, sex, body build, etc. have not been established, however the average normal values of range of motion which can be used as a guide are presented here. As described earlier also, the opposite limb of the individual (if it is present and unimpaired) is perhaps the best norm to which the affected joint or body segment should be restored in rehabilitation.

Summary of Normal Range of Motion (ROM) of Major Joints

Shoulder: *Flexion* 0°–180°
Extension 0°, *hyperextension* 0°–45°
Abduction 0°–180°
Medial rotation 0°–90°, *lateral rotation* 0°–90°
Elbow: *Flexion* 0°–145°
Extension 0°
Forearm: *Pronation* from midposition 0°–80°
Supination from midposition 0°–80°
Wrist: *Flexion* 0°–80°
Extension 0°–70°
Radial flexion 0°–120°
Ulnar flexion 0°–30°
Hip: *Flexion* 0°–120°
Extension 0°, *hyperextension* 0°–20°
Abduction 0°–45°
Medial rotation 35°, *lateral rotation* 45°

Knee: *Flexion 0°–135°*
Extension 0°
Ankle: *Dorsiflexion 0°–20°*
Plantarflexion 0°–45°
Foot: *Inversion 0°–30°*
Eversion 0°–15°

MUSCULAR SYSTEM

The muscles of the body are the machines by which chemically stored energy is converted into mechanical work. **According to the location and distribution, the muscles of the body are classified into three types; (i) Smooth muscles, (ii) Cardiac muscles, and (iii) Skeletal muscles.** All three types of muscles are similar in certain characteristics, however each of these three types has some distinguishing characteristics of structure, innervation, and function. On the basis of presence of cross-striation, the muscles are classified as *striated*, and *nonstriated* (plain or smooth) muscles. Similarly, upon the basis of nature of control, the muscles are classified as *voluntary* and *involuntary* muscles. The skeletal muscles are both striated and voluntary. Cardiac muscles are striated but involuntary, and the smooth or visceral muscles are non-striated and involuntary.

Smooth Muscles

The smooth muscles form the walls of the hollow viscera, such as the stomach and bladder, and of various systems of tubes, such as are found in the circulatory system, the alimentary tract, the respiratory system, and the reproductive organs. These muscle-cells possess myofibrils, but they do not have cross-striations. Therefore, these muscles are nonstriated, and have only one nucleus. These muscles are also known as involuntary muscles since the motions caused by them are not under the control of our will.

Cardiac Muscles

Cardiac muscles form the heart which is the most vital organ of the body. The cardiac muscles display structural and functional resemblances to both skeletal and smooth muscles. The cardiac muscle cells are transversely striated and are involuntary. The most characteristic features of the cardiac muscle fibers are that they give off branches which furnish a means of communication between adjacent fibers for the conduction of the impulse for contraction. Therefore, the cardiac muscle is often considered a *functional syncytium* as its whole muscle tissue acts electrically as though it were a single cell (though actually the muscle cells are separated structurally).

Skeletal Muscles

The skeletal muscles are named because of being attached to the skeleton. These are also called as *striated*; and *voluntary* muscles (as the movements produced by them are under our conscious control). The principal functions of these muscles are the body movements and maintenance of posture. Since this type of muscles are directly responsible for motor activity and movement, this is the type with which we in kinesiology are mostly concerned.

Functions of the Muscular Tissues

The specialized function of the muscular tissue is to contract and thereby produce motion. The functions of the muscular tissues are summarized below:
- The muscles cause movement of the bones at the joints
    - Locomotion of the body from one place to another (ambulation)
    - Changes in position of the body (i.e. standing, lying down, moving of eyeballs, tongue and extremities etc.)
    - Development of muscular skills involved in our activities and in our speech.
- The muscles help in maintaining the posture of the body, the formation of walls of body cavities and the support of the organs within cavities.
- The muscles contribute to production of body heat and fluid balance to the whole body.
- The muscles protect the blood vessels and assist in maintaining the circulation of the body by forceful contraction of the heart.
- The muscles through respiration contribute to providing oxygen and eliminating carbon dioxide maintaining the vital acid-base balance of the body.
- The muscles are agents of the brain by which we maintain our independence and give expression to our inner thoughts and feelings.
- The muscles also assist in reacting to the threats of danger to our well-being.

Skeletal Muscles

As described above, the skeletal muscles are responsible for movement of the body and all of its joints. Muscle contraction produces the force that causes joint movement in the human body. There are approximately 434 voluntary muscles in the human body, making up 40–45% of the body weight of most adults. These muscles are distributed in about 217 pairs on the right and left sides of the body. However, only about 75 pairs are involved in the general posture and movement of the body, which concern the kinesiology most. Other than 75 pairs, the other muscles are smaller and are concerned with such minute mechanisms as those controlling the voice, facial expression, and the act of swallowing.

These pairs of muscles usually work in co-operation with each other to perform opposite actions at the joints they cross. In most cases, muscles work in groups rather than independently to achieve a given joint motion.

Skeletal muscles most usually have two parts;
1. **Muscle belly,** the soft fleshy, thick and the central part of the muscle containing predominantly the contractile cells, and
2. **Tendon,** composed of densely packed nonelastic white fibrous tissue. The tendons are strong yet flexible structures. At the end of a muscle belly, the contractile cells disappear but the connective tissue (i.e., epimysium and perimysium) of the muscle belly continue as tendon in order to attach the muscles to the bones, cartilage or fascia.

Properties or Functional Characteristics of Skeletal Muscles

The human skeletal muscle tissue possesses the remarkable four properties related to its ability to produce force and movement about joints. These properties or characteristics of skeletal muscles are, (i) *irritability* (or excitability), (ii) *contractility*, (iii) *extensibility*, and (iv) *elasticity*. No other tissue in the body has all of these properties or characteristics.

To better understand these properties, it will be helpful to know that muscles have a *normal resting length* which is defined as the length of a muscle when it is unstimulated, that is, when there are no forces or stresses placed upon it.

- **Irritability** refers to the ability of a muscle to respond or react to a stimulus. In other words it refers to the muscle property of being sensitive or responsive to chemical, electrical or mechanical stimuli. This property enables a muscle to receive stimuli and then respond to them.
- **Contractility** refers to the ability of muscle to contract and develop tension or internal force against resistance when stimulated. When a muscle receives an adequate stimulation, it contracts and changes its shape, and may become shorter, stays the same length or lengthen. This property is very unique and highly developed in the muscle-tissue only.
- **Extensibility** refers to the ability of a muscle of being stretched or extended in length when a force is applied to it.
- **Elasticity** refers to the ability of muscle to readily return to its original or normal resting length after the stretching or shortening force is removed.

The two properties of muscle, i.e. extensibility and elasticity enable a muscle to be stretched like an elastic band and, when the stretching force is removed or discontinued, to return again to its normal resting length. Tendons which are simply continuations of the muscle's connective tissue, also possess these two properties. The unique properly of contractility is possessed by muscle tissue alone. *Researchers have found out that the average muscle fiber can shorten to approximately one-half its resting length, and can be stretched to about one and one-half times its resting length.* The ability of a muscle for its elongation varies proportionately with the length of the muscle fiber, and inversely with its cross section.

The properties of a muscle are summarized as follows; stretch a muscle, and it will lengthen (*extensibility*). Remove the stretch, and it will return to its normal resting position (*elasticity*). Stimulate a muscle, and it will respond (*irritability*) by shortening (*contractility*), then remove the stimulus and it will return to its normal resting position (*elasticity*).

Gross and Microscopic Structure of a Skeletal Muscle

Muscle tissue is composed of cells, and intercellular substance, which consists of small amount of cement that holds the cells together. A single skeletal muscle cell is termed a *muscle fiber* because of its thread-like shape which is specially adopted for the contractile function. **A muscle fiber is also termed the anatomical or structural unit of the skeletal muscle.** Each skeletal muscle consists of many thousands or lacs of muscle fibers. The skeletal muscle fibers are multinucleated, cylindrical, elongated structures, each about one to several inches in length, and about 0.01 mm (10 μ) to 0.1 mm (100 μ) in width, and having a clear display of longitudinal and cross-striations.

Each muscle fiber is enclosed within a thin and delicate connective tissue sheath called *endomysium*. The muscle fibers are bound into bundles within bundles. Each individual bundle called a "*fasciculus*", contains about 100-150 muscle fibers and is enclosed in a fibrous tissue sheath called *perimysium*; and there are about 20-100 bundles or the fasciculi that constitute a complete muscle, which is encased within a tougher sheath of connective tissue called *epimysium*. In other words, this epimysium is the outer most covering for each whole skeletal muscle. As described earlier, at the end of the soft, fleshy central belly of a muscle (which contains the contractile cells predominantly), the connective tissue (the perimysium and epimysium) continue as tendon in order to attach the muscle to the bone. The muscles vary in length from about few cms. to many inches. The gross structure of skeletal muscle showing series of connective tissue membranes is shown in **Figure 2.23**.

About 20% of the weight of a muscle fiber is represented by proteins; the rest is water, plus a small amount of salts and other substances utilized in metabolism.

The connective tissue content of different muscles varies widely. The proportion of connective tissue is highest in the muscles responsible for the fine and precise movements. A blood capillary is also seen near each muscle fiber alongwith the endomysium. Around each muscle fiber is an electrically polarized membrane the inside of which is about a 10th of a volt negative with respect to the outside.

The individual muscle fibers show variation in length, width, number of nuclei, and relative amount of myofibrils

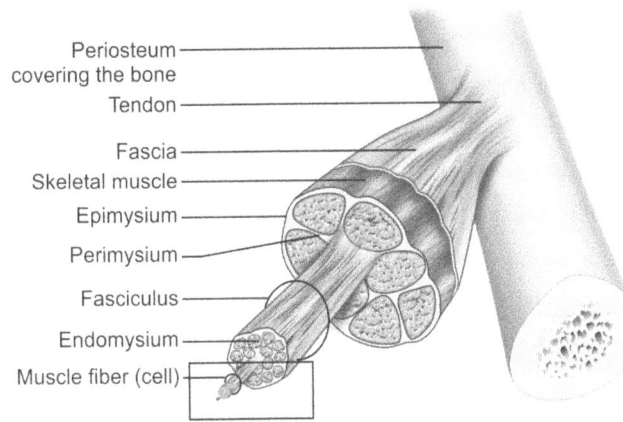

Fig. 2.23: Gross structure of skeletal muscle showing series of connective tissue membranes

and sarcoplasm present in them. According to one estimate there are about 270 million striated muscle fibers in the body. The biceps brachii muscle of the arm is reported to have some 6 lacs muscle fibers. The muscle fibers are of various sizes in the same muscle. The extent of muscle fibers in the muscle-belly may be from one end to the other, or from one end to somewhere at midway, or both the ends within the muscle may not have any attachment on either side.

The thickness of muscle fibers also varies with the degree of nourishment of the individual muscle fibers. The growth of muscle on systematic exercise (hypertrophy) is due to increase in total fiber volume due to increased blood supply, *and not due to increase of muscle fiber number.* It is generally believed that the total number of muscle fibers in a muscle is not increased due to the exercise or training, however some of the researches now show that greatly enlarged muscle fibers can split down the middle along their entire length to form entirely new fibers, thus increasing the numbers of muscle fibers as well.

The muscle fibers are innervated by the cranial or spinal nerves and are under voluntary control. The skeletal muscle contains both pain endings and proprioceptors.

Microscopic examination has revealed that each muscle fiber contains several hundreds myofibrils (1–2 μ or micron diameter), which lie parallel to one another (lengthwise) through the entire fiber **(Fig. 2.24)**. The fluid material called 'sarcoplasm' surrounds the myofibrils. Both the myofibrils and sarcoplasm are enclosed by the *sarcolemma*. The endomysium is the covering on the outside of the sarcolemma. *The myofibrils, which are the contractile elements, are arranged in parallel formation within the muscle fiber and are made up of alternating dark and light bands which give the muscle fibers their striated appearance.* These 'dark' and 'light' bands are called 'A' or 'Anisotropic', and 'I' or 'Isotropic' band respectively.

The electron microscope has revealed the striations to be a repeating pattern of bands and lines due to an interdigitating arrangement of two sets of filaments. *Each myofibril is composed of some 400 to 2,000 tiny filaments arranged parallel to the length of myofibril, and each consists of light and dark bands.* In the center of each 'A' band is found a slightly less refractile region called 'H' band. Similarly, in the center of each *I band* is found a narrow line of highly refractile material which looks dark and named as 'Z line'. Another line, the *M band* lies transversely across the middle of the H band which interconnects adjacent myosin filaments. The portion enclosed by two adjacent Z lines of a myofibril is named as *sarcomere* and considered to be the functional contractile unit of skeletal muscle. The sarcomere extends 2–3 μ in length. *Thus, each myofibril is divided into a series of several sarcomeres.*

The fine structure of a muscle fiber also reveals that myofibrils contain two main filaments of contractile proteins, the *actin* and *myosin*, which when stimulated, slide past each other. The myosin filaments are *thicker* ones (about 100 Å in diameter) and are only confined to the A-band. The length of myosin filament is about 1.5 μ (15000 Å). The actin filaments are *thinner* ones with about 50 Å diameter and stretch from Z line through the I band into the A band, and terminating at H line. The length of actin filament, according to some researchers is about 1 μ.

- When a muscle actively contracts or shortens, the 'Z lines' are drawn in toward the A bands. No change occurs in the width of the A bands, but the I bands narrow. As a result, the A bands come closer together. The H band is obliterated in full muscle contraction, but reappears as an area of lesser density when the length of the muscle increases. If the muscle goes into a state of rigor, forcible attempts to stretch it will result in a tearing of the filaments, usually in the I bands. *As such, according to the theory of sliding filament mechanism, during muscle contraction, the I band and H band diminish in length but the A band remains constant.* As also the relative position of the myofilaments alters during muscular contraction, but neither the actin nor the myosin filaments are shortened themselves.

- During contraction, the actin filaments slide past the myosin filaments and thereby the actin filaments are further extended into the A band causing shortening of the length of the H band and narrowing the sarcomere. In this process, the myosin filament gradually approaches the Z line and the actin filament. At certain stages of contraction, the ends of two adjacent actin filaments may touch each other and the I band is of minimum length.

Protein Constituents of Muscle

A skeletal muscle contains about 80% water, and a great majority of its dry weight consists of proteins. The muscle proteins *actin* and *myosin*, associated with the contractile

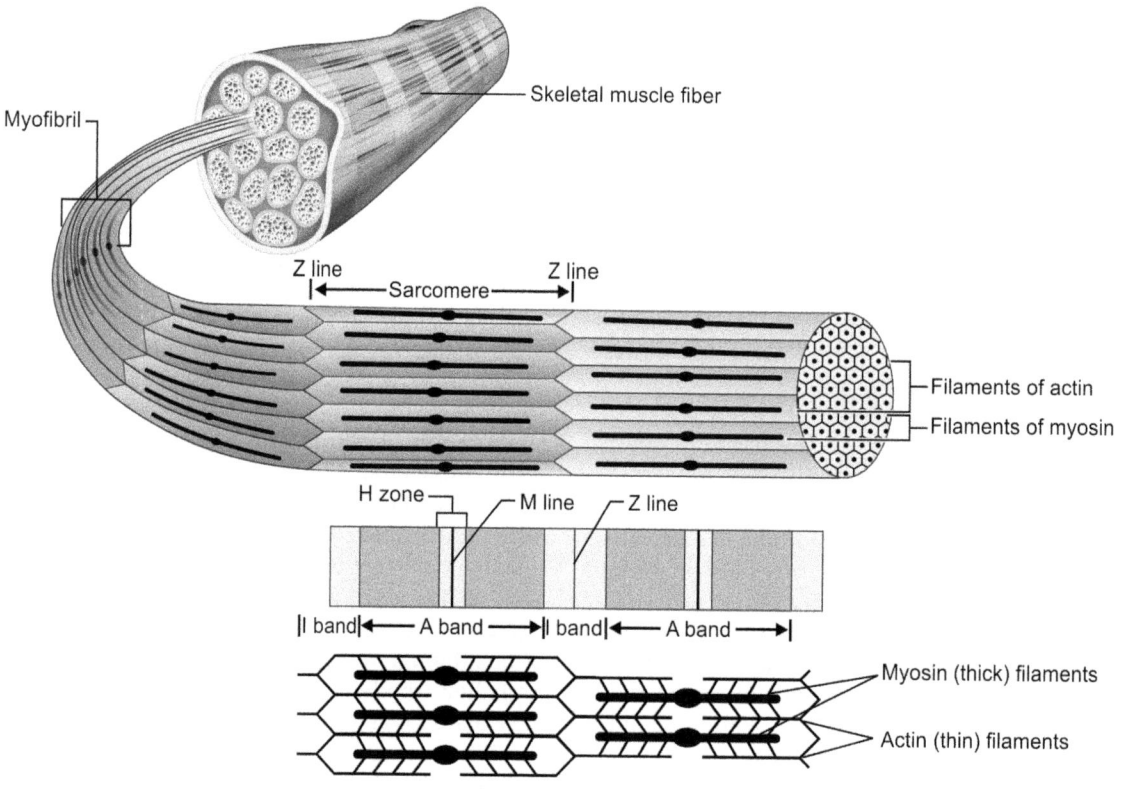

Fig. 2.24: Microscopic structure of a single muscle fiber

mechanism comprise about 60% of this and the remaining 40% of muscle proteins is shared equally by protein-enzymes (as in other cells) and by stroma proteins which help in holding the remaining structure in place.

Slow Twitch (ST) and Fast Twitch (FT) Muscle Fibers

During the first few weeks of life, mammalian skeletal muscle including that of humans develops into two different fiber types:
1. *Slow twitch (ST), red or tonic fibers,*
2. *Fast twitch (FT), white or phasic fibers.*

Most skeletal muscles contain some of each fibers type, but the proportions vary among both different muscles and individuals. The slow twitch (ST) fibers are small and red. Their red appearance is due to the fact that they contain larger amounts of myoglobin, (the muscle hemoglobin), granular material, sarcoplasm and mitochondria per unit area than do white muscle fibers. The myoglobin acts in the transport of oxygen from the blood vessels (capillaries) in the extracellular spaces to the sites of oxidation (mitochondria).

The slow twitch, red fibers depend primarily on oxidative metabolism and probably on fat for fuel. *These fibers tend to be more plentiful in the muscles that are responsible for low tension activities, such as the leg muscles of endurance runners or cyclists.* These fibers are more efficient in maintaining isometric activity, are slower in their contractile action and undergo fatigue less rapidly than do the fast twitch, white fibers. These are well adapted for static, longer and postural contractions and are found more in antigravity muscles. *Since they are slow fatiguing fibers, they are important in any activity requiring endurance.*

The fast twitch (FT), white fibers on the other hand, have larger diameter (about two times larger than slow twitch, red fibers) and therefore occupy a larger proportion of a muscle than their number alone would indicate. These muscle fibers are relatively deficient in myoglobin and depend primarily on the anaerobic glycolytic mechanism. *The fast twitch fibers fatigue more easily, but are better adapted for fast, powerful, rapid, contractions than do the slow twitch, red fibers.* It takes fast twitch fibers only about one-seventh the time required by ST fibers to reach peak tension **(Fig. 2.25)**. These fibers tend to predominate in flexor muscles, used for heavy-strength activities. *They are more evident in the muscles of athletes who engage in high power, short-endurance events.* The fast twitch, white fibers help phasic contractions by which changes in the position of the body or a limb are done.

The proportion of slow twitch (red), and fast twitch (white) fibers must be an important factor in the capacity of a muscle to maintain sustained contractions. *In man, most skeletal muscles contain both FT and ST fibers, in varying*

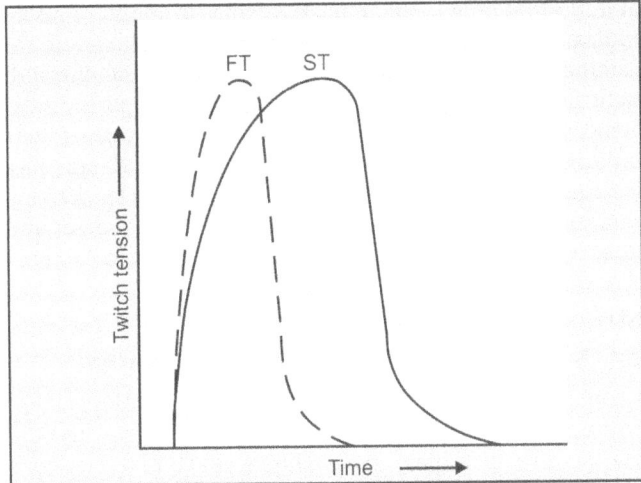

Fig. 2.25: Fast twitch (FT) fibers both reach peak tension and relax more quickly than slow twitch (ST) fibers

amounts in different muscles and individuals. Some persons have considerably more fast twitch, white fibers than slow twitch, red fibers, and others have more slow twitch, red fibers than the fast twitch white fibers.

- Soleus muscle contains predominately the slow twitch, red fibers and is said to be the muscle used to a greater extent for prolonged lower leg muscle activity.
- The gastrocnemius muscle contains predominately the fast twitch, white fibers, which gives it the capability of very forceful and rapid contraction of the type used in jumping. Extensor digitorum longus and semitendinosus muscles also contain fast twitch, white fibers predominantly.
- Whether training or heredity is responsible for the differences in proportion of slow and fast twitch muscle fibers, has not yet been well-established. A world class marathon runner may be such because of being blessed genetically with a large proportion of slow twitch fibers. On the other hand, the runner's success may be due to a training regime that either converted the runner's innate characteristics, or actually caused a modification in the proportion of slow twitch (red), and fast twitch (white) fibers.
- The leg muscles of world class endurance runners contain 75% to 90% slow twitch, red fibers, whereas champion sprinters may have 80% to 90% fast twitch, white fibers. The muscles of the weight lifters appear to be about equally divided between the two types. The weight lifters have significantly larger white fibers than do endurance athletes or untrained men.
- *Individuals genetically endowed with a high percentage of FT, white fibers may choose sports requiring strength, and those with a high percentage of ST fibers may choose endurance sports.*

- Though, athletic training has not been shown to change the relative proportions of fast twitch and slow twitch fibers, however, the fast twitch white fibers have been shown by some to convert into slow twitch, red fibers by training. However, the change of slow twitch, red fibers to fast twitch, white fibers have been less consistent.

Muscular Attachments

Muscles are attached to bone by means of their connective tissue, which continues beyond the muscle belly in the form either of a tendon (a round cord or flat band), or of an aponeurosis (a fibrous sheet). Since muscles are attached to bones and cross at least one joint, so when a muscle contracts, one end of the joint moves toward the other. It is customary to designate the attachments of the two ends of a muscle as '*origin*' and '*insertion*'. The **origin** is usually characterized by stability and closeness of the muscle fibers to the bone. The origin is usually the *more proximal* of the two attachments, over the more stable bone. It also tends to be closer to the trunk. The **insertion**, on the other hand, is usually the distal attachment over the more movable bone. It frequently involves a relatively long tendon, which most usually moves the distal bone to which the muscle's tendon is attached or inserted, and toward the proximal and more stable bone.

For example, when the biceps brachii muscle contracts, the forearm moves toward the humerus, as when bringing a glass toward the mouth. The proximal bone humerus is more stable because it is attached to the axial skeleton at the shoulder joint. The forearm is more movable because it is attached to the hand, which is quite movable. Therefore, the insertion is moving toward the origin, or the more movable end is moving toward the more stable end.

If the site of bony attachment is distant from the belly of the muscle, the extensions of the connective tissue sheaths merge to form either a round cord like *tendon*, or a flat *aponeurosis*. The fibers of the tendon or aponeurosis are folded or interlaced with one another, so that the tension of the muscle is distributed more or less equally to all parts of the attachment to the bone. Since a tendon collects and transmits forces from many different muscle cells onto a small area of bone, the site of the tendinous attachment is normally marked by a rough tubercle on the bone. Likewise, an aponeurosis gives rise to a skeletal line or ridge at its attachment.

The fleshy fibers of some muscles do not give way to tendons at their attachment, but continue almost to the bone, where the individual sheaths of the contractile muscle fibers make the attachment over an area as large as the cross-section of the muscle belly. In these cases, the skeleton will be smooth, as on the surface of the scapula because tensile forces are widely distributed.

- It needs to be understood that the muscle does not pull in one direction or the other. When a muscle contracts, it exerts equal force on both of its attachments and attempts to pull them toward each other. Which bone is to remain stationary and which one is to move depends upon the purpose of the movement. For example, a muscle extending the anterior of a hinge joint tends to draw the two bones toward one another. However, most precision movements require that the proximal bone be stabilized while the distal bone performs the movement.

- It is also emphasized here that a muscle most usually can not pull in a predetermined direction. Actually there are many movements in which the insertion or distal attachment of the muscle remains stationary and the origin or the proximal attachment moves. For example, in the act of chinning oneself with hands holding onto a chinning bar, the movement is of the elbow flexion produced by the biceps brachii, but it is the upper arm that moves toward the forearm, just the reverse of what happens when one lifts a book from the table. The grasp of the hands on the bar during chinning up serves to immobilize the forearm and thus it provides a stable base for the contracting elbow flexors. Since the origin is moving toward the insertion and the proximal bone humerus which is usually more stable has become more movable in this case, many experts refer to this as '*reversal of muscle action*'.

Nomenclature of the Muscles (Muscle Names)

The name of a muscle often tells us a great deal about that muscle. Muscle names tend to fall into one or more of the following categories:
- Location
- Shape
- Action
- Number of heads or divisions
- Attachments = Origin/Insertion
- Direction of the fibers, and
- Size of the muscle.

The tibialis anterior, as its name indicates is located on the anterior surface of the tibia. The rectus (straight) abdominis muscle is a vertical muscle located on the abdomen. Likewise supraspinatus, and intercostal muscles derive the name from their location. The trapezius muscle has a trapezoid shape (quadrilateral of which no two sides are parallel), and the serratus anterior muscle has a serrated (saw) or jagged like attachment anteriorly. Some other muscles indicating the name to their shape are rhomboids, deltoid, etc. The name of the extensor carpi ulnaris muscle tells us that its action is to extend the wrist (carpi) on the ulnar side.

Similarly, the levator scapulae, erector spinae, supinator, and extensor digiti minimii muscles are named indicating their action. The triceps brachii muscle is a three-headed muscle on the arm, and the biceps femoris muscle is a two-headed muscle on the thigh, and similarly, the quadriceps femoris, triceps surae, biceps brachii etc. are named. The sternocleidomastoid muscle attaches on the sternum, clavicle and mastoid bones. The brachioradialias is another muscle indicating the bones to which it is attached. The names of external and internal oblique muscles describe the direction of the muscle fibers and their location to one another. In the same way, the names pectoralis major and pectoralis minor indicate that although these muscles are in the same area, one is larger than the other. Similarly, the other muscles deriving their names as according to their size are, teres major and teres minor; gluteus maximus and gluteus minimus.

Structural Classification of Muscles on the basis of Muscle Fibers Arrangement

Various muscles have different shapes, and the arrangement and orientation of muscle fibers within a muscle, and the manner by which these muscle fibers attach to the tendons vary considerably among the skeletal muscles of the human body. These structural variations form the basis for a classification of the skeletal muscles. The arrangement of muscle fibers determine a role in the muscle's ability to exert force (the strength of muscular contraction) and the range of motion that the muscle can move a body segment.

Essentially, all skeletal muscles may be grouped into two major types of muscle fibers arrangement, (i) parallel, and (ii) pennate; and each type may be further subdivided into several categories according to the shape. In a parallel fiber arrangement, the muscle fibers are arranged in parallel to the length of the muscle (longitudinal axis of the muscle). *Generally, parallel muscles will produce a greater range of movement than similar sized muscles with a pennate arrangement of muscle fibers.*

Parallel muscles are categorized into the following shapes;

- **Longitudinal:** This is a long strap like muscle whose fibers lie parallel to its long axis, running the whole length of the muscle. Such muscles are more uniform in diameter **(Fig. 2.26A)**. This arrangement enables a focusing of power onto small, bony targets. The sartorius muscle which slants across the front of the thigh is an excellent example of this type.
- **Quadrate or Quadrilateral:** Muscles of this type are usually thin, flat, and are four sided. They consist of parallel fibers which originate from broad, fibrous, sheet-like aponeuroses that allow them to spread their forces over a broader area **(Fig. 2.26B)**. The examples include rhomboid (muscle between the spine and the scapula), pronator quadratus (on the front of the wrist). The other examples are rectus abdominis and external abdominal oblique muscle.
- **Fusiform or Spindle Shaped:** These are usually a rounded type of muscles with a central belly which tapers to tendons on either end **(Fig. 2.26C)**. The muscles of this type may be long or short, large or small. This fiber arrangement allows

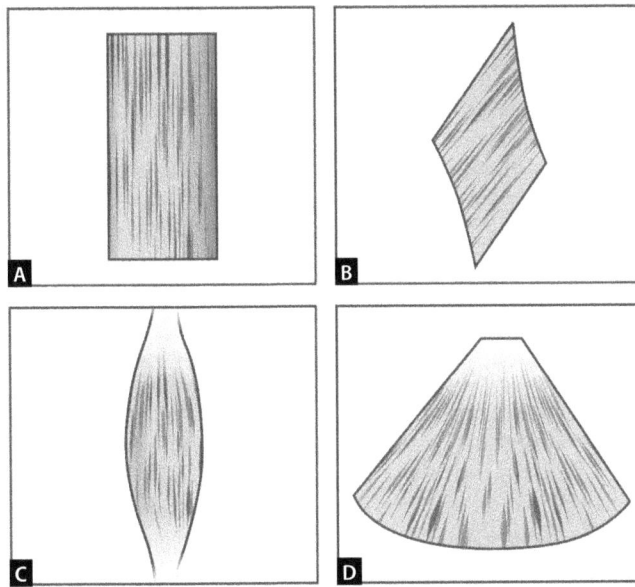

Figs. 2.26A to D: Parallel muscle—fibers arrangements. (A) Longitudinal; (B) Quadrate or quadrilateral; (C) Fusiform (spindle); (D) Triangular (fan-shaped)

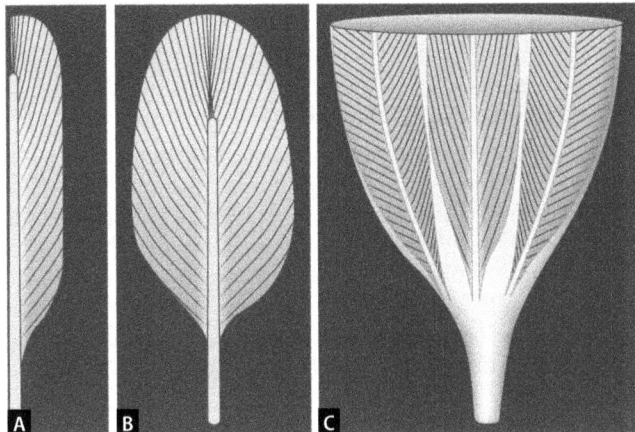

Figs. 2.27A to C: Pennate muscle fibers arrangements. (A) Unipennate; (B) Bipennate; (c) Multipennate

the muscles to focus their power onto small, bony targets. Examples are the brachialis, biceps brachii, and brachioradialis muscles of the upper extremity.

- **Triangular or Fan-shaped:** These are also a relatively flat type of muscles whose fibers originate from broad aponeuroses and converge onto a narrow tendon **(Fig. 2.26D)**. In other words, the muscle fibers at one end have a broad attachment and on the other end a narrow attachment. A good example of this type is pectoralis major muscle on the front of the chest. The trapezius is another example.

In the *pennate* muscles, the muscle fibers are shorter and arranged obliquely or diagonally to attach to a central tendon at an angle to the muscle's longitudinal axis, in a structure similar to that of a feather. Each fiber in a pennate muscle attaches to one or more tendons some of which extend the entire length of the muscle. This arrangement increases the cross-sectional area of the muscle, thereby increasing the power, however these muscles tire quickly.

Pennate muscles are categorized as following based upon the exact arrangement between the fibers and the tendon.

- **Unipennate:** In this type of muscle, a series of short parallel, feather like muscle fibers runs obliquely or diagonally from a long tendon on one side only, giving the muscle as a whole the appearance of a wing feather **(Fig. 2.27A)**. Examples are seen in the biceps femoris, extensor digitorum longus, and tibialis posterior muscles of the leg.
- **Bipennate:** This is double penniform muscle in which the muscle fibers run obliquely or diagonally in pairs on both sides of a long central tendon **(Fig. 2.27B)**. It resembles a symmetrical tail feather. Examples are rectus femoris and flexor hallucis longus muscles of thigh and leg respectively.
- **Multipennate:** These muscles have several tendons with the muscle fibers running diagonally between them **(Fig. 2.27C)**. The middle portion of the deltoid muscle is a prime example of this type of muscle.

Effect of Muscle Structure on Force and Range of Motion

The force a muscle can exert is proportional to its total physiological cross-section which is a measure that accounts for the diameter of every muscle fiber and whose size depends upon the number and thickness of the fibers. **Keeping all other factors constant, a muscle with a greater cross-section diameter will be able to exert a greater force**. *A broad thick, longitudinal muscle will exert more force than a thin muscle. However, in a pennate muscle with oblique fiber arrangement, the total cross-sectional area of the muscle is much larger than that of a parallel muscle and as such, the force or power produced by the muscle is significantly greater. Pennate muscles are the most common type of skeletal muscle and predominate when forceful movements are needed.*

The range through which a muscle shortens to move a joint depends upon the length of its fibers. *Experimentally it has been shown that an average muscle fiber can shorten to about one-half of its resting length, and can be stretched until it is about one and half times of its original length. Generally, the longer muscles with parallel, longitudinally arranged muscle fibers such as sartorius can shorten and exert force through a greater range and are more effective in moving joints through large ranges of motion than the pennate muscles having the shorter muscle fibers producing a smaller ranges of motion.* The pennate muscles, with their oblique fiber arrangement and short fiber length, can exert greater force through only a short range of motion.

Types of Muscle Contraction

The word 'contract' literally means to 'draw together' or to shorten. As such, the term muscular contraction appears to create some confusion initially. A muscle contraction occurs 'whenever there is development of tension within the muscle', or 'the muscle fibers generate tension in themselves'. As such, a muscle contraction may not necessarily change the length, rather it may shorten, or lengthen a muscle depending upon the nature of the work demanded during the muscle contraction.

Muscle contractions are mostly classified as:
- *Isometric*, or *Static contraction*, and
- *Isotonic contraction*.

A third type of muscle contraction, i.e. *isokinetic* contraction is a less common type of muscle contraction. This is referred by some experts as a type of exercise rather than a muscle contraction.

Isometric or Static Contraction

The Greek term 'isometric' composed of two parts, 'iso' (means same) and 'metric' (means length) altogether means '*same length*'. As such, an isometric contraction occurs when tension is developed within the muscle, without any appreciable change in the length of the muscle. The muscle may contract partially or completely producing force within it, but the joint angle remains the same. Therefore, an isometric contraction is often referred to as a static contraction as a significant amount of tension developed in the muscle maintains the joint angle, more or less in a static or stable position.

An isometric type of muscle contraction is likely to occur under two different conditions;
1. The muscles that are antagonistic to each other contract with equal strength, thus balancing or counteracting each other. The body part is held tensely in place without moving. Tensing the biceps brachii to show of its bulge is an example of this. The contraction of triceps brachii prevents the elbow from further flexing.
2. A muscle is held in either partial or maximal contraction against another force such as the pull of gravity, or an external mechanical or muscular force. Examples of this are holding a book with outstretched arm, a tug of war between two equally stronger opponents, and attempting to move a too heavy object.
    - The force or tension developed in the muscle is either insufficient or not desired to move the body part against a given resistance. The force developed by the muscle either can not move the given resistance or is equal to the given resistance. And, since the muscle does not move at all, technically no work is accomplished by it. The energy which would normally be displaced as mechanical work is dissipated as heat.
    - *According to some kinesiologists, actually no muscle contraction is perfectly isometric or static, as even under the most rigid laboratory conditions, the contractile elements within a muscle shorten by about 3% of their length by stretching the elastic components.*
    - Posture is largely maintained by isometric contractions of certain muscles of the back and the legs, where muscular tension is required to counter-balance the effects of gravity upon the body.
    - *According to an estimation, on the average, the isometric strength of the muscles of the upper extremities of women is about 55% that of men; that of the muscles of the lower extremities, about 70%, and that of the trunk muscles, about 66% of men.*

Isotonic Contraction

The term 'isotonic' originating from the Greek means '*same tone or tension*'. Isotonic contraction is a type of contraction in which the tension developed in the muscle remains the same, but with definite change in the length of the muscle or the joint angle. Due to causing change in the length of the muscle or the joint angle, isotonic contractions are sometimes also referred by some as dynamic contractions. The force developed by the muscles is either greater or lesser than a given resistance and these types of muscle contractions either produce or control the active joint movements.

During an isotonic contraction, the muscle either shortens or lengthens. The 'shortening' and 'lengthening' type of isotonic contractions are usually referred to as 'concentric' and 'eccentric' contractions respectively, though technically these terms do not have the same meanings.

The terms 'concentric' and 'eccentric' do not indicate the degree of tension, rather they merely indicate a decrease or increase in length of the muscle. Similarly, *the use of term 'isotonic' is also not without criticism because it is felt that the tension developed within a muscle does not remain the same throughout its range.*

A described above, the isotonic type of muscle contraction is further classified into;
- *Concentric*, and
- *Eccentric contraction*.

Concentric

A concentric contraction involves development of tension as a muscle shortens. The attachments of the muscle (origin and insertion) move closer toward each other. The muscle develops enough force to overcome the applied resistance and there is a joint movement in the direction of applied muscle force. As such, in the concentric contraction, force applied by the muscle is greater than the resistance and the body part moves against the gravity or the external force. Lifting a weight held in the hand, while flexing the elbow to bring the weight up toward the shoulder is an example of a concentric contraction of the biceps brachii muscle.

The concentric contractions are also referred as *'positive'* contractions as the work is being done in these type of contractions.

Eccentric

An eccentric contraction involves lengthening of the muscle under tension while the attachments of the muscle (origin and insertion) are drawing apart. When either the given weight or the resistance overcomes the force developed by the muscle, or when a controlled movement is desired, the muscle gradually lessens in tension. The movement of the body part occurs in the direction of the gravity or the external weight or the resistance. It is however emphasized that though the weight or resistance is overcoming the force of muscle contraction, however, it is not to the point that the muscle cannot control the descending movement. Lowering the weight held in the hand downward is an example of eccentric contraction. Though the joint motion is elbow extension, and it appears that the triceps brachii is contracting, actually the biceps brachii is contracting eccentrically to lower the weight.

Eccentric contractions are often referred to as *'negative'* contractions as the antagonists are losing energy and the movements of the gravity or the resistance are being controlled by them.

In the slow, controlled movements toward the gravity, the flexor muscle groups, in the upper limb, and the extensor muscle groups in the lower limb usually contract or work eccentrically. As such the effect or force of gravity should always be considered while analyzing the muscle work.

The examples of isometric, concentric and eccentric contractions are shown in **Figures 2.28A to C**.

Isokinetic Contraction

While there are some experts who consider isokinetic contraction a modification of isotonic contraction rather than a separate type of muscle contraction, on the other hand there are certain other experts who relate this more to a type of dynamic exercise rather than a type of muscle contraction.

Whether a type of muscle contraction or an exercise, 'isokinetic' term means the *'same or constant motion'*. With isokinetic contraction or exercise, which can be done only with special machines, (i.e. Cybex, Orthotron etc.) resistance to the body part varies but the speed of the movement remains the same throughout the range of motion. As such, isokinetic differs from the conventional isotonic contraction in which the resistance remains the constant but the speed of the movement varies. These isokinetic exercises are sometimes also referred to as *'accommodative resistance exercises'* in which the use of machine offers resistance proportional or matching to the muscular force applied by a person throughout the range of movement at a preset speed, during a dynamic exercise to an extremity or the trunk. As such, an isokinetic contraction allows the development of full muscular force throughout the range of motion.

To understand the value and importance of isokinetic machines, the following example should be considered. A person with 5 kgs weight attached to his leg straightens and flexes his knee (isotonic contraction). With this weight or resistance of 5 kgs remaining constant throughout the range of movement, it is easier to move the leg in the middle and at the end of the range rather than at the beginning of the movement because of the angle of pull. In other words, this 5 kgs weight is not as effective in offering maximal resistance throughout the range of motion as according to the *principle of overload*. However, with an isokinetic machine, the speed is preset which remains constant no matter how hard a person pushes, but the resistance will vary. If the person pushes harder, the machine will give more resistance, and if the person does not push as hard, there will be less resistance. **As such, the isokinetic exercises have two major advantages**:

- The amount of resistance given can be varied or altered throughout the range of movement, as according to the strength of the muscles which an isotonic contraction/exercise can not.
- The characteristic of accommodating resistance is also advantageous and important, as during an exercise, if a person suddenly develops pain, he may stop working, and so the machine will also stop working, as no resistance will now be offered by the machine. Or if the person reduces the strength of his muscle contraction, the machine also accordingly offers and adjusts resistance. No isotonic contraction or machine can offer such a response of adjusting the resistance.

Effect of Gravity on Muscle Action

The force of gravity is an important factor in the analysis of any muscle action. The gravity affects the movement to such an extent that it sometimes causes confusion to the novice students.

Movements of the body or its parts may be downward in the direction of gravitational forces, or upward opposing gravity, or horizontal perpendicular to the gravity. Therefore, it is essential to consider the direction as well as the speed of movement when analyzing the nature of muscular involvement during any movement. The muscles may be contracting, either to provide the force for a movement, or the force to resist and control the movement, or they may be completely relaxed.

The gravity is a constant force pulling vertically downward on all objects at all times. Whenever the body or one of its part is moved upward, the force must be sufficient to overcome

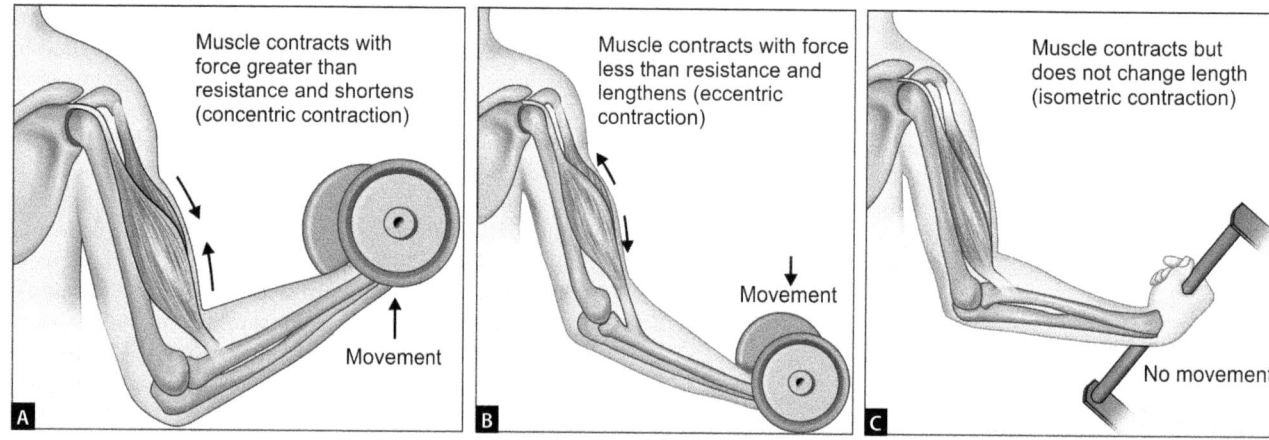

Figs. 2.28A to C: Isotonic and isometric contractions: (A) Concentric contraction; (B) Eccentric contraction; (C) Isometric contraction

the downward pull of gravity. However, on the other hand, whenever gravity can produce the desired movement, we soon learn in the interest of muscular economy to let the muscles relax and allow the gravity to do the work. For example, to perform the elbow flexion in standing position requires the concentric work of elbow flexors to overcome the downward pull of the force of gravity. And, a slow controlled elbow extension in standing position seems to be produced by the force of gravity and elbow extensors, but the elbow flexors, by working eccentrically control the downward, sudden and jerky movement. The forceful elbow extension in this position against the resistance (even toward the force of gravity), however, will require the concentric work of elbow extensors.

Brief Physiology of Muscular Contraction

The muscular contraction is the result of the nervous stimulation which enters the nervous system through any of its receptors. This stimulus, through the receptors and afferent nerve fibers is taken to the central nervous system from where the impulse is transmitted to the muscle through the efferent nerve fibers and then causing the muscle to contract. When a muscle contracts, it results in shortening and thickening without any change in its volume.

As already described, the muscle fiber is a structural unit of a muscle and it is able to contract when a stimulus of sufficient threshold reaches its neuromuscular junction or the motor end plate. A motor end plate is the junction between the terminal branch of a motor nerve (axon) to the muscle fiber, and the point or place where the changes first take place in initiation for contraction.

When an impulse reaches the motor end plate, acetylcholine is released which depolarizes the membrane-sarcolemma by increasing its permeability to the ions like Na, K, etc. (since the outer membrane bears a +ve charge, and the inner side of the membrane a –ve charge). This sets of an action potential (small bio-electrical currents between the +ve and –ve ends) which is propagated along the muscle fiber. This action potential is followed by a wave of contraction. Apart from these bioelectrical reactions, some chemical reactions also take place, all leading to the conversion of the chemical energy to the mechanical energy and work.

When changes occur in the length of a muscle fiber, the actin filaments in the I band slide past the myosin filaments in the A band, i.e. the muscle shortens but its filaments do not.

A muscle has many fibers, and they are stimulated by motor neurons. Each motor neuron may innervate many muscle fibers. Not all the muscle fibers are excited at the same time. The total force of contraction of a muscle depends on how many muscle fibers are stimulated.

In summary, the mechanism of muscle contraction may be expressed in the following scheme:

1. *Contraction*: Membrane depolarization → Ca^{++} released from sarcoplasmic reticulum → myosin ATP-ase is activated → cross-bridges formed → myosin slides along actin → tension is developed.
2. *Relaxation:* Ca^{++} pumped back into sarcoplasmic reticulum → myosin ATP-ase is depressed → cross-bridges broken → myosin is pulled back to its resting site → tension disappears.

NEUROMUSCULAR CONCEPTS

Motor Units

Muscle fibers are organized into functional groups of different sizes. These functional groups of muscle fibers are known as '*motor units*'. **A motor unit consists of a single motor neuron and all the muscle fibers innervated by that motor neuron (Fig. 2.29).** The axon of each motor neuron subdivides many times so that each individual muscle fiber is supplied with a motor end plate. Typically, there is only one motor end plate for each muscle fiber in humans.

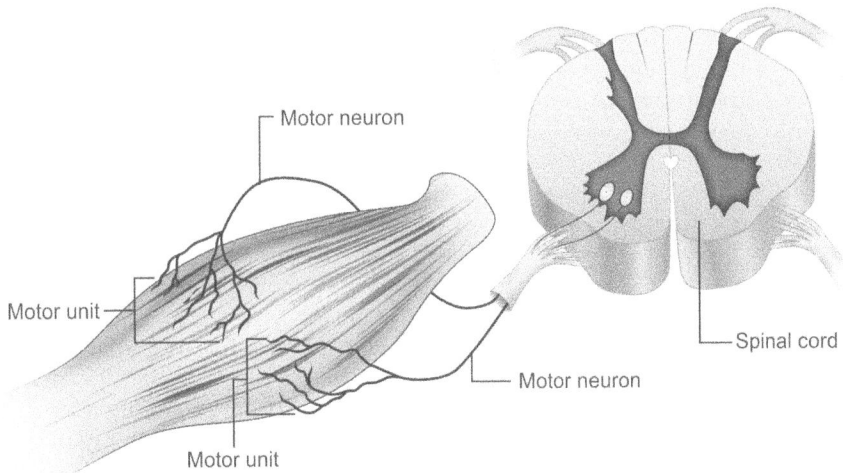

Fig. 2.29: A motor unit consisting of a single motor neuron and all the muscle fibers innervated by that neuron

The muscle fibers of a motor unit may be spread over a several centimeter area and be interspersed with the muscle fibers of other motor units. Motor units are generally confined to a single muscle and are localized within that muscle only.

A single motor unit may contain from less than 100 to nearly 2000 muscle fibers, depending on the type of movements the muscle executes. For example, the movements that are precisely controlled, such as those of the eyes or fingers, are produced by motor units with less number of muscle fibers. However, the gross, forceful movements, such as those produced by the gastrocnemius are usually produced by the activities of large motor units containing greater number of muscle fibers.

All or None Law

This law applies to the contraction of all the muscle fibers in a particular motor unit. When a particular muscle contracts, the contraction actually occurs at the muscle fiber level within a particular motor unit. **According to this law, 'regardless of any other factor, if the stimulus is of threshold value, all of the muscle fibers in a motor unit will contract maximally, or none of the muscle fibers in the motor unit will contract at all if the stimulus is below threshold value'.**

This is again emphasized that this law or principle of muscle contraction applies to the individual motor units, not to the entire muscles. When the stimulus is of threshold value, each muscle fiber in the participating motor unit will contract maximally, however depending upon the contractile state of the muscle fibers, i.e. temperature, nutrients present at that particular moment.

Gradation of Muscle Contraction

Common experience shows us that the same muscles contract with various gradations of strength according to the requirements of the task. For example, the elbow flexors are just able to contract enough to enable the hand to lift a piece of paper from the table, or be able to contract forcefully enough to lift a 10 kg briefcase. In other words, the skeletal muscles have the ability to contract very weakly or very strongly, and an ability to adjust their strength of contraction as according to the requirement of the task before hand.

In a typical muscle contraction, the number of muscle fibers contracting within the muscle may vary significantly from relatively few (as used to lift a piece of paper in the example cited above), to virtually most or all of the muscle fibers (as used to lift a briefcase). The difference between a particular muscle contracting to lift a minimal resistance, and the same muscle contracting to lift a maximal resistance lies in the number of muscle fibers recruited. **There are two major factors that determine the gradation of muscle contraction in a particular act;** (i) *number of motor units* which participate in the act, and (ii) *the frequency of stimulation* (also called the summation).

The greater the number of motor units activated, the greater will be the number of muscle fibers recruited (contained in those motor units) and, as such, stronger will be the strength of muscle contraction, other things being equal. This gradation of muscle contraction is the ability of the central nervous system to send stimuli to a greater or lesser number of motor units, as according to the requirement of a task. When greater tension is needed, more number of motor units are stimulated or recruited. (However, it has already been described under 'All or none law' that the stimulus should be of threshold value and all of the muscle fibers in the motor unit contract maximally).

If the stimuli are discharged at low frequency, the muscle fibers will partially relax between successive stimuli and a weaker muscle contraction will result. However, if the stimuli are discharged at high frequency, the muscle fibers will not have sufficient time to relax and the result is the "*summation*" or maximal muscle contraction.

If these two factors are combined, i.e., if the maximum number of muscle fibers are stimulated with the impulses being discharged at high frequency, the resulting muscle contraction is of maximal strength.

Reciprocal Innervation and Inhibition

'Reciprocal innervation and inhibition' is one of the mechanisms that provides for economical and coordinated movement. This concept was first described by '*Sherrington*', according to which **"when motor neurons are transmitting impulses to the agonist muscles causing them to contract, the motor neurons that supply their antagonist muscles are simultaneously and reciprocally inhibited"**. The antagonist muscles, therefore remain relaxed and lengthened so that the movers or the agonists contract and produce a skilled movement, without opposition. The reciprocal inhibition of antagonists results due to reflex neural activity which occurs automatically in the movements. If the excitation or innervation of the agonists is not accompanied by the corresponding reflex inhibition of the antagonists, an uncoordinated movement results. This can be also demonstrated by comparing the ease at which one can stretch his hamstrings when simultaneously contracting his quadriceps, rather than quadriceps attempting to stretch the hamstrings without the hamstrings relaxing.

However, there are certain experts who are not fully in agreement to the operation of mechanism of reciprocal innervation and inhibition in all volitional movements, as there are certain conditions, activities, or exercises when a rigidity of position or tenseness of movement is required and the muscles antagonistic to each other, (both agonists and antagonists) do contract and are required to contract simultaneously. This phenomenon is called as '*co-contraction*'. However, when the antagonistic muscles do contract, when not required to contract, results in the unskilled performance.

The antagonist muscles which remain relaxed, when the agonist muscles contract during the production of a movement, slight tension in the antagonists elicits a stretch reflex in them, which later leads to an active contraction of these antagonist muscles brought about by the process of successive induction. When these antagonists start to contract, the inhibitory process which was in operation earlier is withdrawn.

Relation of Muscle's Line of Pull to the Joint Axis

The relationship of a muscle's line of pull to the joint axis of motion is also very important in determining the movement that a contracting muscle produces at that joint. It has been observed that some of the muscles that act on triaxial joints may produce a reverse or a different action to their usual or routine function/action, as the muscle's line of pull in relation to the joint axis is reversed or changed in certain positions.

It is very well-established that the type of movements, i.e. flexion, extension, abduction, adduction, or rotation etc. that a contracting muscle produces at a joint is determined by the *type of joint* (i.e. hinge type, ball and socket type, condyloid type etc.) it acts upon, and *the relation of muscle's line of pull to the joint*. For example, contraction of a muscle, i.e. quadriceps whose line of pull is anterior in relation to the axis of knee joint will produce extension movement at the knee. On the other hand, a muscle, i.e. biceps brachii whose line of pull is also anterior to the elbow joint produces flexion at the elbow joint.

Of course, the possible axes of motion of a joint are determined by the structure of the joint itself. For example, a hinge joint (i.e. knee or elbow joint) only permits flexion-extension movements around a single axis, i.e. frontal horizontal axis, whereas a condyloid joint (i.e. wrist) being a biaxial joint, allows flexion-extension, abduction-adduction movements around two axes i.e. frontal horizontal, and a sagittal horizontal axis. The shoulder and hip are the examples of ball-and-socket type of joints that allow wide variety of movements around three axes, i.e. frontal, sagittal and vertical axis.

- A muscle whose line of pull is lateral to the hip joint is a potential abductor of the thigh, but if there are muscles whose lines of pull lie lateral to the elbow joint cannot produce abduction movement of the forearm as the elbow joint is so formed that there is no provision made for abduction-adduction movements. As described above, the elbow being a uniaxial, hinge type of joint only permits flexion-extension movements.
- The importance of the relation of muscle's line of pull in relation to joint axis of motion is especially seen in some muscles that act on triaxial joints. For example the clavicular part of the pectoralis major muscle is primarily a flexor, but also acts as an adductor of the humerus at the shoulder. However, when the arm is elevated in sideward position, slightly above the shoulder level, the line of pull of this muscle shifts from below to above, in relation to the sagittal horizontal axis of the shoulder joint, and the contraction of clavicular part of the pectoralis major muscle in this position contributes to the abduction of the humerus, rather than usually and customarily performed adduction movement of the humerus at the shoulder **(Figs. 2.30A and B)**. Similarly, several other muscles, or parts of muscles, particularly of the hip joint (i.e. adductor longus) are also reported to perform reversed function to their customary function, as their line of pull changes in relation to the joint axis of motion. The adductor longus muscle usually causes adduction movement of the femur at the hip joint, but it also flexes the hip upto 70° hip flexion. However, beyond 70° of hip flexion, the adductor

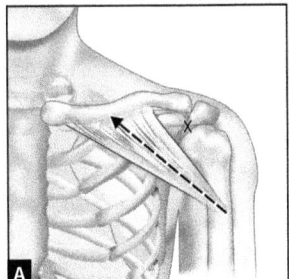 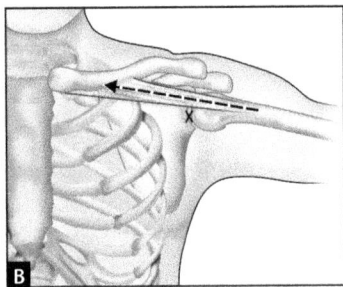

Figs. 2.30A and B: The clavicular portion of pectoralis major muscle reversing its customary function. (A) The line of pull is below the center of the shoulder joint; (B) The line of pull is above the center of the shoulder joint

longus helps to extend the femur rather than customarily flexing the femur at the hip, due to its changed line of pull in relation to the axis of hip joint.

Angle of Pull

Angle of pull of a muscle is one of the several aspects upon which the action of an applied muscular force depends. In considering muscular force, the angle between the muscle's line of pull and the axis of the bone to be moved determines the angle at which the force is applied. **The angle of pull may be defined as the angle between the line of pull of the muscle and the bone on which it is inserted.**

It may be clarified that the angle of pull is actually the angle of muscle's attachment to the bone, facing away from the joint, and not the angle on the side of the joint. With every degree of joint movement, the angle of pull changes. The pull of muscle results in the rotation of the bone which constantly changes the angle of pull. Joint movements and insertion angles involve mostly small angles of pull. The angle of pull decreases as the bone moves away from its anatomical position through the contraction of the local muscle group. This range of movement depends on the type of joint and bony structure.

The angle of pull of a muscle has a direct bearing on the effectiveness of the muscle's pulling force in moving the bony lever. *Most muscles in the body work, at a small angle of pull, generally less than 50°.*

- *When the angle of pull of a muscle is 90°* (**Fig. 2.31B**), all of the muscular force is completely rotatory force and 100% of the muscular force is contributory and effective in producing the desired movement of the bone. And, all of the muscular force is used to rotate the bony lever about its axis, i.e. the joint.
- As with any vector, a force may be resolved into a vertical and a horizontal component, and the relative size of these components depends upon the angle at which the force is applied. The vertical component of the muscular force, also referred to as the '*rotatory component*' always acts perpendicular to the long axis of the bone (lever); and it is that part of the muscular force that moves the bony lever. The horizontal component of the muscular force, also referred to as the '*non-rotatory component*' acts parallel to the bony lever, and does not contribute to the movement of the bony lever.
- *The greater the angle of pull between 0° and 90°*, the greater the vertical or rotatory component and the lesser the horizontal or non-rotatory component. The horizontal or non-rotatory component of muscular force either has a stabilizing or a dislocating effect, depending on whether the angle of pull is less than or greater than 90°.

The angle of pull of most muscles in the resting position is less than a right angle and it usually remains so throughout the movement. This means that the non-rotatory component of force is directed toward the joint axis which gives it a stabilizing effect. By pulling the bone lengthwise toward the joint, it helps to maintain the integrity of the joint. Under most circumstances, therefore, the muscular force has two simultaneous functions, i.e. movement and stabilization. The muscles perform this stabilizing function efficiently during the movements of body segments when the integrity of the joint may be threatened.

- *When the angle of pull is less than 90°* (**Fig. 2.31A**), only a portion of the muscular force is effective in producing the desired movement, as much of the muscle pull particularly, the horizontal or non-rotatory component directs the bone toward the joint axis, and acts as a stabilizing force. Thus, all of the muscular force is not used to produce the movement. The extensor muscles cross the outside of a joint, and since their pull is in line with the axis of the bone, their angle of pull is very small which exerts considerable stabilizing force.
- *Occasionally when the angle of pull is greater than 90°* (**Fig. 2.31C**), then also only a small portion of muscular force is effective in moving the bony lever and producing the desired movement, as considerable non-rotatory horizontal component of the muscular force is directed to pull the bone away from the joint axis and is therefore dislocating. *The angle of pull greater than 90° does not occur in many instances.* However, when it does occur, the mechanical efficiency of a muscle is reduced, as the muscle is in the shortened range, and the joint is less stable in this position, and some of the muscular force is used to separate the joint surfaces. When the angle of pull is closer to 90°, the rotatory component is greater.
- *When the angle of pull is 45°*, the rotatory and stabilizing components are equal. Since the angle of pull usually remains less than 45°, more of the muscle's force serves to stabilize the joint than to move the bony lever. In fact, there are some muscles whose angles of pull are always so small

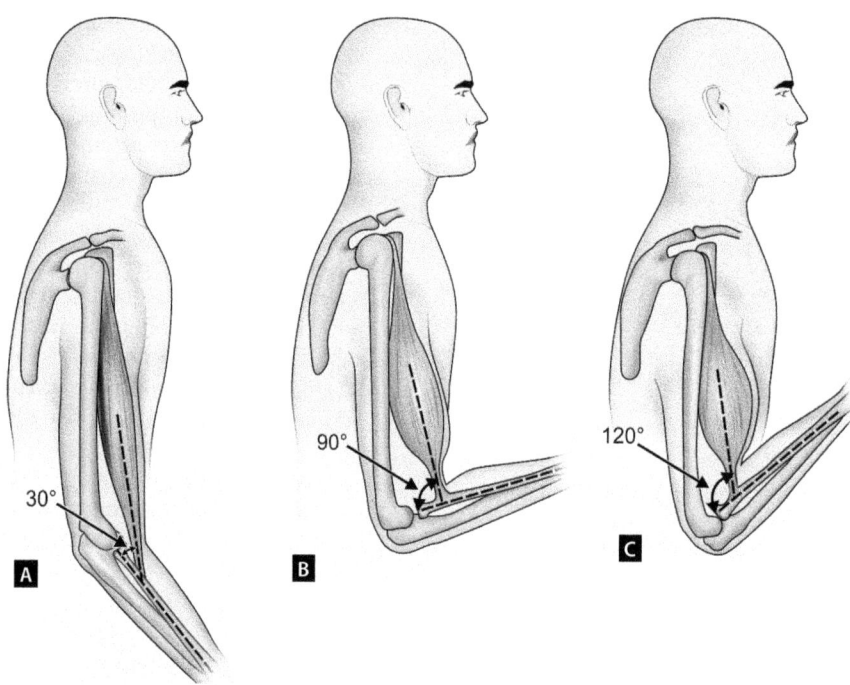

Figs. 2.31A to C: Angles of muscle pull. (A) An angle less than 45°; (B) An angle of 90°; (C) An angle greater than 90°

that their contribution to motion seems to be negligible. The examples are coracobrachialis and the subclavius muscles, which while performing stabilizing function, prevent dislocation of shoulder and sternoclavicular joints during violent, powerful movements of the upper extremities.
- In some activities, it is desirable that a person begin a movement when the angle of pull is at 90°. For example, to do a chin up (pull up) at 90° elbow flexion makes it easier because of the more advantageous angle of pull for elbow flexor muscles, and this fact can be applied when sufficient muscle-strength is lacking in an individual. However, a greater muscle-strength is required to do chin up efficiently at disadvantageous angles which are greater or lesser than 90°.

Two Joint Muscles

Mostly there are 'single joint' or 'uniarticular muscles' in the body that extend across and act at one joint only. However, there are many 'two joint' or 'biarticular muscles' in the body that cross, and act on two different joints.

The examples of two joint muscles in lower limb are rectus femoris, hamstrings, sartorius and gastrocnemius; and in the upper limb, are the biceps brachii, long head of triceps etc. Rectus femoris muscle flexes the hip and extends the knee. Hamstrings (comprising the semitendinosus, the semimembranosus, and the biceps femoris) flex the knee and extend the hip. Sartorius flexes both hip and knee joints. The gastrocnemius helps to flex the knee though its primary function is to plantarflex the ankle. Similarly, the biceps brachii is primarily a flexor of the elbow and contributes to the supination of the forearm. However, since its both long head and short head cross the shoulder joint anteriorly, it does assist in shoulder flexion. The triceps brachii muscle is mainly an elbow extensor, however since its long head crosses the shoulder joint posteriorly, it assists in adduction and extension of the shoulder joint.
- Depending on a variety of factors, a two joint muscle may contract and cause motion at either one or both of the joints it crosses. Similarly, there are 'multiarticular' muscles in the body which act on three or more joints due to the line of pull between their origin and insertion crossing multiple joints. Flexors and extensor muscles of the fingers are the examples of multijoint muscles since they pass over and act over the wrist, and at least two joints of fingers, i.e. metacarpophalangeal (MP) joint and the proximal interphalangeal (PIP) joint. The other examples of multiarticular muscles are tibialis posterior in the foot, erector spinae (longissimus dorsi and iliocostalis portions) in the back. The principles of action, as applied to the two joint muscles also apply in a similar fashion to the multiarticular muscles.
- One of the main *characteristics* of the two joint or multijoint muscles, whether they act on joints that flex in the same direction as in the case of the wrist and fingers, or in the

opposite direction, as in the case of the knee and hip (i.e. rectus femoris, or hamstrings) is that *they are not long enough to permit complete range of movement in both the joints they act upon, the same time*. This is because when a two joint muscle acts simultaneously on both the joints, it loses its tension soon. However, the loss of tension is gradual if it acts on one joint only at a time. This results in the tension of one muscle being transmitted to the other, in much the same manner that a downward pull on a rope which passes through an overhead pulley is transmitted in the form of a pull in reverse direction to the rope on the other side of the pulley. Thus, if the hamstrings contract to help extend the hip, tension in the form of a stretch is placed on the rectus femoris causing it to extend the knee. Or, if the rectus femoris contracts to flex the hip, tension in the form of a stretch is placed on the hamstrings causing them to flex the knee.

- **Two joint or biarticular muscles have two advantages over uniarticular or single joint muscles**. The two joint muscles can cause and/or control movement at more than one joint, and they may be able to maintain a relatively constant length due to 'shortening' at one joint and 'lengthening' at another joint. The muscle does not actually shorten at one joint and lengthen at the other; rather the concentric contraction or shortening of the muscle to move one joint is compensated by motion of the other joint which moves its attachment of the muscle farther away (stretching of the muscle). This maintenance of a relatively constant length results in the muscle's ability to continue its exertion of force.
- A single joint muscle usually has a sufficient excursion to allow the joint to move through the joints entire range. However, as stated above, a muscle extending over two or more joints may not have sufficient excursion to allow the joint to move through the combined range of all the joints it crosses.

One of the factors determining the amount of tension in a muscle is its length. According to a physiological principle, *a muscle loses its tension when it contracts or shortens and develops tension when it is under stretch (elongation)*. It has been demonstrated that a muscle is strongest if put on a stretch prior to its contracting. For example, prior to kicking a football, the player hyperextends his hip to put his hip flexors on a stretch before contracting them forcefully. As such, the two joint muscles have an advantage of maintaining greater contractile force through a wider range. While contracting over one joint, the two joint muscle is being elongated over the other joint. This can be further demonstrated by the action of the hamstrings muscles during stair climbing **(Figs. 2.32A and B)**. Hamstrings function to extend the hip and flex the knee. As one goes up the stairs, he starts by flexing the hip and knee. This elongates the hamstrings over the hip and shortens them over the knee. Next as his hip goes into extension (shortening the muscle), while his knee also goes into extension (elongating the muscle). In other words, the hamstrings muscles are being shortened over the hip while they are being elongated over the knee. Therefore, they are able to maintain an optimal length-tension relationship throughout the range.

As such, the two major **disadvantages** associated with the functioning of two joint and multijoint muscles are that, (i) they are incapable of shortening to the extent required to produce a full range of motion at all joints crossed, simultaneously. This limitation is also termed by some as '*active insufficiency*' **(Figs. 2.33A and B)**. For example, the finger flexors cannot produce a tight fist when the wrist is held in flexion; (ii) In most people they can not be stretched to the extent required for full range of motion in the opposite directions at all joints crossed. This problem is referred by some as '*passive insufficiency*' **(Figs. 2.34A and B)**. For example, a larger range of hyperextension is possible at the wrist when the fingers are not fully extended.

- *The two joint muscles have two different patterns of action*, described as '*concurrent*' and '*countercurrent*' movements. An example of concurrent movement is seen in the simultaneous extension of the hip and knee, and also in the simultaneous flexion of these joints. As the muscles contract, they act on each other in such a way that they do not lose length; thus their tension is retained.

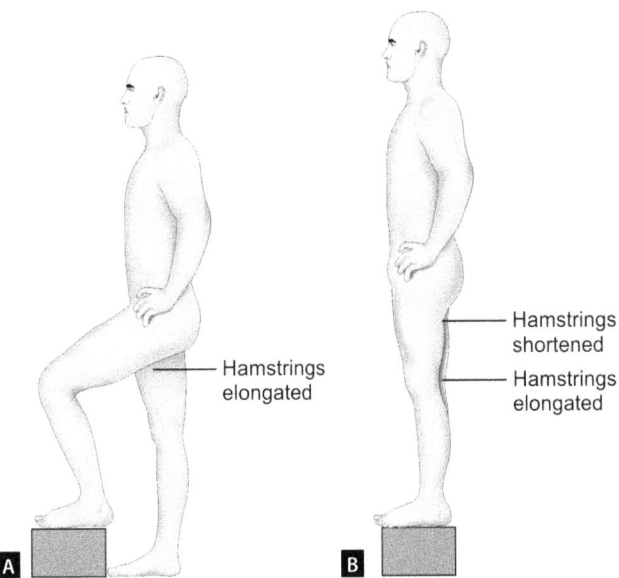

Figs. 2.32A and B: Optimal length-tension relationship of hamstrings when going up stairs; (A) When the foot is placed on the step, the hamstrings are being stretched over the hip while shortened over the knee; (B) Stepping up requires the hip to extend (hamstrings shortening) and the knee to extend (hamstrings being stretched)

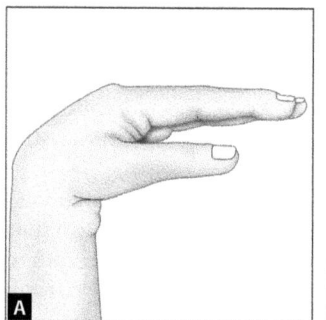 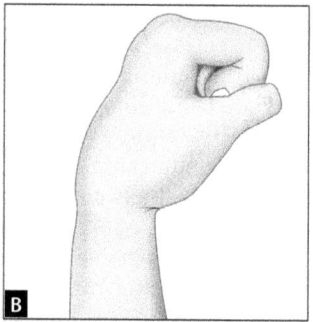

Figs. 2.33A and B: Active insufficiency: (A) The finger flexors unable to produce a tight fist when the wrist is fully flexed. (B) A tight fist is formed as the wrist is extended to a more neutral position

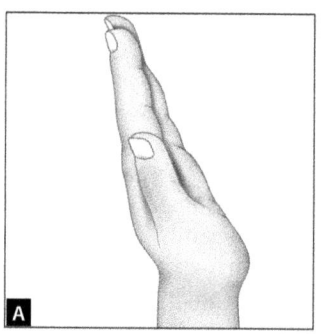 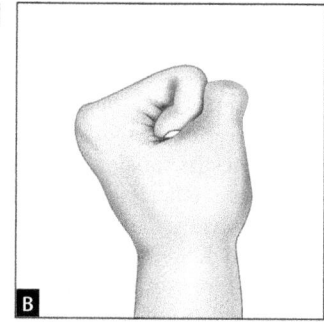

Figs. 2.34A and B: Passive insufficiency: The larger range of wrist extension is possible when the fingers are flexed (B), than when the fingers are extended (A)

The two joint or biarticular muscles of the hip and knee provide excellent examples of concurrent and countercurrent patterns of action. An example of a *'concurrent'* movement pattern occurs when both the knee and hip extend at the same time. In simultaneous extension of the hip and knee, the rectus femoris's loss of tension at the distal end, i.e. knee is balanced by a gain in tension at the proximal or hip end. If only the knee were to extend, the rectus femoris (alongwith other three portions of quadriceps muscles) would shorten and lose tension, but its relative length and subsequent tension may be maintained due to its relative lengthening at the hip joint during extension. Similarly, the hamstrings, which are losing tension at their proximal end are gaining it at their distal end.

The *'countercurrent'* movement pattern presents a different picture. In this type of movement, while one of the two joint muscles shortens rapidly at both joints, its antagonist lengthens correspondingly and thereby gains tension at both ends. An example of this movement pattern is seen in the rapid loss of tension in the rectus femoris muscle while kicking a football, and corresponding gain in the tension in the hamstrings when the hip is flexed and the knee is extended simultaneously.

These movements when combined increase the tension or stretch on the hamstring muscles at both the knee and the hip. In the two joint countercurrent movements of hip flexion and knee extension, the rectus femoris has experimentally been shown to be strongly active; similarly also, the medial hamstrings during hip extension and knee flexion. However, when the motion was concurrent, the activity of the rectus femoris and hamstrings were inhibited when they were antagonists. *Therefore, in both the patterns of movement, the two joint muscles appear to supplement each other in producing smooth, coordinated and efficient movements.*

When a two joint muscle contracts, it acts on both joints it crosses. However, if action is desired in only one joint, the other joint must be stabilized by another muscle or by some external force. For example, gastrocnemius (which is a plantarflexor at the ankle, and a flexor at the knee), is desired only to act as a plantarflexor at the ankle, the knee should be held in extension where the gastrocnemius will be under some stretch/tension. However, if the knee is held in flexion (the action which gastrocnemius performs), the gastrocnemius will no more function as a plantarflexor at the ankle as this muscle is now on a slack over the knee, and the other one joint muscle i.e. soleus can be easily stretched at the ankle by dorsiflexing the ankle. In other words, to stretch a one joint muscle, i.e. soleus, it is necessary to put any two-joint muscles, i.e. gastrocnemius on a slack over the joint, i.e. knee, not crossed or acted upon by the one joint muscle, i.e, soleus. Therefore, to stretch the soleus, which crosses the ankle only, the gastrocnemius muscle (which crosses the ankle and the knee) must be put on a slack over the knee. This can be accomplished by flexing the knee while dorsiflexing the ankle.

The above discussion should not mean that the single joint or uniarticular muscles are less important. Single joint muscles are necessary in order to provide a single joint movement, separately from any other joint action. However, it has been proved experimentally that a single joint muscle in performance of an act requires more expenditure of energy than when the same act is performed by a two joint muscle and therefore, two joint muscles are much more energy efficient when some definite combination of movements at two different joints simultaneously is required. However, this gain in efficiency is achieved at the cost of greater vulnerability to injury. When a two joint muscle is maximally stretched or elongated simultaneously over both the joints it crosses, the resulting tension in the muscle may be great enough to tear some of its muscle fibers. A common example of this is seen in 'tennis leg'; when the knee is fully extended and the ankle is also dorsiflexed the same time, a sudden strain may result in transverse rupture of the medial gastrocnemius at the musculotendinous junction. Similarly, the biceps femoris and semimembranosus components of hamstrings, and rectus femoris muscles are also shown to be susceptible to injury because of their two joint nature.

Roles of Muscles

Muscles assume different roles during joint motion depending on such variables, as the motion being performed, the direction of the motion, and the amount of resistance the muscle must overcome. If any of these variables change, the muscle's role may also change.

An activated muscle rarely acts in isolation, rather an effective, purposeful movement of a body part involves considerable activity of many muscles in addition to the muscles which are directly responsible for the desired movement itself. First, the muscles causing the movement must have a stable base. This means that the bone or bones not involved in the movement but providing attachment for one end of main muscles producing the movement, must be stabilized by some other muscles. Secondly, in some movements, such as those in which the hands are used at a high-level, the upper arms may need to be maintained in an elevated position. This requires contraction of shoulder muscles to support the weight of the arms. Similarly, there are many muscles, particularly the two joint muscles which can cause movements at both the joints they act upon, however at a given time, only one of the actions of one joint alone may be needed for the desired movement. For example, biceps brachii causes elbow flexion, and supination of the forearm, and an assistant role in some shoulder joint actions. However, if at a particular time, when only elbow flexion is desired of biceps brachii, the other actions of the muscle on forearm and shoulder need to be neutralized by some other muscles. Similarly, there are muscles which tend to produce more than one movement in a joint. And, at a particular time, it may necessitate only the desired movement to be produced by the muscle and not the other movement which is not desired. Then some other muscles must contract to prevent the unwanted movement of the muscle. For example, the lower part of the trapezius when it contracts, it tends to produce adduction, depression and upward rotation of the scapula. However, at a time when only adduction movement of scapula is desired of the trapezius, the unwanted actions, i.e. depression and upward rotation of scapula will need to be neutralized by some other muscles, to effect a purposeful and coordinated movement.

Thus, even seeming to be a simple movement may require the cooperative action of a relatively large number of muscles, each performing its own particular task in producing a single, well coordinated movement. Therefore, the muscles have various roles; and their particular role in a given movement depends upon the requirements of that movement. **In nutshell, the various roles of the muscles are described as following:**

Movers or Agonists

The movers or agonists are the muscles that are directly responsible for effecting a movement. In the majority of the movements, there are several movers or agonist muscles, some of which are of greater importance than the others. These muscles are called as the *prime-movers*. The prime-movers are the most effective muscles in causing the desired movement. These muscles contract concentrically to produce joint movement through a specified plane of motion.

The '*assistant movers*' are the muscles that are of less importance and are not as effective, but do assist in producing the movement under certain circumstances. For example, the biceps brachii and brachialis are the prime movers in the elbow flexion, but the pronator teres, and brachioradialis muscles are the assistant movers which are called upon to contract when an extra force is needed to perform elbow flexion against resistance. Similarly, in hip flexion, the iliopsoas are the prime movers but sartorius, rectus femoris and tensor-fascia-lata are the muscles that work as assistant movers.

Sometimes, as assistant mover is also termed as emergency muscle, e.g. the long head of biceps brachii does not ordinarily contract in shoulder abduction. However, it is reported to assist shoulder abduction when there is paralysis of deltoid and supraspinatus muscles, the prime movers of shoulder abduction.

Antagonists

The antagonists are the muscles that while contracting concentrically, tend to produce an opposite joint action to that produced by the agonists or the mover muscles. These muscles are located on the opposite side of the agonist muscles. The elbow extensors, i.e. triceps brachii, localed on the posterior arm are antagonistic to the elbow flexors located on the anterior arm.

The antagonists work in co-operation with agonists (**Figs. 2.35A and B**). The antagonist muscles have the potential to oppose the agonists but are usually relaxed reciprocally, as agonists are contracting to permit a smooth movement. For example, during elbow flexion, the triceps brachii is antagonist to the elbow flexors, i.e. biceps brachii and brachialis muscles. Similarly, during elbow extension, the biceps brachii and brachialis are the antagonists to the triceps brachii, the elbow extensor.

The antagonists often provide controlling or braking actions particularly at the end of fast, forceful movements. When movement of a body part is effected by the force of gravity and resisted by muscle force, the antagonists contract eccentrically to allow a smooth movement. For example, when a person runs down a hill, the quadriceps functions eccentrically as an antagonist to control the amount of knee flexion occurring.

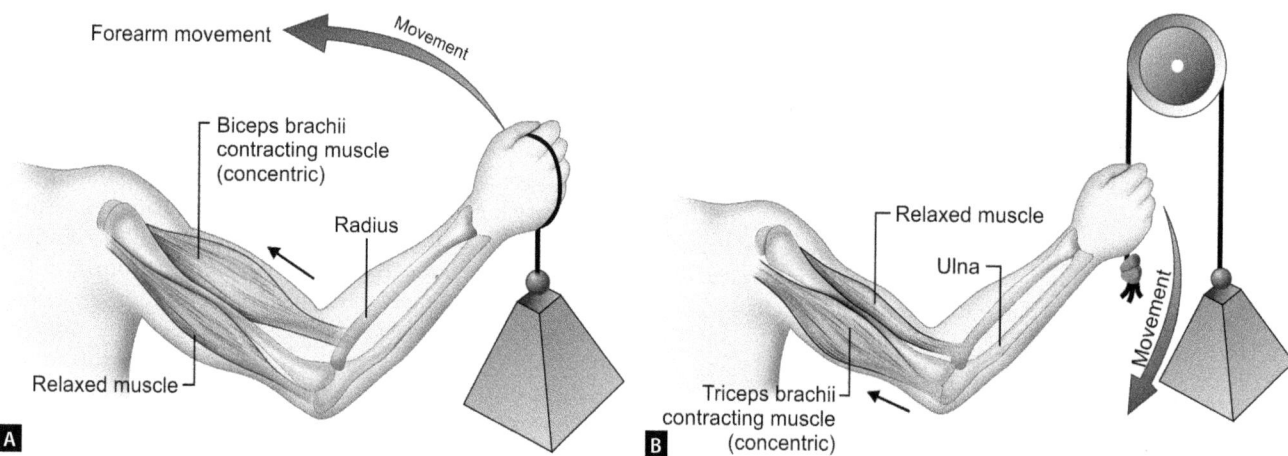

Figs. 2.35A and B: (A) In flexing the elbow, as the biceps (agonist) contracts, the triceps (antagonist) relaxes; (B) In elbow extension, as the triceps (agonist) contracts, the biceps (antagonist) relaxes

Fixators or Stabilizers

These are the muscles that mostly contract statically to steady or support the bone or some part of the body, allowing agonists to work more efficiently. These muscles contract unconsciously to fixate or stabilize the bone to which a contracting agonist muscle is attached. It is only the stabilizing of one of the attachments that the agonist muscle is able to cause an effective movement of the bone at its other attachment. As such, these muscles are essential for establishing relatively a firm base to the proximal bones or body parts to which the agonists are attached, so that the agonist muscles can act upon more distally, at other attachment. For example, during push-ups, the elbow extensors are the agonist muscles, however the abdominal muscles (the trunk flexors) act as stabilizers to keep the trunk straight and prevent sagging of the hip and trunk while the arms move the trunk up and down. Similarly, to open the door, the elbow flexion may be needed, and if the scapula is not stabilized, the contraction of agonist (biceps) may cause a pulling forward of the shoulder girdle rather than an opening of the door. Therefore, the scapula is stabilized by the fixator or stabilizer muscles, i.e. the rhomboids in this case.

Synergists (Syn-with)

These are the muscles that work along with other muscles and assist in the action of the agonists to make it stronger and to facilitate a smooth desired movement. These muscles are not necessarily the prime movers for the movement, rather they act as guiding muscles to assist in the refined movement, and to prevent the undesired action of the agonists or to control the joints not involved in the movement. For example, while doing sit-ups exercise from supine lying position, several abdominal muscles cooperate in producing spine flexion. But, for the moment, let us consider only the right and left external oblique muscles. Both of these muscles cooperate in flexing the spine since each is a prime mover for this movement. But the right external oblique muscle is also a prime mover for right lateral spine flexion and for left spine rotation. Similarly, the left external oblique is a prime mover for left lateral spine flexion and for right spine rotation. The lateral flexion and rotation tendencies in opposite directions, are mutually counteracted or neutralized, and the resultant motion is pure spine flexion. Thus, both of these muscles are the synergists to each other in the spine flexion. These muscles are therefore movers, the mutual synergists and neutralizers the same time at the spine.

The other example is when the fist is clenched, the extensor muscles of the wrist act as synergists. While the fist is clenched, if the wrist were not held extended, then the long flexors of the fingers would produce wrist flexion as well as finger flexion. Thus, the flexion of the wrist added to the flexion of the fingers stretches the tendons of the long finger extensors, which will ultimately lead to the opening out of the fingers, and cause the fingers grip to slacken. Here the wrist flexion of the long finger flexors (multijoint muscles) is counteracted or neutralized by the wrist extensors which are the synergists in this case.

Neutralizers

These are the muscles which contract to prevent or neutralize the unwanted action of the agonists or the prime mover muscles. For example, if a particular muscle both flexes and abducts at a joint, but only flexion is desired in the movement, an adductor muscle contracts as neutralizer to eliminate the abduction component of the agonist muscle. Similarly, if the biceps muscle can flex the elbow and supinate the forearm, and if only elbow flexion is wanted, the supination component must be neutralized by the contraction of the pronator teres muscle. Then only the pure elbow flexion would occur.

Since there seems to be an overlapping in the action of the synergists and neutralizers, many experts advise them to study together. These muscles working together facilitate a smooth and purposeful movement, prevent the undesired action of the agonists, control the joints not involved in the movement (of two joint muscles), and modify the direction of the movement as desired.

As described earlier, the performance of human movements typically involves the cooperative actions of many muscle groups acting sequentially and together. For example, even the simple task of lifting a glass of water from a table requires several different muscle groups to function cooperatively though performing different roles. In this case, the stabilizing roles are performed by the scapular muscles, and both the flexors and extensor muscles of the wrist. The agonist function is performed by the flexor muscles of the fingers, elbow, and the shoulder. Since the anterior deltoid and pectoralis major muscles acting as the main shoulder flexors also produce horizontal adduction of the shoulder (which is not desired here), the horizontal abductors such as the middle deltoid and supraspinatus muscles act as neutralizers to counterbalance the horizontal adduction. The speed of the movement during lifting of the glass may be partially controlled by the antagonistic role of the elbow extensors. When the glass of water is returned back to the table, the gravity is the main force producing the movement. However, the flexors of the elbow and shoulders now work eccentrically, controlling the speed of downward movement.

NERVOUS CONTROL OF VOLUNTARY MOVEMENTS

As human beings, we make demands on our neuromuscular system which enable us to fulfil our daily activities and as such, the movement is our fundamental characteristic by which we adapt ourselves to various demands made by the environment in which we live. Out of all the motor activities, the most important are the voluntary movements which have purpose are consciously initiated, goal directed, and are learned motor acts. By learning process, the voluntary movements may become more automatic and require less conscious thought as our skill improves, e.g. playing the piano.

When we discuss muscular activity, we should really state it as a neuromuscular activity since muscle cannot be active without nervous innervation. All voluntary movement is a result of both the muscular and nervous systems working together. All muscle contraction occurs as a result of stimulation from the nervous system. Ultimately, every muscle fiber is innervated by a somatic motor neuron, which, when an appropriate stimulus is provided, results in

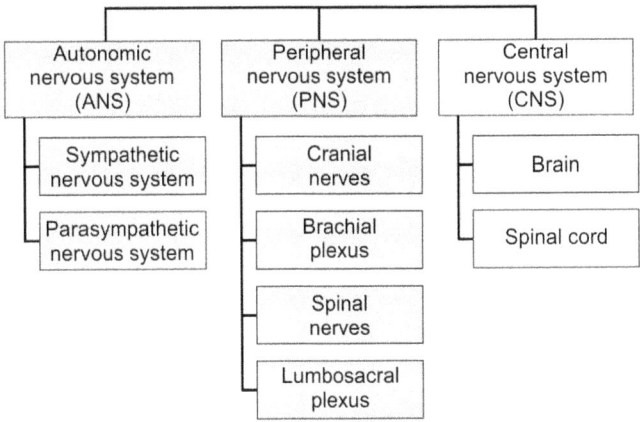

Flowchart 2.1: Divisions of nervous system

a muscle-contraction. Depending upon a variety of factors, this stimulus may be processed in varying degrees at different levels of the Central Nervous System (CNS). Various divisions of the nervous system are shown in **Flowchart 2.1.**

All movement involves the integrated activity of several motor control systems which are organized *hierarchically* and in *parallel*. Necessarily, the hierarchical model involves the integration and interpretation of sensory information in the *association areas of the cortex* which determines the current position of body parts relative to each other and also to the environment. The information received is shared between the *cortex* and the *basal ganglia* (which represent the *highest levels* in the hierarchy) to develop motor strategies for action. The *middle level* in hierarchy is represented by the *motor cortex* and *cerebellum* which both determine the tactics for the action to achieve the strategy. The *lower levels* in the hierarchy is represented by *brainstem* and *spinal cord* which execute the movements through activation of motor neuronal pools for selective movements and postural adjustment.

Parallel processing also takes place, whereby the same information received (input) is shared between several areas of the brain at the same time. For example, the cerebellum and basal ganglia process the information received from the cortex before sending signals to motor cortex for final resultant action.

The role of different levels of the central nervous system in the order from the most general level of control and the most superiorly located to the most specific level of control and the most inferiorly located, including the cerebral cortex, the basal ganglia, the cerebellum, the brainstem, and the spinal cord are briefly described. The pathway from the brain's motor cortex to the spinal cord (corticospinal tracts) is shown in **Figure 2.36.** The role of peripheral nervous system and of the kinesthetic sensations in coordination of movements are also described toward the end of this chapter.

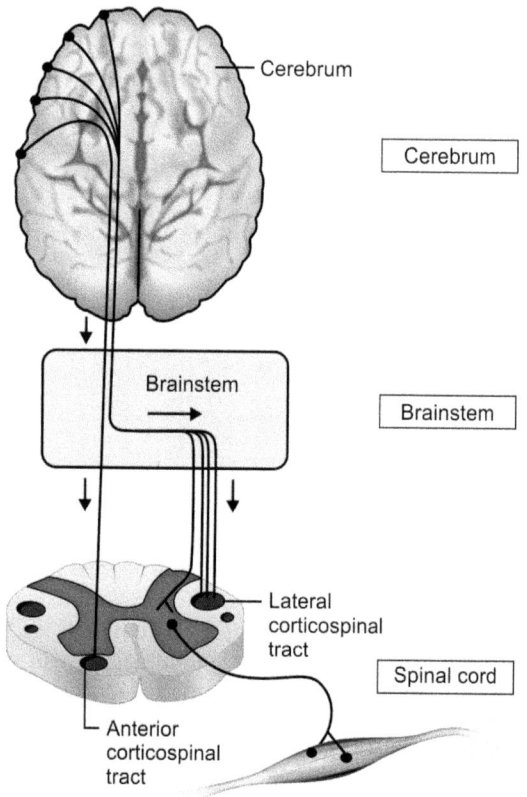

Fig. 2.36: Pathway from the brain's motor cortex to the spinal cord (corticospinal tracts)

Role of Cerebral Cortex in Movement Control

The cerebral cortex is the highest center for control of movement. It provides for the creation of voluntary movement as aggregate muscle action but not as specific muscle activity. It receives sensory information from the body, eyes and ears and then plans it to initiate movements. It is also related with memory, emotions and intellectual ability which may have some bearing on its motor functions.

Voluntary movement is usually initiated in response to some sensory stimulus (pain, touch, temperature, proprioception etc.). It is believed that an initiation center exists in the brainstem which alerts the cerebral cortex for planning the pattern of movements. The planning of movements is based on memories of patterns used on previous occasions.

Role of Basal Ganglia in Movement Control

The basal ganglia is the next level. It controls the maintenance of postures and equilibrium and learned movements such as driving a car. Sensory integration for balance and rhythmic activities is controlled here. The basal ganglia, (comprising of four bilateral structures lying just below the cerebral cortex, and one substantia nigra lying in the midbrain), apart from performing the cognitive and behavioral function, also performs a role in the motor control. The basal ganglia receives input from the cerebral cortex and sends signals to the motor cortex to activate the descending systems to the spinal cord for execution of the movements. It is also believed to play a role in motor programing which includes the planning and execution of complex movement strategies.

Role of Cerebellum in Movement Control

The cerebellum is an important center for movement control. It is a major integrator of sensory impulses and providing feedback relative to motion. It controls the timing and intensity of muscle activity to assist in the refinement of movements.

The cerebellum is a receiving station of information which reaches it by the afferent pathway conveying impulses of kinesthetic sensation from the periphery, and from other parts of the brain (cerebral cortex, brainstem, vestibular nucleus etc.). After receiving the information, it processes it and makes delicate adjustments ensuring coordinated interaction of various groups of muscles as required in the specific pattern of movement, and finally conveys for action to the anterior horn cells of the spinal cord (either by extra pyramidal tracts or other descending pathways).

Role of Brainstem in Movement Control

The brainstem integrates all central nervous system activity through excitation and inhibition of desired neuromuscular actions and functions in arousal or maintaining a wakeful state. One of the important functions the brainstem performs is that it acts as a channel for ascending and descending pathways concerned with the control of movement. The ascending ones include all somatic afferent pathways (medial lemniscus and spinothalamic tracts) and the descending motor tracts include the corticospinal tracts. Certain motor tracts like reticulospinal tracts and vestibulospinal tracts also originate in the brainstem which play important roles in controlling certain muscle tone and movements.

Brainstem is also the site of reflex activity. The reflexes operating are vestibular and neck reflexes which stabilize the head and neck and align the body relative to gravity. Changing head position will elicit vestibular reflexes, and bending or turning the head will elicit neck reflexes (symmetric and asymmetric tonic neck reflexes).

Role of Spinal Cord in Movement Control

The spinal cord is the common pathway between the central nervous system and the peripheral nervous system which contains all the remaining nerves throughout the body. It has the most specific control and integrates various simple and complex spinal reflexes as well as cortical and basal ganglia activity.

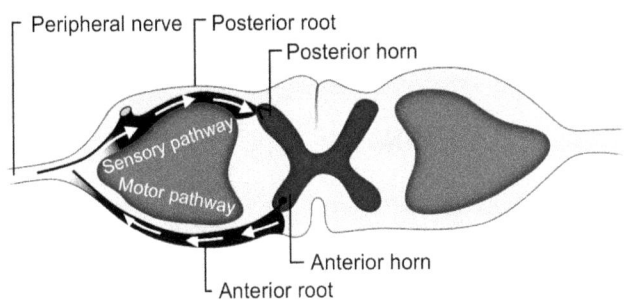

Fig. 2.37: Sensory and motor pathways within the spinal cord (cross section)

The spinal cord is the lowest level in the hierarchy for movement control. It receives somatosensory information from muscles, tendons, joints and skin which it processes and acts upon. The spinal cord also transmits information to higher centers in the brain. The spinal cord is believed to be mainly concerned with reflex activity which is moderated by descending influences from higher centers to produce movements as are required for the task.

The spinal cord segments and spinal nerves are shown in **Figure 2.38.**

Role of Peripheral Nervous System

Functionally, the peripheral nervous system (PNS) can be divided into sensory and motor divisions. The sensory or afferent nerves bring impulses from receptors in the skin, joints, muscles and other peripheral aspects of the body to the CNS, while the motor or efferent nerves carry impulses to the outlaying regions of the body. The sensory and motor pathways within the spinal cord (cross section) are shown in **Figure 2.37.**

The *spinal nerves* provide both motor and sensory function for their respective portions of the body, and are named for the locations from which they exit the vertebral column. From each side of the spinal column, there are 8 cervical nerves, 12 thoracic nerves, 5 lumbar, 5 sacral, and 1 coccygeal nerve. Cervical nerves from 1st to 4th form *cervical plexus* which is generally responsible for sensation from the upper part of the shoulders to the back of the head and front of the neck. The cervical plexus supplies motor innervation to several muscles of the neck. Cervical nerves from 5th to 8th, along with 1st thoracic nerve form the *brachial plexus* which supplies motor and sensory function to the upper extremity and most of the scapula. Thoracic nerves from 2nd to 12th run directly to specific anatomical locations in the thorax.

All of the lumbar, sacral, and coccygeal nerves form the *lumbosacral plexus* which supplies sensation and motor function to the lower trunk and the entire lower extremity and perineum.

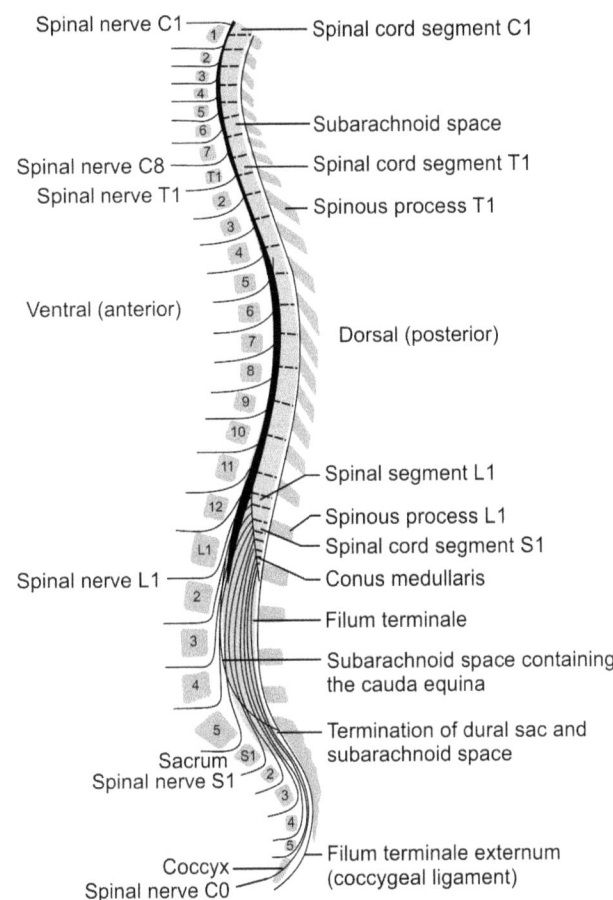

Fig. 2.38: Spinal cord segments and spinal nerves

One aspect of the sensory function of spinal nerves is to provide feedback to the CNS regarding skin sensation. A defined area of skin supplied by a specific spinal nerve is known as a **'dermatome'**. The dermatomes (segmental areas of innervation of the skin) of anterior and posterior aspect of the body are shown in **Figures 2.40A and B.** Regarding motor function of spinal nerves, a **'myotome'** is defined as a muscle or group of muscles supplied by a specific spinal nerve. The myotomes (spinal cord level and muscle innervation) are shown in **Figure 2.39.**

The basic functional units of the nervous system responsible for generating and transmitting impulses are nerve cells known as *'neurons'*. Neurons consist of a *'neuron cell body'*, one or more branching projections known as *'dendrites'* which transmit impulses to the neuron and cell body, and an *'axon'* which is an elongated projection that transmits impulses away from neuron cell bodies. These neurons are classified into three types, according to the direction in which they transmit impulses. *'Sensory neurons'* transmit impulses to the spinal cord and brain from all parts of the body, whereas *'motor neurons'* transmit impulses from

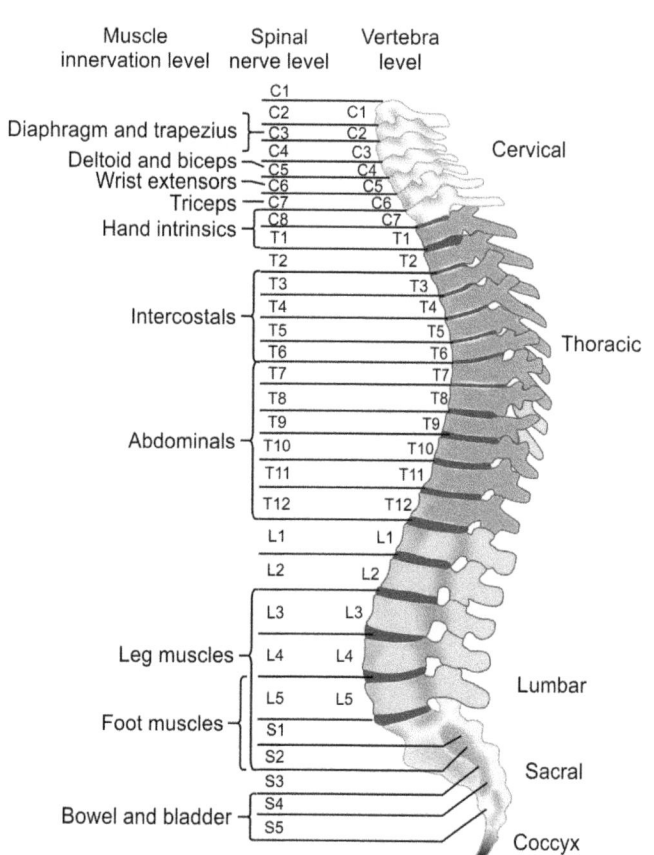

Fig. 2.39: Myotomes: Spinal cord level and muscle innervation. Also shown is the position of a spinal nerve exiting the vertebral column, and its corresponding vertebral level

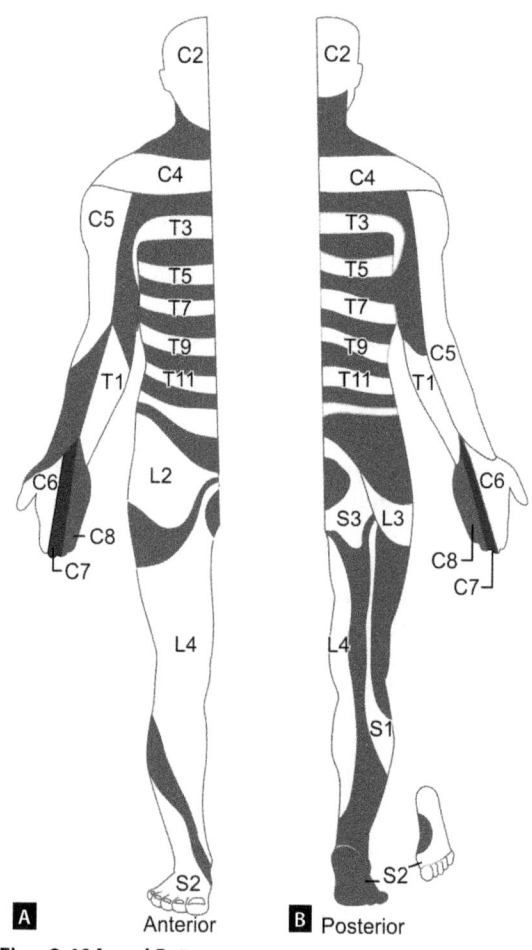

Figs. 2.40A and B: Dermatomes: (A) Anterior; (B) Posterior segmental areas of innervation of the skin

the brain and spinal cord to muscle and glandular tissue. 'Interneurons' are central or connecting neurons that conduct impulses from sensory neurons to motor neurons.

A **motor unit**, as described earlier is the functional unit of the neuromuscular system which initiates and achieves movement in response to a demand for activity. The anterior horn of the spinal cord houses the alpha and gamma motor neurons via which the movements and postural adaptations are executed. *Each alpha motor neuron along with the several skeletal muscle fibers it supplies form a motor unit.* Each muscle comprises several motor units and movements are graded by recruitment of many or few motor units as per the requirement of the task to be performed. The anterior horn cells of the spinal cord are influenced and stimulated by impulses which reach them from many sources in the central nervous system and other parts of the body (as already mentioned) which in turn discharge impulses to the muscle fibers (via motor units) causing them to contract. The impulses reaching the anterior horn cells may be *excitatory* or *inhibitory* and it is the predominance of one or the other type at any one time which determines their effect.

Role of Kinesthetic Sensations in Coordination of Movements

The performance of various activities is significantly dependent upon neurological feedback from the body. Very simply, we use five special senses, i.e. sight, hearing, taste, smell, and touch to determine a response to our environment. We are also aware of other sensations such as pain, pressure, heat, and cold, but we often ignore the sixth sense referred to as 'kinesthesis' which is associated with neuromuscular activity through 'proprioception'.

The kinesthetic sensations arise from proprioceptors situated in muscles, tendons and joints and record contraction or stretching of muscle, knowledge of movement and position of the limbs. These afferent impulses may reach the level of consciousness but most of them end in the spinal cord and cerebellum where they are processed and ultimately help in the coordination of the movements.

The **proprioceptors** are internal receptors located in the skin, joints, muscles and tendons which provide us feedback relative to the tension, length, and contraction state of muscle, and the position of the body and limbs, and movements of the joints. These proprioceptors in combination with other sense organs of the body are vital in 'kinesthesis' which is the awareness of the position and movement of the body in space.

The proprioceptors are of several types. Proprioceptors specific to the muscles are *muscle spindles* and *Golgi Tendon Organs (GTO)*, whereas *Meissner's Corpuscles, Ruffini's Corpuscles, Pacinian Corpuscles*, and *Krause's end bulbs* are proprioceptors specific to the joints and skin.

- The *muscle spindles* are highly specialized sensory receptors often referred to as '*intrafusal fibers*' that consist of bundle of smaller cells, and are concentrated primarily in the muscle belly between the muscle fibers. The muscle spindles provide feedback from the muscle-cells to the spinal cord for regulating the degree of muscle-contraction. These muscle spindles are sensitive to stretch and rate of stretch. As a result, when stretch occurs, an impulse is sent to the CNS. The CNS then activates the motor neurons of the muscle and causes it to contract. The knee jerk or patella tendon reflex can be used as an example. When the reflex hammer strikes the patella tendon, it causes a quick stretch to the musculo-tendinous unit of the quadriceps. In response, the quadriceps fires and the knee extends. To an extent, the more sudden the tap of the hammer, the more significant the reflexive contraction. The stretch reflex provided by the muscle-spindle may be utilized to facilitate greater response, as in the case of a quick short squat before attempting a jump. The quick stretch placed upon the muscles in the squat enables the same muscles to generate more force in subsequently jumping off the floor.
- The *Golgi Tendon Organ (GTO)*, found in the tendon close to the muscle-tendon junction is sensitive to both muscle tension and active contraction. However, GTO is much less sensitive to stretch than muscle spindles are, and requires a greater stretch to be activated. Tension in tendons and consequently in the GTO increases as the muscle contracts which in turn activates the GTO. When the GTO stretch threshold is reached, an impulse is sent to the CNS which in turn causes the muscle to relax and facilitates activation of the antagonists as a protective mechanism. Thus, the Golgi tendon organ protects us from an excessive muscle contraction by causing the muscle to relax.
- *Pacinian Corpuscles* are proprioceptive endings concentrated around joint capsules, ligaments, tendon sheaths, subcutaneous fascia and articular cartilages. These corpuscles are activated by rapid changes in the joint angle and by pressure changes affecting the joint capsule. This activation lasts only briefly. *The Pacinian corpuscles are helpful in providing feedback regarding the location of a body part in space following quick movements such as running or jumping.*
- *Ruffini's Corpuscles* are located in deep layers of the skin and the joint capsule and are activated by strong and sudden joint movements as well as pressure changes. Though their reaction to pressure changes is slower to develop, however their activation is continued as long as the pressure is maintained. *These corpuscles are essential in detecting even minute joint position changes and providing information as to the exact joint angle.*
- *Meissner's Corpuscles* and *Krause's end bulbs* are located in the skin and in subcutaneous tissues. While they are important in receiving stimuli from touch, they are not much relevant to kinesthesis.

The quality of movement and how we react to the position change is significantly dependent upon the proprioceptive feedback from the muscles, tendons and joints. Like other factors involving body movement, proprioception may also be enhanced through specific training.

CHAPTER 3

Basic Biomechanical Concepts

The human body, in many respects can be referred to as a living machine. This implies that both the body and the implants it employs must follow the conventional laws of mechanics. While learning about the kinesiology, it is very important also to learn how various forces placed on the body cause movements. Therefore, in order to study kinesiology effectively and meaningfully, the basic underlying principles and concepts of the mechanics which are frequently applicable must be known.

GRAVITY

Gravity is a force of mutual attraction which is acting constantly on all objects at all times. This force is always exerted in a vertically downward direction, toward the center of the earth. *Newton concluded from the experiments and observations that a force of mutual attraction existed between all the material objects in the universe within the gravitational field of the earth, and that the magnitude of this attraction was directly proportional to the mass of each body/object, and inversely proportional to the square of the distance between them.* However, since the total mass of the earth is so great and its proximity to the objects is so close that practically the total gravitational force on the objects is only directed toward the center of the earth, and therefore all the objects are directed downward toward the center of the earth, practically toward the ground.

This gravitational force of the earth is also continuously acting upon the human body and, if unopposed, the human body will fall to the ground. The effects of gravity can be counterbalanced when a force equal and opposite to it is employed, such as the support of a plinth, the buoyancy of water, or a static muscular contraction. However, if the gravity is opposed by a force which is greater, the movement will occur in the direction of that force. In other words, to move a body part upward, the force must be sufficient to overcome the downward pull of the gravity.

CENTER OF GRAVITY

Center of gravity (COG) may be defined as an imaginary point about which a body balances, or as the point at which the weight of the body can be considered to be concentrated. Therefore, the center of gravity represents the balance point or the weight center of a body about which the body would balance without a tendency to rotate.

A rigid body will balance only when it is supported at its center of gravity. The COG of a rigid symmetric body with uniform density is at its geometric center. As the density of a rigid body varies, the COG does not remain at the geometric center, but is shifted toward the more heavier section. The location of the COG of any object remains fixed as long as the body does not change shape. However, if the shape or position of an object changes, the location of its COG will also change.

The position of the center of gravity in man is important for maintaining balance and for his many forms of locomotion. The location of the center of gravity of a human being in the normal standing position varies with body-build, age and sex. *In an adult human in the anatomic standing position the COG is found to be slightly anterior to the level of the second sacral vertebra.* Because the body proportions and weight distribution change with age, the COG of a child is higher than that of an adult as the head and thorax of a child are much larger in proportion than his legs, arms and rest of the body.

The shape of an object may be such that its COG is actually outside the object itself. Similarly, the human body may assume many positions in which its COG is outside the body itself. In general, when two parts of the body are at an angle, the center of gravity for the total is on a line which joins the centers of gravity of the two parts or segments.

The COG in an adult male in the normal standing position is slightly higher, i.e. approximately at 56–57% of his total height from the floor, due to his broader shoulders than that of an average female where it is slightly lower, i.e. at about 55% of her standing height from the floor, due to her heavier pelvis and thighs, and shorter legs.

The human body is multi-segmented structure, which is capable of several positions, and the location of its COG also changes accordingly. The COG of the body as a whole is the sum of the centers of gravity of individual segments of the body. The researchers have found out the location of COG of major segments of the body in an adult male of average body weight.

The center of gravity of the extended upper extremity is just above the elbow joint, and that of the extended lower extremity it is just above the knee joint. The arm, forearm, thigh, and leg are larger proximally, and thus their individual centers of gravity lie closer to the proximal ends. The COG of the extended body segments, i.e. limbs is at about 4/7th of the distance from their distal ends, or at about 4/9th (45%) of the length of the segment, measured from their proximal ends.

The location of COG may also change as the anatomic position changes. If the arms are folded on the chest or raised overhead, the COG rises. If the subject bends the head, trunk and hips, the COG moves toward the feet. Similarly, wearing high heel shoes, pregnancy and other circumstances also cause the COG to shift in proportion to the percentage of the body's mass which is moved. Even in the normal adult male, the rhythmic upward and downward motion of the body during walking is reported to displace the COG of the body about 1.8 inches during the cycle from one heel strike to the subsequent striking of the same heel, in an effort to conserve body's energy.

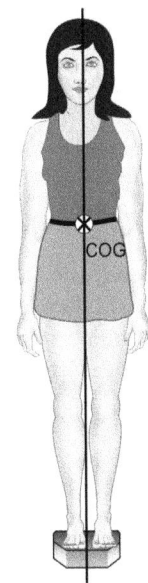

Fig. 3.1: Line of gravity (LOG) passing through the COG, and falling on the base of support (BOS)

LINE OF GRAVITY

Line of gravity (LOG) is defined as an imaginary vertical line passing through the center of gravity toward the center of the earth. Hence, the location of LOG depends on the position of the center of gravity, which changes with every shift of the body's position.

When the human body is in the normal standing position, the line of gravity passes from the vertex, through the second sacral vertebra (the COG of the body) to just infront of the ankle joints and between both the feet **(Fig. 3.1)**.

However, the relationship of various body structures to the line of gravity is subject to considerable variation according to the individual differences in posture and anatomic structure. It is estimated that on an average when the posture is good, the LOG passes through the mid-cervical and mid-lumbar vertebrae, and infront of the thoracic vertebrae. The external ear and the point of shoulder are in the same frontal plane and lie lateral to the line. The central axis of the knee joint as well as the ankle joint lie posterolateral to this line.

As long as the line of gravity falls within the base of support, and through the COG, the body remains stable and balanced. However, if this line falls outside the base, the balance is lost. This is one of very important principles related to body mechanics.

BASE OF SUPPORT

Base of support is the area by which a body is supported. Base of support (BOS) is that part of a body which is in contact with the supporting surface. In case of human body, the base of support in standing position comprises the area of the feet and the space between them. In supine lying position, the posterior aspect of the whole body forms the base. As such, the base of support in standing is lesser than the supine lying position. In case of a chair, the base is formed by the area covered by the lines joining the legs of the chair.

Larger the supporting base, the more likely is the line of gravity falling within the base, and as such, the stability will be more. On the other hand, a smaller base of support decreases the stability.

AXES AND PLANES

When studying the various joints of the body and analyzing their movements, it is helpful to characterize them according to specific axes and planes of motion. Whenever a movement is performed, the body part moves through the surface or the space, and to record its location in space a reference point is required.

An axis is an imaginary point or line through a joint about which a movement takes place.

A plane is an imaginary surface or the space which lies at right angle to the axis, and in which the movement takes place. A plane of motion has also been defined as an imaginary two-dimensional surface through which a limb or body segment is moved.

Various movements of the body take place at the joints and therefore the axes pass through the joints, and the body part moved is in the plane which is always at right angles or perpendicular to the axis of movement. A particular motion will always occur in the same plane and around the same axis. For example, flexion/extension will always occur in the sagittal plane around the frontal axis. Abduction/adduction will always occur in the frontal plane around the sagittal axis. Similar motions, such as radial and ulnar deviation of the wrist will also occur in the frontal plane around the sagittal axis. However, the thumb is the exception because the flexion/extension, and abduction/adduction movements do not occur in these traditional planes.

To describe various movements of the body, generally three traditional axes and three planes are described, with the human body in the anatomical position (standing erect with the head, toes and the palms of the hands facing forward and with the elbows fully extended and the fingers extended) **(Fig. 3.2)**. The three traditional planes correspond to the three dimensions of space and each plane is perpendicular to each of the other two. Similarly, there are three axes of motion, and each axis is perpendicular to the plane in which the motion occurs.

Axes of Movements

As described above, there are three following axes:

1. **Sagittal (anteroposterior) axis: This is a point or line that runs horizontally through a joint from front to the back.** It lies parallel to the sagittal suture of the skull, and runs in the anteroposterior direction. The movements around the sagittal axis occur in the frontal plane. Abduction and adduction at the hip or the shoulder are the examples of the movements occurring around the sagittal axis and in the frontal plane **(Fig. 3.3)**.

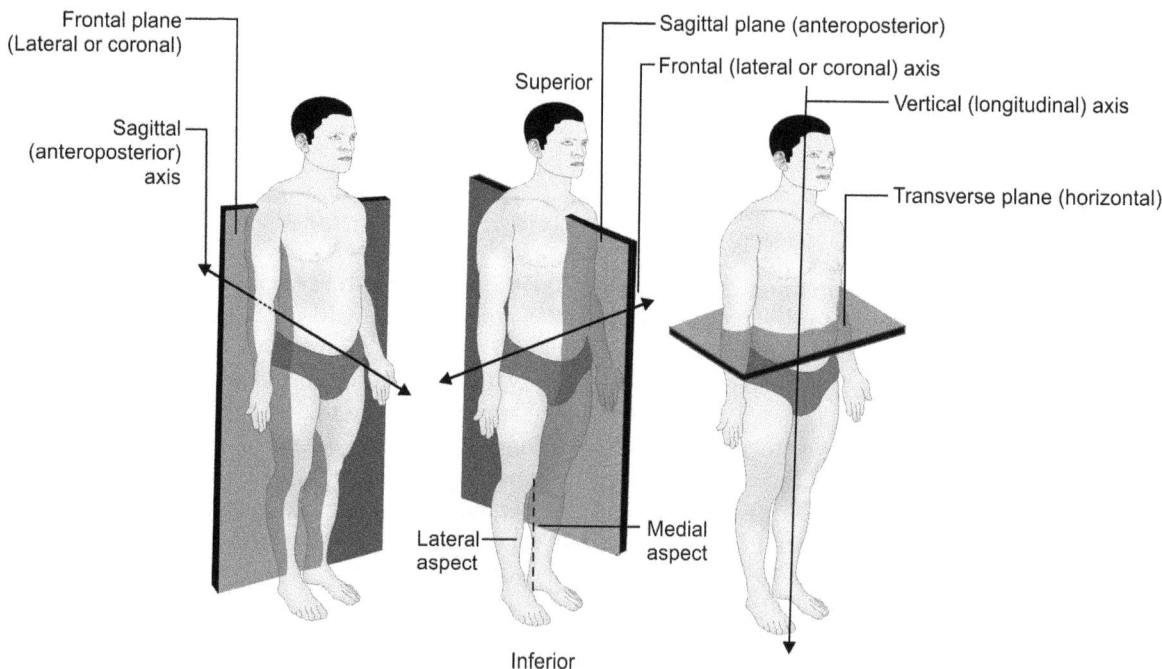

Fig. 3.2: Three axes and three planes of motion

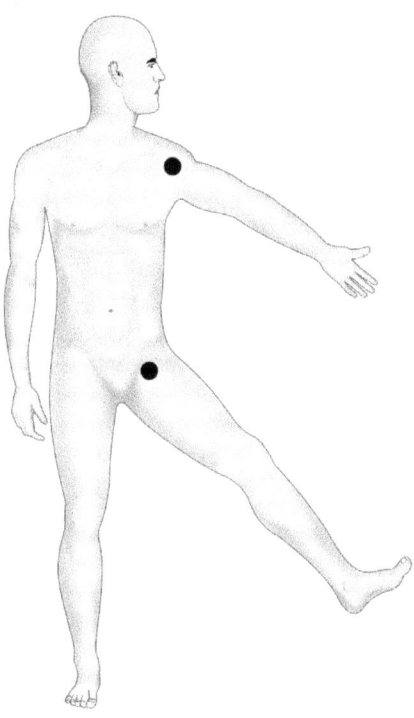

Fig. 3.3: Abduction movement at the shoulder and hip joints in the frontal plane around the sagittal axis

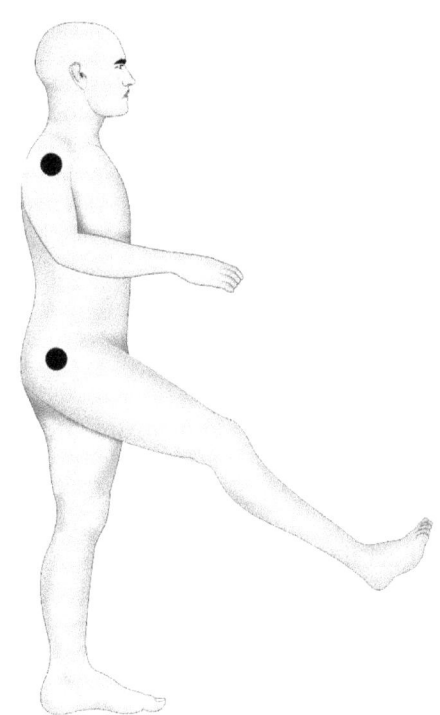

Fig. 3.4: Flexion movement at the shoulder and hip joints in the sagittal plane around a frontal axis

2. **Frontal (lateral, or coronal) axis: This axis is a point or line that runs horizontally through a joint from side to side.** It lies parallel to the frontal suture of the skull. The movements around the frontal axis occur in the sagittal plane. Flexion and extension at the hip or the shoulder are the examples of the movements that occur around the frontal axis and in the sagittal plane **(Fig. 3.4)**. The flexion/extension movement at the elbow (as when performing a biceps curl) is also an example of movement occurring around the frontal axis.
3. **Vertical (longitudinal) axis: This is a point or line that runs straight down through a joint (longitudinally) from top of the head to the bottom.** It lies parallel to the line of gravity and is always perpendicular to the ground. The movements around the vertical axis occur in the transverse or horizontal plane. Internal and external rotation at the hip or the shoulder are the examples of the movements that occur around the vertical axis and in the transverse or horizontal plane. The other examples of movements around the vertical axis include the rotation or turning of the head and spine to the right and left **(Fig. 3.5)**.

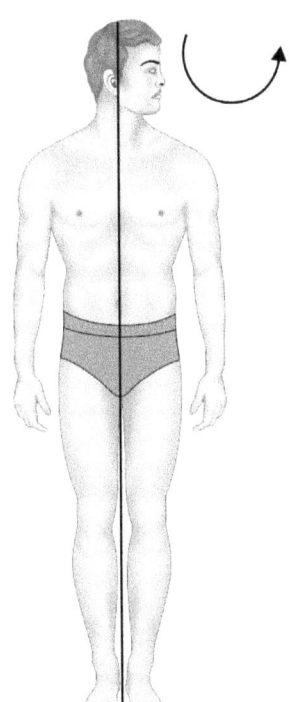

Fig. 3.5: Movement of head rotation in the horizontal plane around a vertical axis

Planes of Movements

Like axes, there are three following planes, and each plane is at right angles or perpendicular to the other two planes.

1. **Sagittal (anteroposterior) plane: This plane is a vertical plane that passes through the body from front to back, and divides the body into right and left symmetrical halves.** The movements that occur in the sagittal plane are the flexion and extension at the hip or the shoulder, around the frontal or coronal axis. The other examples of the flexion-extension movements occurring in this plane are biceps curls (elbow flexion), knee extensions, and sit-ups.

2. **Frontal (lateral, or coronal) plane: This plane is a vertical plane that passes through the body from side to side, and divides the body into the anterior and posterior halves (front and back parts).** The movements that occur in the frontal plane are the abduction and adduction at the hip or the shoulder, around the sagittal or the anteroposterior axis. The sideward or lateral flexion of the spine is the other example of the movement occurring in the frontal palne.

3. **Transverse (horizontal) plane: This plane is a plane that passes through the body horizontally and divides the body into upper and lower halves (top and bottom parts).** Generally, rotational movements that occur in the transverse or horizontal plane are the internal and external rotation at the hip or the shoulder, around the vertical axis. Pronation/supination of the forearm and the spinal rotation in the anatomical position are the other examples of the movements that occur in this plane.

Summary of three axes and three planes of movements is given in **Table 3.1**.

Cardinal Planes

There are only three specific or cardinal planes of motion in which the various joint movements can be classified. The specific three planes (sagittal, frontal, or transverse) that divide the body exactly into two equal halves and the movement occurring in the plane passing through the center of gravity are often referred to as 'cardinal planes'. **The point where these three cardinal planes intersect each other in the anatomical position is the center of gravity, which in the human body lies at the level slightly anterior to the second sacral vertebra (Fig. 3.6).**

Thus, it should be clear that there are only three cardinal planes but an infinite number of vertical and horizontal planes parallel to the cardinal planes, within each half. When describing a movement in terms of a plane, it does not necessarily imply that the movement occurs in a plane passing through the center of gravity. The forward-backward nodding of the head is a movement occurring in the cardinal sagittal plane, but movement of either leg or arm forward-upward (i.e. flexion) is movement occurring in the sagittal plane. It means that the movement of flexion of leg or arm occurs in a plane parallel to the sagittal plane, and not in a cardinal sagittal plane.

Sit-ups that involves the spine are the other example of the movements occurring in the cardinal sagittal plane (also known as mid-sagittal plane), but the movements of biceps curls, and knee-extensions occur in the parasagittal planes, which are parallel to the mid-sagittal plane. Even though, these examples are not in the cardinal sagittal plane, they are thought of as movements in the sagittal plane.

Although each specific joint movement can be classified as being in one of the three planes of motion described above, **our movements are usually not truly and totally**

Table 3.1: Summary of axes and planes of movements

Axis	Plane	Movements
Sagittal (anteroposterior) • Runs from front to the back direction.	*Frontal (lateral or coronal)* • Divides the body into anterior and posterior halves.	• Abduction/adduction at hip and shoulder • Radial/ulnar deviation at the wrist • Eversion/inversion at the foot • Sideward/lateral flexion of the trunk to right/left.
Frontal (lateral or coronal) • Runs from side-to-side direction.	*Sagittal (anteroposterior)* • Divides the body into right and left halves.	• Flexion/extension at the hip and shoulder • Tipping the head forward/backward • Flexion/extension at the elbow.
Vertical (longitudinal) • Runs in the longitudinal direction from top to the bottom.	*Transverse (horizontal)* • Divides the body into upper and lower halves.	• Internal/external rotation of thigh and arm • Supination/pronation of the forearm • Rotation of head/spine to the right/left.

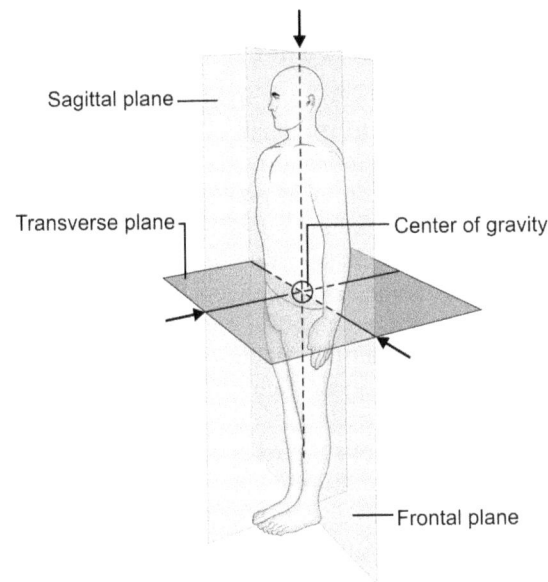

Fig. 3.6: Three cardinal planes intersecting at the center of gravity of the body

in one specific plane, but occur as a combination of motions from more than one plane. These movements from the combined planes may be described as occurring in diagonal or oblique planes of motion.

A diagonal or oblique plane is a combination of more than one plane. In reality, many motions used in our daily life activity and sports do not follow the traditional planes of motion, rather they occur in a diagonal plane (which is not parallel or perpendicular to the other planes). However, whatever the degree of obliquity of the plane, the axis for a movement is always perpendicular to the corresponding plane.

In our body, there are various types of joints such as uniaxial, biaxial, and triaxial (or polyaxial) which determines the degree of freedom or number of planes in which they can move. For example, a **uniaxial joint** (i.e. elbow or knee) permits movements of flexion-extension around one axis (frontal axis) and in one plane (sagittal plane). A **biaxial joint** of which the wrist joint is the best example allows movements around two axes and in two planes, i.e. flexion/extension around the frontal axis and in the sagittal plane; and the radial/ulnar deviations occur around the sagittal axis and in the frontal plane. A **triaxial joint** of which the shoulder and hip are the examples can permit movements maximally around three axes and in three planes:

1. Flexion/extension movement occurs about a frontal axis and in a sagittal plane.
2. Abduction/adduction movement occurs around the sagittal axis and in a frontal plane, and
3. Internal/external rotation occurs around the vertical axis and in a horizontal plane.

EQUILIBRIUM AND STABILITY

Equilibrium

All the objects at rest are in equilibrium. **A state of equilibrium results when all the forces (internal and external) acting upon a body are perfectly balanced and the body remains at the rest.** If the body is moving, there is no change in its speed or direction. However, all objects at rest are not equally stable.

Stability

The stability refers to the resistance to a disturbance of the body's equilibrium. The stability may be enhanced by determining the body's center of gravity and changing it appropriately.

Balance

Balance is the ability to control equilibrium, either static or dynamic. Balance is very important for the resting body, as well as for the moving body. Generally, the physical education and sports activities demand stability and balance, but sometimes there are circumstances in which the movement is improved when the body tends to be unbalanced. Therefore, an understanding of the mechanical principles and factors involved can help us enhancing equilibrium, maximizing stability and ultimately achieving balance.

Types of Equilibrium

Basically, there are three states or types of equilibrium:
1. **Stable equilibrium:** It occurs when sum of all the applied forces as well as inertial forces acting upon a body is nil. The body is at rest or completely motionless in this type of equilibrium. If a body is displaced, the forces acting upon the body tend to restore it to its original position and the body is said to be in stable equilibrium.

   The most stable position of equilibrium is when the COG is as low as possible and the line of gravity falls near the center of a large supporting base, as in supine lying position. A body becomes progressively less stable as its COG is raised and the line of gravity falls near the margin of the base.

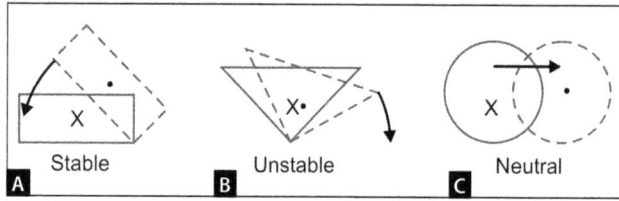

Figs. 3.7A to C: Types of equilibrium, (A) Stable; (B) Unstable; (C) Neutral

The wrestler and defensive lineman both know the value of shifting body position to increase stability by lowering the COG. However, for any reason, if the equilibrium is not very stable, assuming a crouching, kneeling or sitting position will lower the COG and increase the stability.

2. **Unstable equilibrium:** It occurs when only a slight force is needed to disturb the equilibrium of an object. In an unstable equilibrium, after an initial displacement, the forces acting upon the body tend to increase the initial displacement. This occurs when the COG is as high as possible and the base of support is very small. Even a slight push or displacement causes the line of gravity to fall outside the base, and the body will fall down. Balancing a pencil on the pointed end, walking on a tightrope, or standing on one leg are good examples of unstable equilibrium.

   Sprint runners at the start of their race, gymnasts performing on balance beam, or toe dancers are all in unstable equilibrium.

3. **Neutral equilibrium:** This equilibrium exists when in spite of displacement or change of position of a body by the force exerted on it, the height and position of center of gravity of the body remains the same in relation to its base. In other words, the COG of the body is neither raised nor lowered when it is disturbed. Objects in neutral equilibrium will come to rest in any position without a change in the level of the COG. A good example is a ball, which when rolls across the floor, its COG remains the same.

All three types of equilibrium are shown in **Figures 3.7A to C**.

Factors Affecting Stability

Since the humans ordinarily hold themselves in an upright position and the effect of gravity is always in operation on this earth, the problems of stability are ever present. Probably the only time the human body is not adjusting itself in response to gravitational force is when it is completely resting in supine lying position. Consciously or unconsciously, humans spend most of their waking hours adjusting their body positions best suited to the task.

The following factors affect the stability:

- **Size and shape of the base of support:** A larger or wider base of support adds to the greater stability of an object. The center of gravity can move a greater distance without falling outside the base. An increase in the size of the base, while the height of the object remains the same, usually lowers the center of gravity. Likewise, a decrease in the base raises the COG, all other things remaining equal.

   While walking on a balance beam, rail road track, or tight rope, much of the difficulty is experienced due to the narrow base of support apart from a higher COG. In all of these situations, there is a problem to keep the COG over the base of support, which is a requisite for maintaining equilibrium. The wider base makes it easier to keep the COG over the base.

   In addition to the size of base of support, its shape is also a factor in stability. In making the base wider, the base should be enlarged in the direction of the moving or opposing force to allow for a decided shift of the center of the gravity without the line of gravity falling outside the base. To resist lateral or sideward external forces, the base should be widened in the lateral direction. Similarly, when the external forces are known to be coming from a forward-backward direction, the base should be widened in the forward-backward direction. For example, when catching a fast moving ball at the front, a forward-backward stance is recommended, as it widens the base in line with the direction of the force. The similar adjustment and widening of the stance is also made when one stands in a bus or a subway train, in the direction that the vehicle is moving, i.e. forward-backward direction in relation to the vehicle.

- **Height of the center of gravity:** The lower center of gravity increases the stability of the object and vice-versa. When the supporting base of two objects is same, the object having a higher COG is more unstable than the object having a lower COG, as it would take less force to disturb and throw the COG from over the base of support when the COG is higher (**Figs. 3.8A and B**).

   Balancing a weight on the head shifts the COG higher and therefore it is much difficult to maintain the equilibrium.

   To increase the stability, a performer on a balance beam should quickly squat when he feels as if balance is being lost. Similarly, a wrestler should try to remain as stable as possible by lowering his center of gravity.

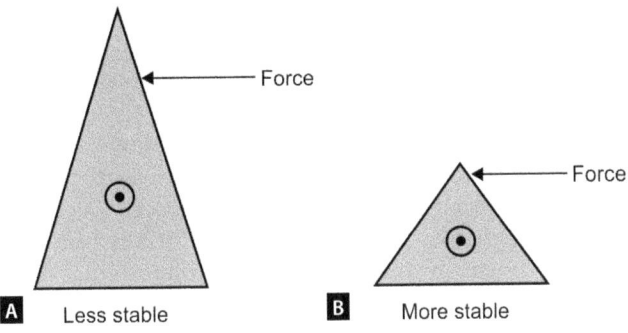

Figs. 3.8A and B: Relationship of height of COG to stability. (A) Less stable; (B) More stable

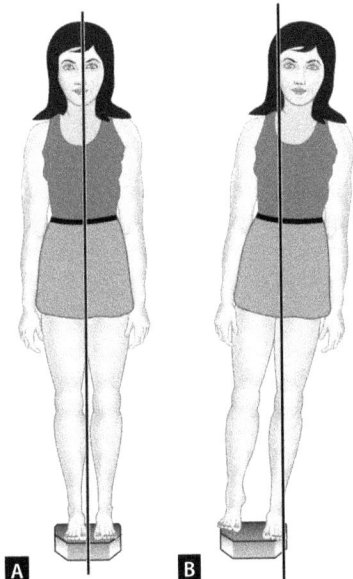

Figs. 3.9A and B: Relationship of LOG with respect to the base of support: (A) More stable as LOG is falling on the middle of the BOS; (B) Less stable as the LOG is falling near the edge of BOS

Both the factors, i.e. size of the base of support and the height of the center of gravity are often dependent upon each other. In general, the larger the base the lower the center of gravity and the greater stability of the object, conversely a small base and high center of gravity reduce stability.

- **Location of line of gravity with respect to the base of support:** When the line of gravity falls near to the center of supporting base, the object is more stable. As the line of gravity shifts toward the periphery of the supporting base, there is a greater tendency for the object to rotate around the edge and to fall **(Figs. 3.9A and B)**. And, if the LOG falls outside the base of support, there is still much greater tendency for the object to fall unless there is some other force holding the object in place.

  Stability can be increased by a football player knowing he will be pushed from in front, by leaning forward so that he can 'give' in a backward direction without losing his balance.

- **Mass or weight of the object:** When the mass or weight of an object is greater, all other things being equal, there is greater stability. This is largely due to the inertia of the object particularly when the object is subjected to some external force or some motion. A light weight object such as an empty card board carton can be easily blown over by a little push or wind than when it is heavier, filled with canned goods.

  In all sports involving physical contact, the heavy, solid sports person stands a better chance of keeping his footing than does the light weight player. This concept can also be observed by looking at the size of players on a football team. Linemen are traditionally heavier, and thus harder to push over, than the halfbacks who are much lighter as their job is to run with the ball. Therefore, it is also said that what is gained in stability is lost in speed and vice-versa.

- **Friction:** A body is more stable when there is greater friction between the body and the surface it contacts. Friction has even greater influence when the body is in motion or is being acted upon by an external force. Walking on an icy pavement is quite slippery as there is essentially no friction between the ice and the shoe. When the supporting surface presents insufficient friction, the person or the player should wear the specially made shoes or footgear, i.e. rubber-soled shoes, 'creepers' on the shoes etc. appropriate to the specific sport, so as to increase the stability in positions held momentarily between quick or forceful movements, as in basketball, football, hockey etc.

  However, having a surface with a great deal of friction is not always desirable. For example, pushing a wheel chair across a hardwood floor is much easier than pushing it across a carpeted floor, as the carpet creates more friction, making it harder to push the wheel chair.

- **Alignment of body segments:** In an object or a body, like human body which consists of multisegments, placed one above the other, the greater stability is assured when the centers of gravity of all the weight bearing segments lie in the same vertical line that is centered over the base of support. For example, a column of children's building blocks will stand easily, if the blocks are placed squarely one over the other. However, if the blocks deviate to one side or are placed

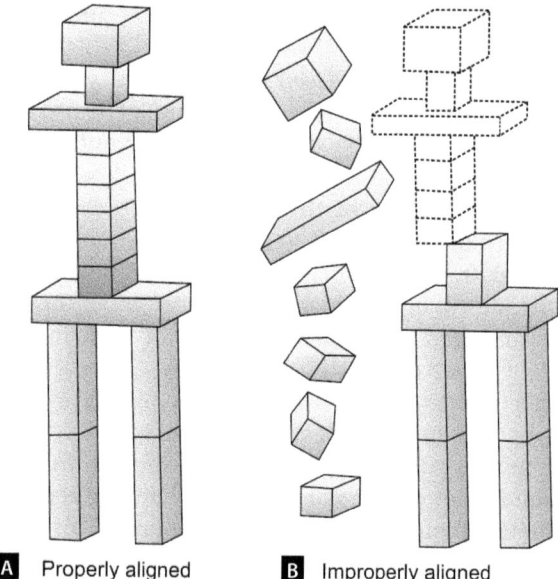

Figs. 3.10A and B: Alignment of human body segments: (A) Properly aligned; (B) Improperly aligned

in a zig-zag alignment, the column becomes unsteady. In the human body which is a multisegmented, united yet flexible structure, the proper alignment of major weight-bearing segments, i.e. head, thorax and pelvis, one over the other in standing posture is not only pleasing in appearance, but there is less likelihood of strain to the joints and muscles. When any one of these segments gets out of line, there is usually a compensatory disalignment of another segment in order to maintain a balanced position of the body as a whole, resulting in fatigue or strain **(Figs. 3.10A and B)**.

This aspect should be taken care in pyramid building and other balance stunts in which one person supports the weight of another person or persons, and thus aligning or balancing the several centers of gravity over the center of the base of support.

- **Sensory organs:** Many sensory organs are important in the maintenance of balance and stability, including the organs of *vision* (eye), the organs of the internal *ear* (the semicircular canals), the organs of *touch*, and the end organs of the *kinesthetic sense* (the proprioceptors in the muscles, tendons, and joints). All these factors though known, are not usually explained with respect to the stability or equilibrium. For example, role of vision is quite evidently experienced when crossing a swirling river on a high foot bridge, or walking close to an unprotected edge high above the ground. Even when the supporting surface is entirely adequate, the sense of balance may be disturbed. The eyes give a point of reference and therefore, are important in the maintenance of body balance.
  - The semicircular canals of the internal ear, touch (pressure), and kinesthetic senses all provide balance information to the performer.
  - Whenever the head is rotated for any period of time as in rolling, the fluid in the internal ear is also put into motion. Because of the inertia, this motion of fluid continues even after the movement of the head is stopped and gives the individual a sense of continued motion although the body is actually still. It is believed that with training the adjustment to this can be learnt. For example, a beginner after executing one forward roll may feel considerable dizziness and has difficulty controlling his movements, however with effective training, the same gymnast can execute several rolls in succession and maintain control of his subsequent movements.
  - The proprioceptors of the kinesthetic sense are extremely important particularly when the muscles and tendons are stretched. The proprioceptors present in the muscles, tendons and joints particularly of the feet and legs are stimulated and the information as to the position and movement is carried to the brain, which enables body adjustments leading to the maintenance of balance.

FORCE

Motion occurs only when sufficient force is applied to an object. Force produces or stops the motion. It may increase or decrease the speed, or cause objects to change direction.

Force is one of those concepts that everyone understands, but is difficult to define. Force may be a push, or a pull to cause motion, or produce a net effect so that bodies remain stationary. Movement occurs if one side pushes or pulls harder than the other.

Force is defined as that which produces push or pull on an object in an attempt to affect motion or shape. This pushing or pulling can be affected through direct mechanical contact or through the force of gravity. It is the effect that one body has on another.

Force may also be defined as the cause, which changes the state of rest or uniform motion of a body. When forces are applied to the human body, either a push (compression) or a pull (tension) is caused. If a body is at rest, we have to pull or push it to set it in motion. To push or pull an object, we have to apply muscular force and our muscles experience a tension due to this push or pull.

Though no motion can occur without force, it is not necessary that application of force will always cause motion. Sometimes the motion can result due to the invisible force, i.e. gravitational force of the earth.

The action of a force may be either internal or external. *Internal forces* in kinesiology refer to those forces that are produced by the body itself and usually act on various structures of the body and cause differences in body shape. *External forces* are those forces which act from outside the body, i.e. weight, friction, gravity, another person, some machine, wind or water resistance etc.

In the human body, muscles are the main source of force that produces or changes movement or motion of a body segment or the whole body or an object thrown, struck, or stopped. Strong muscles can produce more force than the weak muscles.

Force is the product of mass times acceleration.

Force = mass × acceleration

$F = m \times a$

The mass of a body segment or the entire body times the speed of acceleration determines the force.

The standard unit of force in metric system is *Newton (N)*. In British or English system, the unit of force is *pound (lb)*. *A Newton is the amount of force which when acting on 1 kg mass produces an acceleration of 1 meter per square second (m/s^2)*.

Forces are vector quantities. A vector quantity describes both magnitude and direction. A vector force can be shown graphically by a straight line of appropriate length and direction.

In order to describe the force fully, the following three characteristics of force must be considered:

1. **Magnitude of force:** Since force is a vector quantity, it has both magnitude and direction. The force must be sufficient to overcome the inertia of an object if movement is to result. The greater the mass of an object, greater its inertia and the more force necessary to move it. The magnitude of force depends on the mass and speed of the object imparting the force. The greater the mass of the object imparting the force and the faster that it is moving, the greater the force imparted, all other things being equal.

   Graphically, the force is represented by an arrow. The length of arrow (or the vector line) represents the magnitude of the force. The head of the arrow indicates the direction of applied force, which is always represented as a pull away from the point of application of force. The tail of the arrow indicates the point of application of force where the force vector starts.

   a. •⎯⎯⎯⎯→ Head of arrow

   b. Tail •⎯⎯Shaft⎯⎯→

   Here, the magnitude of applied force is greater in case of 'b' than 'a'.

   In case of muscles, more force is available from strong than from weak muscles and all muscles are not equal in potential maximal force and, therefore are not equally adapted to the task of supplying force. **The magnitude of muscular force is directly proportional to the number and size of muscle-fibers in the muscle that are contracting**. However, since normally the muscles act in group, their force or strength is measured collectively, rather than the force exerted by an individual muscle, measured in a given movement.

   To determine the amount of force exerted by a single muscle in the living body, the *cross section of the muscle* is mainly used. Since there is a variety of arrangement of muscle-fibers in different muscles (i.e. longitudinal, fan-like, spindle-like, penniform etc.), physiological cross-section of the muscle that cuts across every fiber in the muscle measures the true cross-section of a muscle. And the potential force or strength a muscle is calculated on the basis of force exerted by per square unit of muscle's cross-section.

2. **Point of application of force:** The point or application of a force is that point at which the force is applied to an object. *Graphically, the point of application of force is the point where the force vector starts*, and the direction of the force is always represented as a pull away from the point of application of force. Where gravity is concerned, the point of application of force is always through the center of gravity of an object. For practical purposes, it may be assumed that the point of application of muscular force is the center of the muscle's attachment to the bony lever. This usually corresponds to the muscle's insertion or distal attachment.

   – Force applied through the center of gravity when the object is free to move, results in its linear motion, with relatively less force. However, when the object is not free to move as when there is great deal of friction, rotatory motion results.

   – If one point of the object is fixed, the object rotates no matter where the force is applied.

- If the force is applied farther from the object's center of gravity, the object will tend to rotate (turn or tip).
- If the force is applied toward one end of an object, the total object will move in the direction of the force and the same time it will also rotate.

In other words, unless there is considerable friction, both ends of the object will move but the end nearest the point of force application will move forward faster. An object tends to rotate if force is applied above or below, or on either side of the center of gravity instead of through the center of gravity. The force applied farther from the center of gravity requires less force to rotate the object.

3. **Direction of application of force:** Usually an object moves in the direction of the force applied to it, unless it is hindered by some other force such as gravity, friction or an obstacle in the pathway. Since the force of gravity pulls all objects toward the earth's center, the direction of gravitational force is always vertically downward.

The direction of an applied force is along its action line from the point of force application through the center of gravity of the object. The direction of muscular force is represented by the direction of the muscle's line of pull (i.e. muscle's angle of pull formed between the mechanical axis of the bone and the muscle's line of pull).

In walking, running or jumping, the direction of the force applied by the 'push off' of the toes is determined by the relationship of the center of gravity to the toes at the moment of take off. Since in order to move itself, the body is dependent upon the reaction force from an outside medium (ground, water), *the force produced by the body must be applied in the direction opposite to the desired movement.*

Resolution of Forces

Many a times when an object is acted upon by two forces the same time, the movement of the object will be in the direction of the resultant of these two forces. This can be illustrated diagrammatically by drawing lines in the directions that the two forces are acting and the length of lines representing the relative magnitude of the two forces. If the lines parallel to the other force are constructed at the end of each force line, the point of intersection indicates both the direction and magnitude of the resultant, or the actual force acting on the object.

- **Linear forces applied along the same action line in the same direction:** When two or more forces are applied in the same direction along the same action

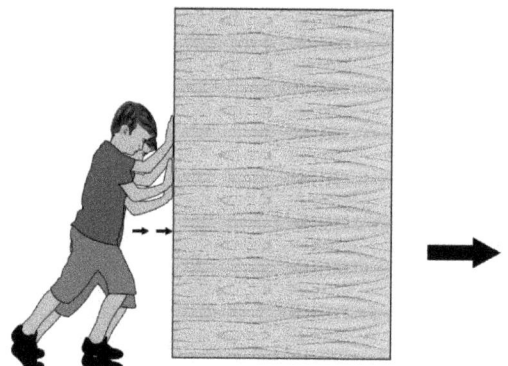

Fig. 3.11: Two linear forces being applied along the same action line in the same direction

line, then the resultant force can be obtained by adding the value of two forces (**Fig. 3.11**). For example, if a horizontal push is applied to a piece of furniture and the push is in line with the object's center of gravity, the object will move forward in a horizontal direction, providing there is no additional resisting force. If another force is applied to the furniture in line with and in the same direction as the first force, the resultant of the two forces, i.e. 'c' in the following example will have a value equal to the sum of the two forces 'a' and 'b' (a + b = c).

$$\xrightarrow{a} + \xrightarrow{b} = \xrightarrow{c}$$

Though there are rare examples in the body where two or more forces are acting at the same point and the same direction, however, there are two examples, i.e. gastrocnemius and soleus acting at the ankle joint and psoas and iliacus acting at the hip joint, and in each of both cases, the two muscles have a common tendon for distal attachment.

- **Linear forces applied along the same action line in opposite directions:** If the two forces are acting along the same line, but in opposite directions (i.e. tug of war), the resultant force can be obtained by subtracting the forces and the movement will be in the direction of the greater force [a + (–b) = c].

$$\xrightarrow{a} + \xleftarrow{b} \underset{a}{\overset{b}{\xleftarrow{}}} = \xrightarrow{c}$$

However, when two forces of equal magnitude are acting along the same line but in opposite directions, it results in equilibrium and no motion occurs (**Fig. 3.12**).

- **Concurrent forces:** Sometimes two or more forces act on the object at one point, but at different angles, such forces are called as *concurrent forces*. For example,

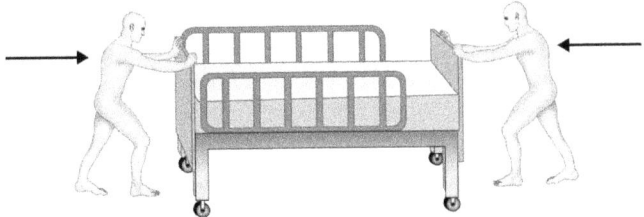

Fig. 3.12: Two linear forces being applied along the same line, in opposite directions

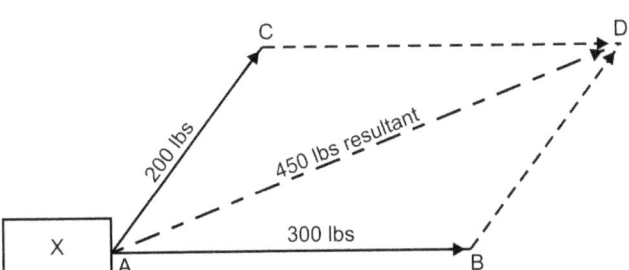

Fig. 3.13: Two concurrent forces AB and AC acting on an object 'X' and the resultant force AD

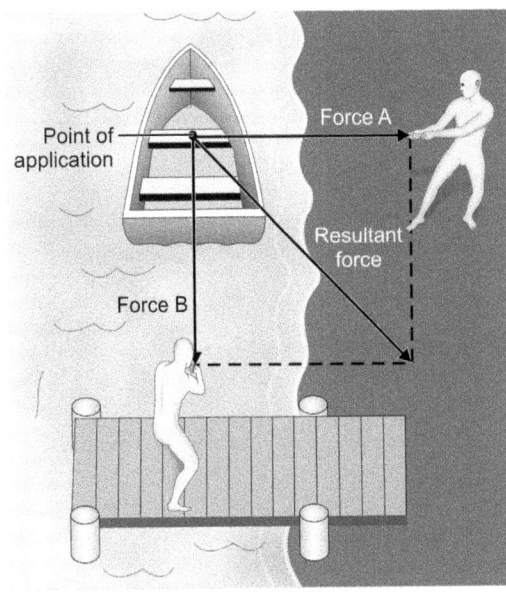

Fig. 3.14: A parallelogram shows graphically the resultant force of two concurrent forces acting on the boat

several football players opposing each other in a blocking situation apply forces at the same point from different or opposing angles.

However, it is important to remember that the resultant magnitude of the two or more concurrent forces is not their arithmetic sum and the resultant direction of two concurrent forces is not the halfway between them unless two forces are of equal magnitude. The resultant of two or more concurrent forces depends upon (i) magnitude of each force, and (ii) angle of application, i.e. the direction of each force. The resultant outcome of the blocking in the following example can be found using the combination of vectors method. The forces *AB* and *AC* acting on object *X* would move in the direction *AD*, and the magnitude of this resultant force is indicated by the relative length of line *AD*. The resultant force *AD* lies some where in between the forces *AB* and *AC*, usually more toward the side of stronger force **(Fig. 3.13)**.

Two concurrent forces acting at one common point, but acting in different angles/directions, as two people pulling on the boat is shown graphically **(Figure 3.14)** using the *parallelogram* method. The two solid lines represent the two concurrent forces. The dotted lines drawn complete the parallelogram. The diagonal of the parallelogram drawn as the middle line represents the resultant force. However, the resultant force being represented by the diagonal is moved toward the side of the stronger force if both the concurrent forces are of unequal magnitudes.

- **Parallel forces:** Sometimes, a situation may exist when two or more forces act upon a body, in which the forces are not in the same action line but are parallel to each other and act at different points on the body. The parallel forces are applied in the same plane.

The resultant effect the parallel forces will have on the body they act upon depends on (i) the magnitude of each force, (ii) direction, and (iii) application point of each force.

  – The parallel forces may act in the same or opposite directions. They may be balanced and cause no motion or they may cause linear or rotatory motion.
  – When parallel forces act on an object their relationship to the object's fixed axis or to its center of gravity, (if it can move freely), determines the resultant action. In the case of the weight (mats) held at opposite ends being carried by two persons **(Fig. 3.15)**, each person is exerting an upward force on an end while the weight of the mats exerts a downward force between them. As such, there are three parallel forces here, two in one direction and the third in opposite direction. All three forces are acting on the same object, but at different points. The mats will remain balanced and motionless if all the forces about the center of

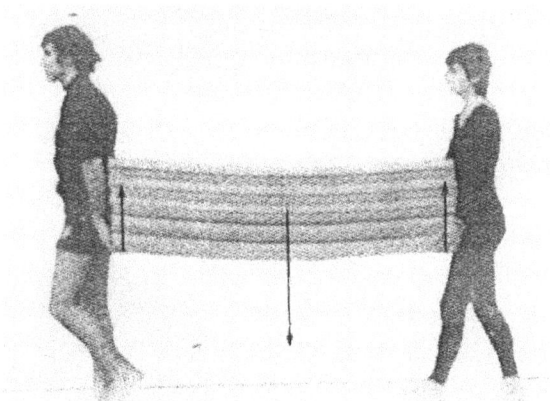

Fig. 3.15: An example of parallel forces, the mats exerting a downward force while both persons exert parallel forces upward

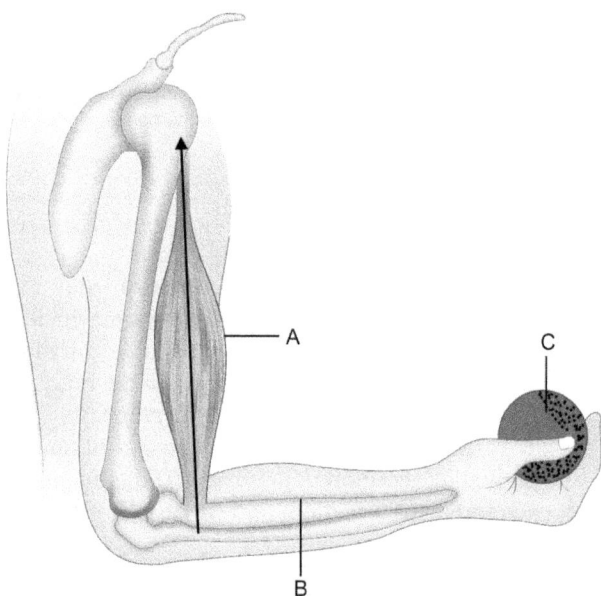

Fig. 3.16: An example of parallel forces in equilibrium. The force of biceps (A) balancing the opposite force of gravity acting on the forearm at its center of gravity (B), and on the weight held in the hand (C)

gravity are balanced. The mats will move upward in a linear fashion if both the persons exert equal parallel forces upward, that are greater than the downward force of the weight. However, if both the persons apply equal and parallel forces in the opposite directions (one in upward, the other in downward direction), the mats will move in a rotatory fashion about their center of gravity.

- The another example of parallel forces acting at different points in opposite directions is shown in **Figure 3.16**, where say a 10 lb weight is held in the hand when the forearm is flexed so that the angle of pull of the biceps is 90°. The force of biceps (A) acting in upward direction is balancing the force of gravity acting in opposite downward direction at two different points, i.e. forearm's center of gravity (B) and the weight held in the hand (C). The force of biceps is pulling the forearm up, while the force of gravity acting at two different points is pushing the forearm and the weight downward. Here, all three forces are parallel to each other but acting at different points.

In applying force, whether to move the body or some object, the objective of the movement must be considered. It is not necessary to apply maximum force and thus increase the momentum of a ball or an object being struck in all situations. It is necessary to consider whether the desired outcome calls for a maximum production of force available in the body or a controlled application of force. An all-out effort and therefore maximum force, is required if the purpose is to throw as far or as fast as possible or to jump as far or as high as possible. However, in skillfull performance, regulation of amount of force is necessary. Many a putt has been missed in golf because too much force was used and many a catch has been missed because the ball came to the receiver from a short distance too forcefully. Therefore, judgment as to the amount of force required to throw a softball a given distance, or hit a tennis ball across the net and into the court is all important.

- In activities involving movement of various joints, as in throwing a ball or putting a shot, there should be summation of forces from the beginning of movement in the lower segment of the body to the twisting of the trunk and movement at the shoulder, elbow, and wrist joints. The speed at which a golf club strikes the ball is the result of a summation of forces of the trunk, shoulders, arms, and wrists. Shot-putting, discus and javelin throwing are other good examples that show that summation of forces is essential.

MOTION

Motion is happening all around us—people walking, cars traveling on highways, airplanes flying in the air, water flowing in rivers, balls being thrown, and so on.

Motion can be defined as an act or process of changing place or position with respect to some reference point.

Whether or not a body is at rest or in motion depends totally upon the reference. Motion involves direction and speed.

Since motion is a change in position, it must be a term defining change relative to something else. For example, the fingers flex on the palm, the elbow flexes on the upper arm, etc. The pedestrian moves with respect to the side walk. Likewise, motion in space without a point of reference may not be perceived readily.

When walking down the street or riding a bicycle or serving a tennis ball, it seems obvious that movement or motion is involved. However, the motion status of a sleeping passenger in a smoothly flying plane is less obvious with reference to the earth. On the other hand, the person riding a bicycle is at rest with respect to the bicycle and the sleeping passenger in the plane is at rest with respect to anything in the plane. Therefore, rest is also a relative term. An object is said to be at rest when its position with respect to some point, line, or surface remains unchanged. However, it is also possible to be at rest and in motion at the same time, relative to different reference points. The sleeping passenger is at rest relative to the plane, and in motion relative to the earth. Similarly, a person may sit quietly in a chair—a state of rest with respect to the earth, but at the same time he could be typing rapidly—which denotes a state of motion of hands relative to the arm and the rest of his body. Therefore, the human motion varies from very precise, little movements to a wide range of movements involving the whole body.

The relative motion of two bodies depends entirely upon their relative velocities through space. Two athletes running at 8 km/hr in the same direction are at rest with respect to each other. However, if one jogs at 8 km/hr and the other at 10 km/hr, the former athlete would be considered to be at rest with respect to the later athlete, but the later athlete would be in motion both with respect to the slower runner and to the earth.

Motion is fundamental in physical education and sports activity. Motion of a body or object or any part thereof, is generally produced or at least started when a force of sufficient magnitude is applied to overcome the inertia of a body or object. As such, motion can not occur without a force and the muscular system is the source of force in the human body. Thus, development of the muscular system is very essential for causing movement or motion.

Types of Motion

Basically, there are two types of motion;
1. *Linear,* or *translatory motion,* and
2. *Angular,* or *rotatory motion.*

Either an object moves entirely from one place to another, or it turns about a center of motion. Sometimes, an object does both types of motion simultaneously.

Linear or Translatory Motion

This type of motion is a motion along a line, directly from one point or location to the another point or location. In other words, the object or the body is carried as a whole in whatever position it may be; for example, a passenger in a car carried forward by the vehicle.

Linear or translatory motion is further classified into two types:
- **Rectilinear motion:** It occurs when the object or body moves along a straight line. In other words, it is the straight line progression of an object as a whole with all its parts moving the same distance in the same direction at a uniform rate of speed. Examples of rectilinear motion are a water-skier being pulled by a boat or a bowling ball moving in a straight path.

- **Curvilinear motion:** It refers to the motion along a curved line. In other words, in the curvilinear motion, the object moves in a curved pathway. The pathway of a ball or any other projectile in the flight, or an athlete running in the sweeping turn of a 400 meter track, or of a somersault dive in aquatics **(Fig. 3.17)** are the examples of curvilinear motion.

Almost the only time the human body experiences pure translatory motion is when it is carried by some means, is pushed or pulled, or glides over a slippery surface. When the whole body is propelled from one place to another, the end result is translatory motion. After a series of steps forward as in walking, the body may appear to have experienced translatory motion. Actually this is accomplished by a series of movements of the body segments, each movement being rotatory.

Angular or Rotatory Motion

This type of motion involves rotation about an axis. This type of motion is typical of levers, and of wheels and axles. *This motion is characterized by a movement around an axis with all parts of the object or body moving in an arc.* The body or its parts move around a point of contact or around a central point known as the axis of rotation. This point may be within or outside the body.

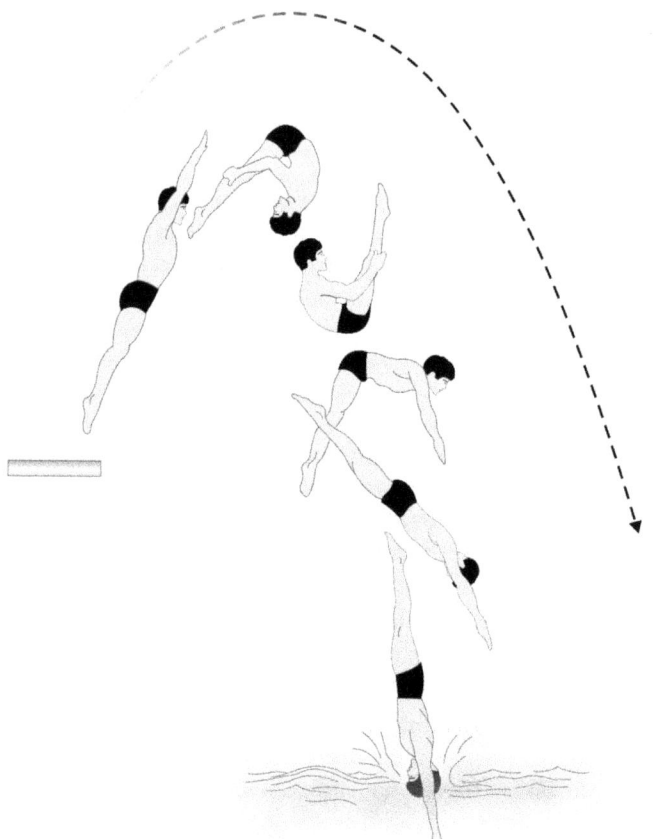

Fig. 3.17: An example of curvilinear motion

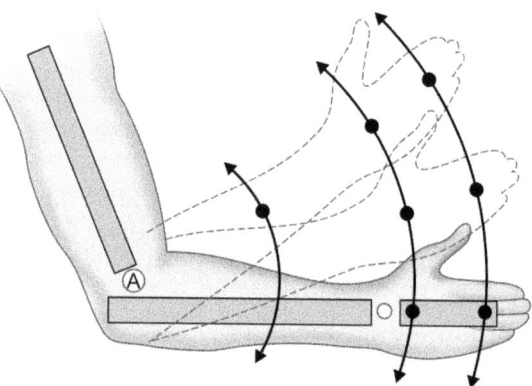

Fig. 3.18: An example of angular motion of the forearm

Rotatory or angular motion occurs when any object acting as a radius moves in a circular path about a fixed point. The distance traveled may be a small arc or a complete circle.

In the human body, the axis of rotation is provided by the various joints **(Fig. 3.18)**. *Most of the human movements are not strictly linear or rotatory, rather a combination of two types. In a sense, these two types of motion are related, since angular motion of the joints can produce the linear motion of walking.*

- In many sports activities, the cumulative angular motion of the joints of the body imparts linear motion to a thrown object (i.e. ball, shot) or to an object struck with an instrument (i.e. bat, racket).
- The motion of head in the act of indicating 'no', the lower leg in kicking a ball or the hand and forearm in turning a door knob are all examples of rotatory or angular motion. In all these instances, the moving body segment may be likened to the radius of a circle. However, these movements are not to be confused with circular motion. Circular motion describes the motion of any point on the radius, whereas the angular motion involves the motion of the entire radius.
- Rotatory or angular motion may take place around either of two axes, i.e. the longitudinal or the transverse axis. The longitudinal axis runs the length of the object and the transverse axis runs across the object at any level in its entire length. The transverse axis may also extend in any direction. The rotation around the longitudinal axis may be described as clockwise or counterclockwise.
- Various segments of the body are attached to each other, and all movements of these segments are rotations around the articulating point in exactly the same sense that the pendulum rotates about its point of support. Thus, it can be seen that the movements of flexion, extension, abduction or adduction in any joint are rotatory motions around the axis of the joint.

Application of Force Determining the Kind of Motion

Some force is required to produce all motion. The place at which that force is applied to an object and the other conditions and forces affecting the object, i.e. environmental pathways available to the object and the presence or absence of additional external factors, determine the type of motion resulting.

- If the force is applied to the midpoint of a uniform cylindrical body resting on its side, that body will roll or if the surface is very slippery and the friction is reduced to a minimum, the body may slide straight forward. Rolling, the first one is the rotation around the

longitudinal axis and the sliding, the second one is a translatory motion.
- If the force is applied a little nearer one end than the other the movement will no longer be straight ahead but will be in an arc with a long radius. As the force is moved nearer the end, this rotation becomes more marked and the radius shortens.
- When the force is applied at one end, rotation will be around the center of the body.

Thus, the farther the force is applied from the center of the object, the greater the effectiveness of that force for rotation about the center.

Factors Modifying Motion

Motion is usually modified by a number of external factors such as friction, air resistance, and water resistance. Whether these factors are a help or a hindrance depends upon the circumstances and the nature of the motion. The same factor may facilitate one type of motion, yet hinder another type of motion. For example, friction is a great help to the runner because maximum effort may be exerted without danger of slipping; however, on the other hand friction hinders the rolling of a ball, as in hockey or golf. Similarly, the wind or air-resistance is of great help to the speed and motion of sailboat, and if it is not a tailwind, it impedes the runner. However, the records made in track and field events are disallowed if the runner is aided by a tailwind of greater than a said magnitude.

- If the air resistance will assist or impede the motion of the objects, it will depend on the size and weight of the objects and their speed.

   Therefore, it is very important in sports to learn how to take advantage of these factors when they contribute to the desired movement and on the other hand how to minimize them when they are disadvantageous to the desired movement.
- There are also certain anatomical factors that modify the motion of body segments. These include friction in the joints, tension of antagonistic muscles, tension of ligaments and fasciae, and any anomalies of bones and joints etc. These all collectively comprise the internal resistance.

Newton's Laws of Motion

Sir Isaac Newton formulated three laws of motion in 17th century which explain why objects move as they do, and also explain the characteristics of motion which are fundamental to understanding human movement. Newton's laws also indicate many applications of these laws to various physical education and sporting activities. Although all of these three laws can not be proved entirely on earth even in the ideal experimental situations, they are accepted as universal truths to explain the effects of force.

Newton's First Law of Motion

Newton's first law states that **an object which is at rest tends to remain at rest and if it is in motion, it tends to remain at motion at the same speed in the same straight line, unless acted upon by an external force**.

The law is also called as the '*Law of Inertia*'. Inertia is described as the resistance to action or change. Inertia is the tendency of an object to stay at rest or in motion. An object or a body at rest tends to resist being set into motion and if it is set in motion, it will continue in motion until another force stops it.

When the total of all forces acting on an object is zero, the body is in a state of equilibrium. This state changes only when an unbalanced force enters the picture. A projectile like a ball or a missile would travel indefinitely through space in a straight line if there were not forces of gravity, friction, and air-resistance which alter its course or bring it to a halt.

- *In terms of human movement, inertia refers to resistance to acceleration or deceleration.* To demonstrate this law, riding in a car may be considered. If the car moves forward quickly from a starting position, body of person riding pushes against the back of the seat and his neck probably hyperextends. His body was at rest before the car moved and it tended to stay at rest as the car started to move. Once moving, if the car were to stop suddenly, his body would be thrown forward and his neck would go into extreme flexion because his body was in motion and tended to stay in motion when the car stopped. Similarly, when a running horse suddenly stops, the rider falls forward because the lower part of the body of the rider (which was in contact with the body of the horse) stops or comes to rest suddenly whereas his upper body part continues to move forward, due to inertia.
- A force is needed to overcome the inertia of an object or a body. In case of humans, the muscles produce the force necessary to start or stop motion, accelerate or decelerate motion or change the direction of motion. The greater the mass of an object, the greater its inertia. Therefore, the greater the mass, the more force needed to significantly change an object's inertia.
- Numerous examples or Newton's first law are found in physical education and sports activities. For example,

kick a football and it will roll along the grass. If no forces act on it, the ball will roll forward forever. However, the friction of the grass acting on the surface of the ball overcomes the moving inertia and causes the ball to stop.

Similarly, a sprinter in the starting blocks must apply considerable muscular force to overcome resting inertia. A runner on an indoor track must apply considerable force to overcome moving inertia and stop before hitting the wall. Starting, stopping, and changing direction are part of many physical activities that provide a number of other examples of Newton's first law, applied to body motion.

Newton's Second Law of Motion

Newton's second law of motion is also referred to as the '*Law of Acceleration*' or '*Law of Momentum*'. According to this law, **when a body is acted upon by a force, its resulting acceleration is directly proportional to the force and inversely proportional to the mass. And, the change in the acceleration of a body occurs in the same direction as the force that caused it.**

This law is expressed by the formula,

$$a \propto \frac{F}{M}$$

or $F = m \times a$

a = Acceleration of the body
F = Force applied
m = Mass of the object

'*Acceleration*' may be defined as the rate of change in velocity. To attain speed in moving the body, a strong muscular force is generally necessary. '*Mass*', the amount of matter in a body affects the speed and acceleration in physical movements.

In simpler words, a greater force is required to initiate (or stop) the motion if the mass of the body is greater and also that when the mass of a body is certain (fixed one), a greater force is required for affecting greater velocity (or speed) to that body.

Newton's second law can be demonstrated using a football. For example, first when the ball is mildly kicked, notice the distance it travels. Next, when the ball is kicked about twice as hard as the first kick, notice that the ball will now travel approximately twice the distance to the first kick. In this instance, relatively more muscular force was used for the ball to travel greater distance than when the ball traveled lesser distance with the use of less muscular force. Similarly, in hockey also, harder the ball is hit, the faster and farther it will move off across the grass. Thus, the amount of acceleration depends on the strength of the force applied to an object.

- The force required to run at half speed is less than the force required to run at top speed. To impart speed to a ball or an object, it is necessary to rapidly accelerate the part of the body holding the object.
- Acceleration can also deal with a change in direction. Force is needed to change the direction and according to the law, the change in direction of an object depends on the force applied to it.

Another component of Newton's second law deals with the mass of an object. Acceleration is inversely proportional to the mass of an object. If we apply the same amount of force to two objects of differing mass, the object with greater mass will accelerate less (move at lesser speed) than the object with less mass. This can be easily demonstrated by first rolling a football, then rolling a medicine ball with the same amount of force. The heavier medicine ball will not travel as far as the football. Similarly, in throwing events, it would require more force to throw a discus than a javelin because a discus is relatively heavier than a javelin. Football, basketball, track, and hockey are a few sports that demand speed and acceleration.

When the body moves, the supporting surface against which it applies force develops an equal and opposite momentum. The momentum of an object is the product of its mass times its velocity. This is expressed in the following formula,

$M = mV$

whereas, Momentum = mass × velocity
M = Momentum
m = Mass
V = Velocity

Velocity is the distance that the object moves in a given time. It is expressed as a formula:

$$V = \frac{d}{t}, \text{ i.e. Velocity} = \frac{\text{distance}}{\text{time}}$$

Since the momentum is the product of mass times velocity, according to this law, the force or the momentum of one object (mass times velocity) equals the force or the momentum (mass times velocity) of the other object. This can be expressed as following formula:

$M_1 = M_2$ M_1 = Momentum of one object
$m_1 v_1 = m_2 v_2$ M_2 = Momentum of the second object

It shows that the smaller the mass of one object in relation to the other, the faster will be its velocity in relation to the velocity of the other. Thus, when a runner pushes backward against the earth, the equal and opposite force produced by the earth causes him to move forward, because of the tremendous weight of the earth in relation to the weight of the runner.

Newton's second law describes the relationship between applied force, mass, and acceleration.

F = m a
where F = Force
m = Mass
a = Acceleration

Acceleration 'a' can be expressed as the change in velocity per unit of time.

$$a = \frac{v_2 - v_1}{t}$$

here, v_2 = Final velocity
v_1 = Initial velocity
t = Time

Substituting this for 'a' in the foregoing equation, we get,

$$F = \frac{m(v_2 - v_1)}{t} = \frac{mv_2 - mv_1}{t}$$

$$Ft = mv_2 - mv_1$$

Since the product of mass and velocity is called momentum, the force then becomes defined as the change in momentum of the moving object in a given unit of time. In other words, the force is directly proportional to the momentum.

From the foregoing equations it is clear that the racket of a tennis player strikes the ball with a momentum proportional to its mass and the velocity with which it is moving. The player can hit the ball harder by 'getting his body' into the serve, and thus increasing the mass or by increasing the velocity with which the racket moves and hence the velocity which it imparts to the ball. In either case, in order to produce a larger momentum (mv) in the object, a larger Ft must be applied to the object.

Newton's Third Law of Motion

Newton's third law of motion is also known as the '*Law of Reaction*'. According to this law, **for every action there is an equal and opposite reaction.**

The strength or magnitude of the reaction is always equal to the strength or the magnitude of the action and it occurs in the opposite direction. The force applied by a body on other body is known as action, while the effect (of other force) on the first body is called as reaction.

An excellent example is found in the recoil of a gun. The force exerted upon the bullet and the gun is equal and opposite, but the greater weight of the firearm causes it to move relatively less as compared with the lighter bullet.

This law can also be demonstrated by jumping on a trampoline. The action is one jumping down on the trampoline. The reaction is the trampoline pushing back with the same amount of force. This causes the jumper to rebound up in the opposite direction that he jumped. The harder he jumps, the higher he rebounds.

As we walk on a supporting surface, our feet push down and back, while the surface pushes up and forward. We provide the action force, while the surface provides the reaction force (an equal resistance back) in the opposite direction to the soles of our feet. The force of the surface reacting to the force we place on it is referred to as '*ground reaction force*'. It is easier to run on a hard track than on a sandy beach because of the difference in the ground reaction forces of the two surfaces. The track resists the runner's propulsion force and the reaction drives the runner ahead. The sand dissipates the runner's force and the reaction force is correspondingly reduced with the apparent loss in forward force and speed.

A sprinter applies a significant force on the starting blocks which resist with an equal force. When a body is unsupported in the air, as it is in jumping or diving, movement of one part of the body produces a reaction in another part. This occurs because there is no resistive surface to supply a reaction force. The other examples of Newton's third law include, a weight lifter performing a bench press applies force to the barbell to lift it, the barbell in turn pushes down on the hands of the weight lifter. A basketball player dribbling the ball exerts force on it to push it down toward the floor and as he does so, the ball resists his action and exerts force against his hand. This law is also important in fancy diving, in stunts on the trampoline, and tumbling stunts executed in the air.

Thus, an endless number of examples could be cited, where one body exerts a force against another and receives in return a force exerted equally in the opposite direction. This characteristic action of two bodies exerting forces on each other forms the basis of Newton's third law of motion.

The state of motion of a body depends upon the forces acting on that body rather than on the forces it may exert on other bodies. Forces never act in isolation, rather in an interaction to each other, and that too when the two bodies come in contact with each other. Thus, when two bodies come in contact with each other, they exert in equal and opposite force on each other. And, then the motion is caused. Here, it is important to remember that the action and reaction act on different bodies and therefore do not forestall or prevent motion by canceling each other out.

LEVERS

Levers are used many times during the day, whether it is the kitchen, the garden, the workshop or on the sports field, and these levers help us accomplish various activities. Lever system is involved in many activities like using a knife to lift a lid, scissors being used to cut cloth, the head of hammer for pulling out a nail, or when a crowbar is used for lifting a heavy object and so on. All of these implants, though they differ in their shapes and structures, they all are common in that each of them is a rigid bar. When a force is applied to one of them, it turns about a fixed point and it overcomes a resistance.

Functions of Levers

The levers generally serve two important functions:
1. They are used to gain a mechanical advantage so that a force of smaller magnitude (applied over a great distance) overcomes a resistance of larger magnitude (operating over a lesser distance).
2. Speed is gained and since speed and range of motion are linked together, both speed and range of motion are increased through which a resistance can be moved.

When there is no motion, the torque produced by the force and the torque produced by the resistance are equal and the lever system is said to be balanced.
- Some levers are designed to enable using a relatively small force to overcome a relatively large resistance. In these levers, the range of movement of the resistance is relatively small whereas the range of movement of the force or effort is large. For example, the crowbar lifts a rock only a few inches, while the handle moves through a much larger distance. In other words, the power to overcome a considerable resistance is gained at the expense of range of motion.

On the other hand, the striking implements used in sports are the examples of levers that are used for gaining range of motion at the expense of force. For example, the length of golf shaft enables the club head to travel through a large arc of motion, but it is used to overcome the relatively slight resistance of the weight of the club itself. Tennis and squash racquets, hockey sticks and baseball bats are other examples of levers used to gain distance at the expense of force. These levers do not save the strength of the user, but they increase the user's range and speed of movement. For example, by striking a ball with a racquet, the striker can impart more speed to it and send it a greater distance than could be done by striking it with hand. This is because the head of the racquet travels a greater distance and therefore at a greater speed than the hand alone is able to do.
- A different kind of levers is still seen in the seesaw on the playground and the scales in the laboratory. These levers gain neither force nor distance, but provide for a balancing of weights. If the loads are equal, they will balance each other when they are same distance from the axis or fulcrum.

Different types of lever can also be found in the human body though at first sight, it appears difficult for a person to visualize the human body as a system of levers. The anatomical levers of the body cannot be changed, but when the lever system is properly understood, it can be used more efficiently to maximize the muscular efforts of the body. To understand the structure and function of levers, one should be familiar with certain terms.

Definition

A lever is defined as a rigid bar that can turn (or rotate) about a fixed point called an axis when a force is applied to overcome resistance. Thus, the axis (A) or the *fulcrum* is the point of rotation about which the lever moves.

The lever rotates about the axis when a *force (F)* [sometimes also called as *effort (E)*] is applied to it to cause its movement against a *resistance (R)* [sometimes also called as *weight (W)*].

Anatomical Levers

Referring to the definition of a lever, it appears that nearly every bone in the human skeleton can be looked upon as a lever. In the human body the bone itself serves as the rigid bar or the lever, the joint as the axis or fulcrum, and the contracting muscles as the force.

The resistance that must be overcome for motion to occur can include the weight of the body part being moved, gravity, or an external weight. The *force arm (FA)* is the perpendicular distance (or length) between the line of force and the axis, while the *resistance arm (RA)* is the perpendicular distance (or length) between the line of resistance and the axis **(Fig. 3.19)**.
- A large segment of the body, such as the trunk, the upper extremity, or the lower extremity, can likewise act as a single lever if it is used as a rigid unit. For instance,

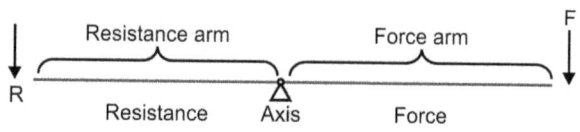

Fig. 3.19: Components of a lever

when the entire arm is raised sideward as in abduction, it is acting as a simple lever. The center of motion in the shoulder joint serves as the axis or fulcrum. The force or effort is supplied mainly by the deltoid muscle and the resistance is the weight of the arm itself. The point at which the force or effort is applied to the lever is approximately the point at which the deltoid inserts into the humerus, and the point at which the resistance is applied is the center of gravity of the extended arm. If a weight is held in the hand, the resistance point is then the center of gravity plus its load and is located closer to the hand than before. While lifting a heavy weight, the weight of the arm for the practical purposes, may be disregarded and the resistance point may be assumed to be the center of the weight's point of contact with the hand.

- Anatomical levers do not necessarily resemble rigid bars. The skull, scapula and vertebrae are the exceptions to the formal definition of the lever. The resistance point, also may be difficult to identify. It is not always easy to say whether the resistance is the weight of the lever itself or is the resistance afforded by the antagonistic muscles, which are put on a steadily increasing stretch as the movement progresses. For example, when the head is turned easily to the left, the resistance point may be regarded as the center of gravity of the head (which can only be guessed as its approximate location). However, if the turning of the head is resisted by the pressure of some one's hand against the left side of the chin, the resistance point is the mid-point of the contact area. And if the head is turned without the external resistance but is forced to the limit of motion, the resistance to the movement is afforded by the antagonistic rotators etc. The resistance point in this case is the mid-point of the area over which these resisting forces act on the head.

Classification of Levers

All lever systems as described above have each of three components, i.e. *axis*, *resistance* and *force* in one of three possible arrangements or classes of levers. The arrangement or location of these three points or components in relation to one another determines the type of lever and the kind of motion for which they are best suited. These points are the axis (A), the point of force (F) application (usually the muscle insertion), and the point of resistance (R) application (sometimes the center of gravity of the lever and sometimes the location of an external resistance).

As such, on the basis of relative arrangement or location of these three points the levers are classified into three types:

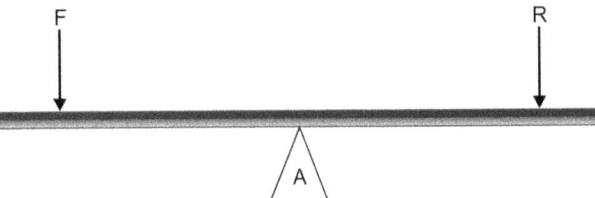

Fig. 3.20: First class lever

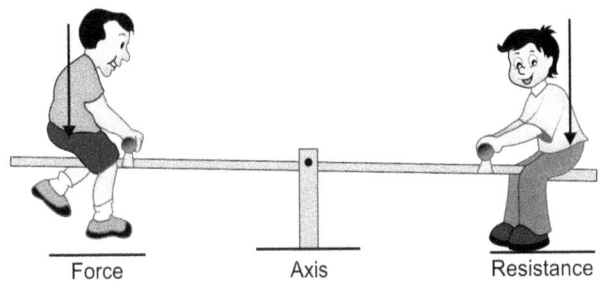

Fig. 3.21: Seesaw—An example of first-class lever
(F = force, A = axis and R = resistance)

1. **First class lever:** *In a first class lever, the axis (A) is located between the force (F) and the resistance (R)* **(Fig. 3.20)**.

   A good example of this class of levers would be a play-ground seesaw **(Fig. 3.21)**. The seesaw (lever arm) rotates on a crossbar (axis), which is located somewhere between a child sitting on one end of the board and pushing down against the ground (force), and the weight of the other child sitting (resistance) at the other end of the board. The other examples of first class lever are crowbar and scissors.

   The first class lever is best designed for balance and often sacrifices force to gain speed. An example in the human body of first class lever is moving up and down (tipping forward and backward) of the head on the 1st cervical vertebra **(Fig. 3.22)**. In this case, the axis of motion would be cervical vertebra in the frontal plane, somewhere between the ears. The resistance to the movement is the weight of one side of the head and the force or effort is supplied by the extensor muscles of the head (which are pulling on the opposite side of the head).

   The another example of this class of lever is the forearm being extended by the triceps muscle against a resistance. The axis or fulcrum is situated at the elbow joint, the force or effort is applied at the olecranon process (where triceps is inserted), and the resistance point is located at the forearm's center of gravity when no external resistance is present and at the hand when it is pushing against an external resistance.

SECTION I: Introduction and Fundamentals

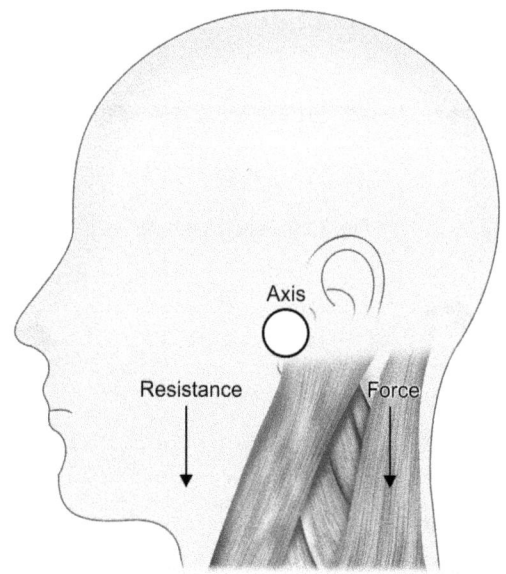

Fig. 3.22: Tipping movement of the head—An example of first class lever in human nody

Fig. 3.24: Wheelbarrow—An example of second class lever

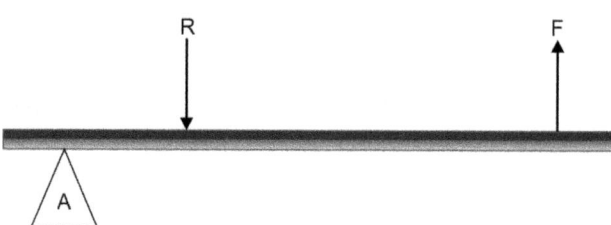

Fig. 3.23: Second-class lever

2. **Second class lever:** *A second class lever has the resistance (R) in the middle and the axis (A) at one end, and the force (F) at the other end* **(Fig. 3.23)**. In other words, the resistance lies between the axis and the force.

The wheelbarrow is an example of a second class lever, in which the wheel at the front end represents the axis, the contents of the wheelbarrow are the resistance, and the person pushing the wheelbarrow is the force **(Fig. 3.24)**. The other example of this class of levers is nutcracker.

The second class lever is best used for power at the sacrifice of the speed and there are only very few examples in the human body. Many kinesiologists believe that there are no pure second class levers in the body in which the force is a muscle contracting concentrically. However, for this class of levers from the body, some give example of the action of the ankle plantarflexor muscles when a person stands on tip

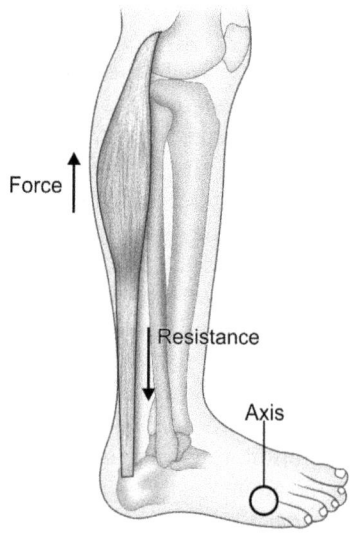

Fig. 3.25: Ankle plantarflexor muscles acting as a person stands on tip toes: An example of second class lever

toes **(Fig. 3.25)**. In this case, the axis or fulcrum is the (metatarsophalangeal) (MTP) joints in the foot at the point of contact with the ground, the resistance is the tibia and the rest of the body weight above it (at the ankle joint where the weight of the body is transferred to the foot). And the force or effort is provided by the plantar flexors at the heel where the tendon of Achilles attaches. As the planter flexors do not have to move the joint very far, however they do have a great deal of weight or resistance to overcome.

Another example of second class lever might be the forearm if it is flexed by the brachioradialis muscle

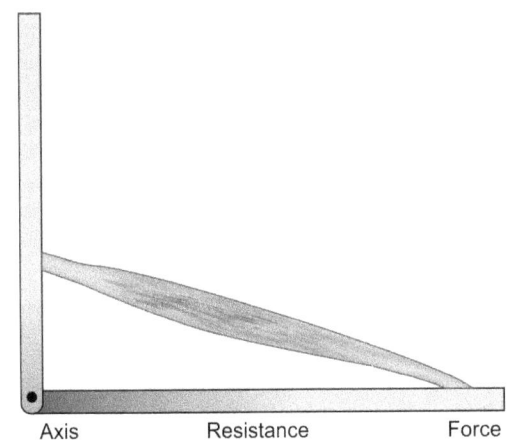

Fig. 3.26: Brachioradialis muscle flexing the elbow—An example of second class lever

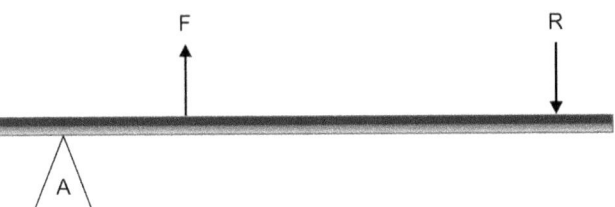

Fig. 3.27: Third class lever

alone **(Fig. 3.26)** (in case when other elbow flexors are paralyzed).

There can be other examples of second class levers of body segments if those actions where gravity supplies the force or effort for the movement and the muscles control or resist the movement through eccentric contraction are included:

- The forearm in slow downward extension; the axis or fulcrum is at the elbow joint, the force or effort may be considered to be applied at the center of gravity of the forearm and the resistance at the insertion of the elbow flexor (the brachialis muscle for this example). The brachialis muscle through its eccentric contraction resists the downward pull of the arm.
- The slow lowering of the leg at the knee from an extended to a flexed position. In this case, the axis or fulcrum is at the knee joint, the force or effort is the lower leg applied at its center of gravity, and the resistance is the eccentric contraction of the quadriceps at its point of insertion on the tibial tuberosity.

- **Third class lever:** *A third-class lever has the axis (A) at one end, the force (F) in the middle, and the resistance (R) at the opposite end;* in other words, the force is placed between the axis and the resistance **(Fig. 3.27)**.

An example of this type of lever would be a door that has a spring attachment in which the axis is represented by the door hinges, the force is the spring that closes the door, and the resistance is the door itself **(Fig. 3.28A)**. The other example of this class of levers is shoveling dirt or snow in which the lifting force is applied to a shovel handle with the lower hand while the upper hand on the shovel handle serves as the axis of rotation **(Fig. 3.28B)**.

The advantage of the third class lever is range of motion (also called speed and distance). This type of levers are the most common in the human body which require a great deal of force to move even a small resistance. A typical example is the biceps muscle flexing the elbow **(Fig. 3.29)**. In this case, the axis or fulcrum is the elbow joint, the force or effort is exerted by the biceps muscle attached to the proximal radius, and the resistance is the weight of the forearm and hand. For the hand to be truly functional, it must be able to move through a wide range of motion. The resistance will vary depending on what (if anything) is in the hand.

The brachialis is also an example of true third class leverage. It pulls on the ulna just below the elbow and since ulna cannot rotate, the pull is direct and true. Other examples include the hamstrings contracting to flex the leg at the knee in a standing position and the iliopsoas flexing the thigh at the hip.

- The presence of so many third class levers in the body (which favor range of motion) than the few second class levers (which favor power) is probably because the advantage gained from the increased range of motion (speed and distance) is more important than the advantage gained from increased power.
- With few exceptions, the levers of the body are of the third class. They have a shorter force arm than resistance arm since the muscles insert close to the joint and the weight is concentrated farther from the joint. The human body, therefore favors speed and range of motion at the cost of force.

On the whole, the levers of the human body are long and therefore, the distal ends can move rapidly. Therefore, wide movements of body can be made with speed, but at the expense of large muscle forces.

The human body does easily those tasks which involve fast movement with light objects, e.g. throwing a ball, for example. However, when heavy work is demanded, the human body must use some type of machine such as a crowbar to gain a force advantage.

- There are many levers which fall into different classes depending upon the way in which they are used. A

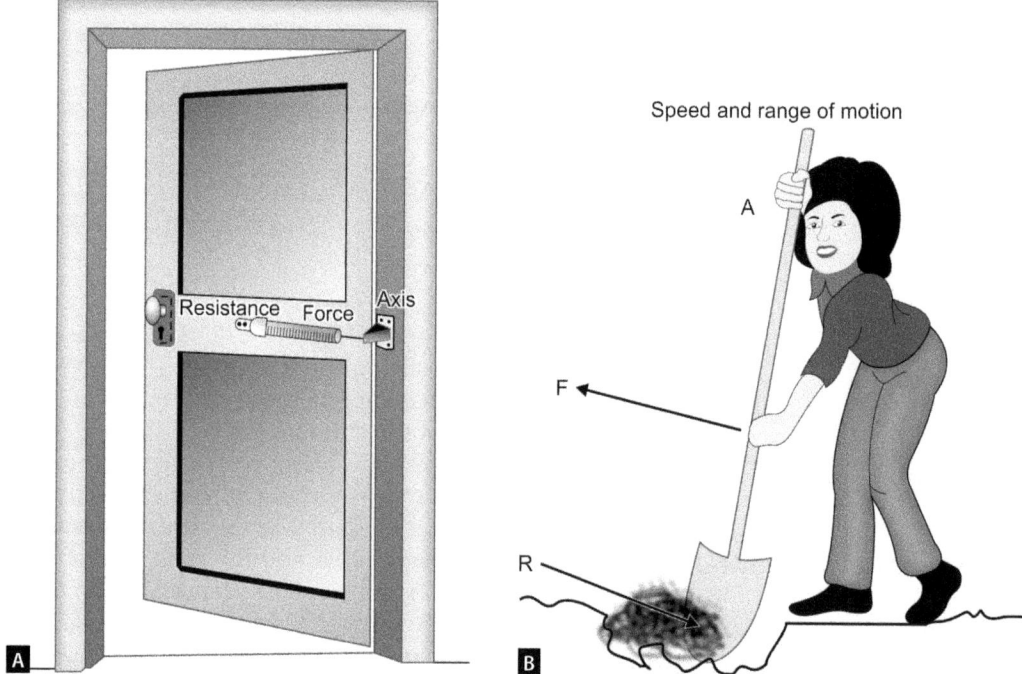

Figs. 3.28A and B: Examples of third class lever: (A) Door with a spring attachment; (B) Shoveling dirt or snow

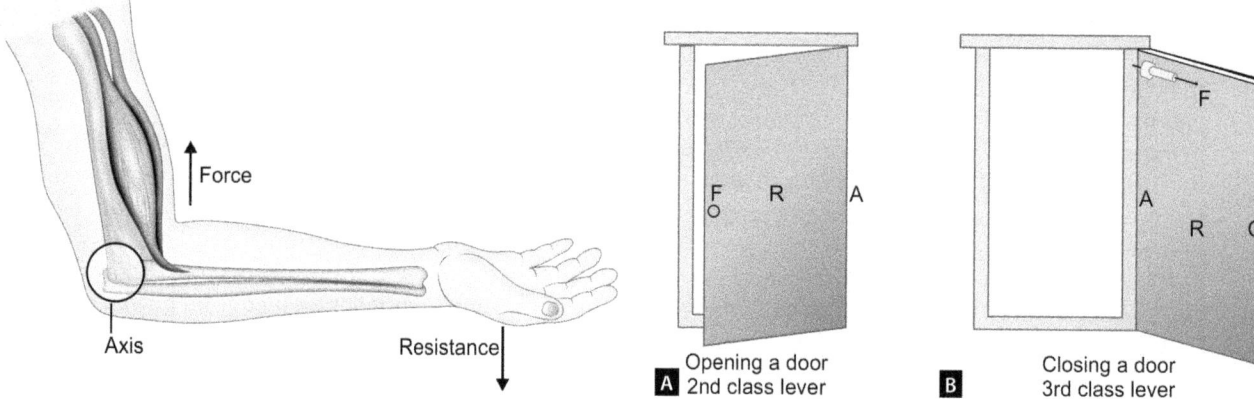

Fig. 3.29: Biceps muscle during elbow flexion—An example of third class lever

Figs. 3.30A and B: Leverage of a door: (A) F: Force applied at the knob; (B) F: Force applied by the spring

door is a second class lever when opened or closed by means of force applied at the knob, and therefore only little force is required to produce the movement. The resistance (the center of gravity of the door) is between the point of application of force (the knob) and the axis (the hinges) **(Fig. 3.30A)**. However, when the door is closed by a mechanical device (spring), it becomes a third class lever since the force is applied between the hinges and the center of the door **(Fig. 3.30B)**.

The summary of classification of levers and characteristics of each class is given in **Table 3.2**.

Factors that Change Class/Type of Levers

- Under certain conditions, a muscle may change from a second class to a third class lever and vice-versa. For example, the brachioradialis has been described as a second-class lever with the weight of the forearm being the main resistance. The weight of the forearm is located between the axis (elbow) and the force (distal attachment of brachioradialis) **(Fig. 3.31A)**. However, if a weight is put in the hand, that weight now becomes the resistance and is located farther from the axis than the force (distal muscle attachment). Therefore, the

Table 3.2: Summary of classification of levers and characteristics of each class

Lever class	Arrangement	Arm movement	Functional design	Relationship to axis	Practical example	Human example
1st	**F–A–R** Axis between force and resistance	Resistance arm and force arm move in opposite directions	Balanced movements	Axis near middle	Seesaw	Erector spinae extending the head on cervical spine
			Speed and range of motion	Axis near force	Scissors	Triceps brachii in extending the elbow
			Force motion	Axis near resistance	Crowbar	
2nd	**A–R–F** Resistance between axis and force	Resistance arm and force arm move in the same direction	Force motion (large resistance can be moved with relatively small force)	Axis near resistance	Wheel barrow, nutcracker	Gastrocnemius and soleus in plantar flexing the foot to raise the body on the toes
3rd	**A–F–R** Force between axis and resistance	Resistance arm and force arm move in the same direction	Speed and range of motion (requires large force to move a relatively small resistance)	Axis near force	Shoveling dirt, catapult	Biceps brachii and brachialis in flexing the elbow

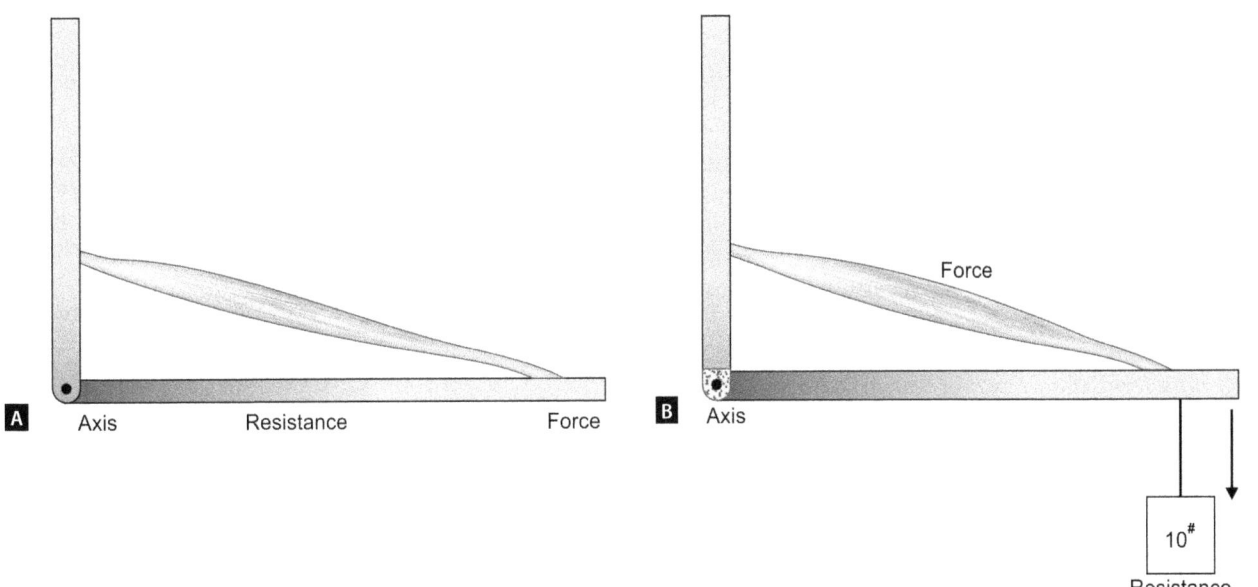

Figs. 3.31A and B: (A) Brachioradialis in elbow flexion as second class lever; (B) Brachioradialis changing to third-class lever with weight in hand

brachioradialis is now working as a third class lever **(Fig. 3.31B)**
- The direction of the movement in relation to gravity is another factor that will affect lever class. For example, the biceps brachii is a third class lever because it contracts concentrically to flex the elbow **(Fig. 3.32A)**. The biceps muscle is the force and the forearm is the resistance. The force is between the axis and resistance. Therefore, it is a third class lever. However, if the biceps muscle contracted eccentrically, it would become a second class lever; as the elbow extends, moving the same direction as the pull of gravity, the biceps must contract eccentrically to slow the pull of gravity **(Fig. 3.32B)**. The gravity and its pull on the forearm becomes the force. The biceps becomes the resistance slowing elbow extension. Now, the resistance lying in the middle between the force and axis, the biceps becomes a second class lever.

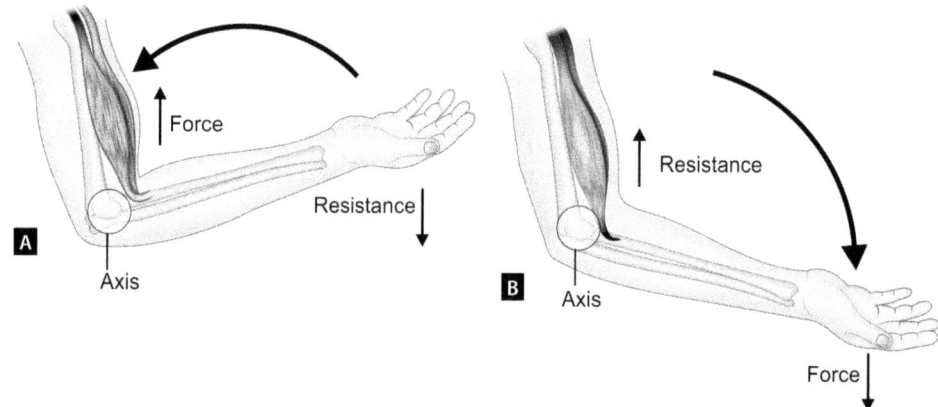

Figs. 3.32A and B: Biceps (A) as third class lever when contracting concentrically to flex the elbow; (B) as second class lever when contracting eccentrically as elbow extends

Lever Arms

The lever arms are commonly defined as the portion of the lever between the axis (fulcrum) and the force points (force, or resistance points). The force arm (FA) is the distance between the axis and the force point and the resistance arm (RA) is the distance between the axis and the resistance point. However, these definitions are only valid when the force and resistance are applied at right angles to the lever. When the force and resistance are applied at some angle other than 90° to the lever, a better definition of a lever arm is synonymous to that of a moment arm. *In fact, a lever arm is a moment arm. As such, as described earlier, in a lever, the perpendicular distance between the axis (fulcrum) and line of force (effort) is the force arm (FA). And, the perpendicular distance between the axis (fulcrum) and the line of resistance force is the resistance arm (RA).*

The Principle of Levers

A lever of any class will balance when the product of the force (effort) and the force arm equals the product of the resistance and the resistance arm. This is known as the principle of levers. It enables us to calculate the amount of force or effort needed to balance a known resistance by means of a known lever or to calculate the point at which to place the axis (fulcrum) in order to balance a known resistance with a given force or effort. If any three of the four values are known, the remaining one can be calculated by using the following equation:

Force (F) × Force Arm (FA) = Resistance (R) × Resistance Arm (RA)

$F \times FA = R \times RA$

(Force times force arm equals resistance times resistance-arm)

The product of force and force arm or resistance and resistance arm is known as the moment of force.

- It is also important to understand the length relationship between the both force arm (FA) and resistance arm (RA). There is an inverse relationship between force and the force arm, as also between the resistance and the resistance arm. **The longer the force arm, the less force is required to move the lever if the resistance and resistance arm remain constant. Similarly, if the force and force arm remain constant, a greater resistance may be moved by shortening the resistance arm.**

- The following example and figures of seesaw **(Figs. 3.33A and B)** illustrates how shortening of force arm (FA) requires application of more force to balance the system. The child *a* and child *b* each weight 20 kg and are both sitting say 5 feet from the crossbar, the seesaw will be balanced **(Fig. 3.33A)**. If child *a* moves one foot closer to the crossbar, the system is no longer balanced and child *a* would go up and child *b* would go down **(Fig. 3.33B)**. Child *a* needs to either add weight (more force) or move back (lengthen force arm) to balance the seesaw again.

- There also exists a proportional relationship between the force components and the resistance components. That is, for movement to occur when either of the resistance components increases, there must be an increase in one or both of the force components. **Figures 3.34A to C** clearly illustrate this relationships.

  Even the slight variations in the location of the force and the resistance are important in determining the effective force of the muscle.

- **As described earlier, the system of leverage in the human body is built for speed and range of**

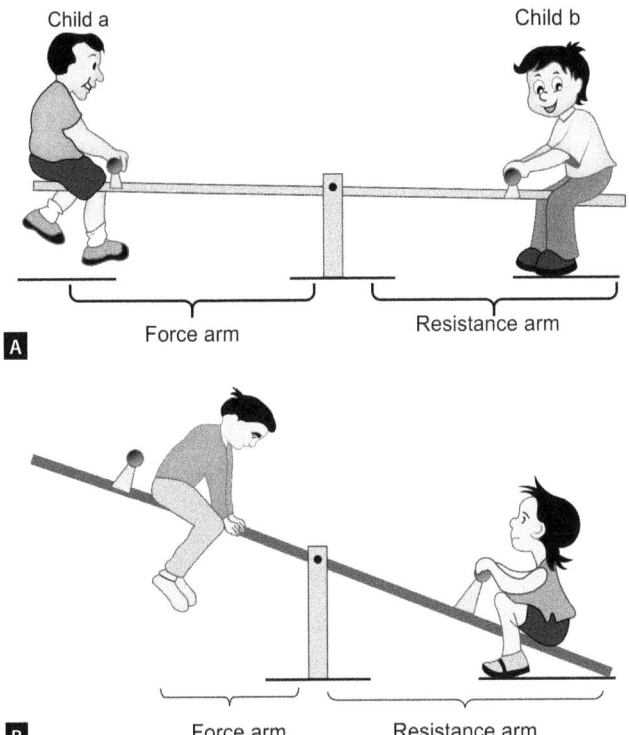

Figs 3.33A and B: In the first figure (A) of seesaw, both force arm and resistance arm are equal and the system is balanced. In the later figure (B), the force arm is shorter than the resistance arm, therefore to balance the system the force arm needs to be lengthened or more weight (force) needs to be added to the force arm

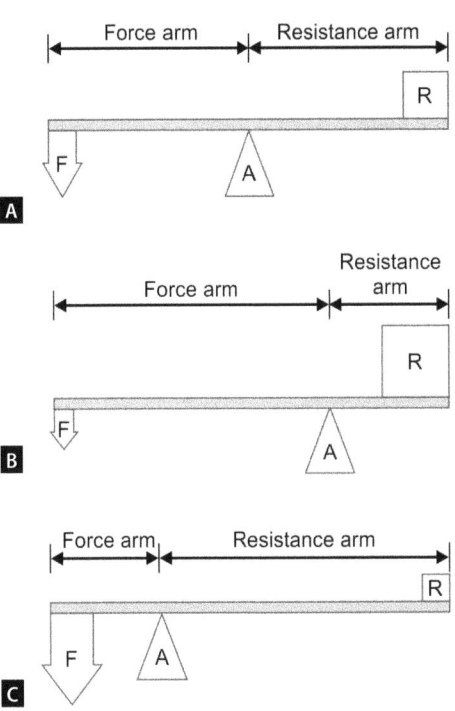

Figs. 3.34A to C: Relationship between force, force-arms, resistance, and resistance arms. (A) Force arm and resistance arm are equal in length, a force equal to the resistance is required to balance it; (B) As the force arm becomes longer, a decreasing amount of force is required to move a relatively larger resistance; (C) As the force arm becomes shorter, an increasing amount of force is required to move a relatively smaller resistance

movement at the expense of force. Short force arms and long resistance arms require great muscular strength to produce movement. In the forearm, the attachments of the biceps brachii and triceps brachii muscles clearly illustrate this point since the force arm of the biceps is 1 to 2 inches and that of the triceps less than 1 inch. Many other similar examples are found all over the body. From a practical point of view, this means that the muscular system should be strong to supply the necessary force for body movements, particularly in strenuous sports activities.

- When speaking of human leverage in relation to sport skills, there are generally many levers. For example, in throwing a ball, there are levers at the shoulder, elbow and wrist joints. In fact, it can be said that there is one long lever from the feet to the hand.

  The longer the lever, the more effective it is in imparting speed or velocity. A tennis player can hit a tennis ball harder with a straight-arm drive than with a bent elbow, because the lever (including the racquet)

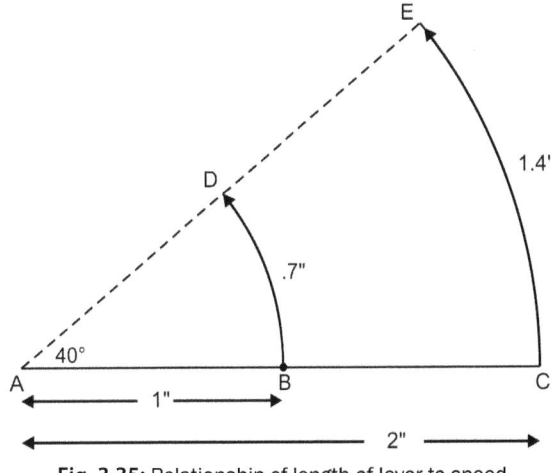

Fig. 3.35: Relationship of length of lever to speed and range of movement

is longer and moves at a faster speed. **Figure 3.35** well indicates how a longer lever travels faster than a shorter lever in traveling the same number of degrees.

In **Figure 3.35**, it is clear that both the levers move through an angle of 40° at the same velocity, the tip of longer lever (AC) travels greater distance (CE), i.e. 1.4" than the tip of the shorter lever (AB) only traveling smaller distance (BD), i.e. 0.7" in the same time. And since distance CE > BD, the speed is gained with longer lever as compared to the speed of shorter lever; (because for the same degree of movement in the same time, greater distance is covered with longer lever than with the shorter lever).

Leverage in Sports

Sports instruments such as a bat or racquet increase the length of body levers further and greatly increase the speed of the object imparting force but their use also adds greatly to the muscular effort required.

- In baseball, hockey and other sports, long levers similarly produce more linear force and thus better performance. However, for quickness of movement, sometimes it is desirable to have a short lever arm, such as when a baseball catcher brings the hand back to the ear to secure a quick throw or when a sprinter shortens the knee lever through flexion so much that the sprinter almost brings his spikes to his gluteal muscles.
- A relatively small difference in the weight of a piece of sports equipment (since this weight is so far from the axis) makes a considerable difference in the demand made on the muscles that are involved in moving it. Therefore, it is important that each individual sportsman use equipment suited to his strength.
- Lever system in the human body rarely, if ever, involves a single body part (a simple lever). Instead, movement results from a system of levers functioning together. Even when movement of a single lever does take place, many other parts of the body must be immobilized.
- When the force produced by the human system of levers is dependent upon speed at the extremity, the levers function in sequence, each coming into action at the time that the one before has reached its maximum speed. However, when many levers are brought into a heavier task such as pushing, they function together.

Mechanical Advantage

Our anatomical lever system in the human body can be used to gain a mechanical advantage that will improve simple or complex physical movements. **Simply speaking, mechanical advantage in a lever system would result when with less input or magnitude of force or effort, a greater output or magnitude of work or resistance is done. Mechanical advantage (MA) is defined as the ratio between the force arm (FA) and the resistance arm (RA).** The mechanical advantage (MA) of a lever is determined by dividing the length of the force arm by the length of the resistance arm.

$$\text{Mechanical advantage (MA)} = \frac{\text{Length of force arm (FA)}}{\text{Length of resistance arm (RA)}}$$

When the force arm (FA) is greater than the resistance arm (RA), as with second class levers, the mechanical advantage (MA) is greater than 1. For example, if a force arm is 2 feet and the resistance arm is 1 feet, the mechanical advantage would be:

$$MA = \frac{FA}{RA} = \frac{2}{1} = 2$$

This means that the force arm has twice the length as the resistance arm. Therefore, it has twice the torque or rotary force. However, if the force-arm is shorter (1 feet) and the resistance arm is longer (2 feet) as with a third class lever, the mechanical advantage would be:

$$MA = \frac{FA}{RA} = \frac{1}{2} = \frac{1}{2}$$

In other words, the force arm has half the length of the resistance arm. Therefore, it has half the torque or the rotary force.

Therefore, it can be summarized that the mechanical advantage will be gained in all the levers of the second class/order, and also in the levers of first class/order when the force-arm (effort arm) is longer than the resistance-arm (weight arm). And a condition of mechanical disadvantage results in all the levers of third class; and also in the levers of first class/order when the force arm (effort arm) is shorter than the resistance arm (weight arm).

It is always true of the lever system in accordance with the law of conservation of energy that what is gained in force is lost in distance and what is lost in force is gained in distance. In other words, to move an object using less force (MA>1) will also require that the force arm (FA) move a greater distance. Conversely, by using more force (MA<1), the force arm will need to move a shorter distance. If the mechanical advantage equals 1, the force arm (FA) and resistance arm (RA) would be equal and the system would be balanced, as in a first-class lever.

SECTION II

Kinesiology of Body Regions

- Shoulder Girdle
- Shoulder
- Elbow and Radioulnar Joints
- Wrist and Hand
- Pelvic Girdle
- Hip
- Knee
- Ankle and Foot
- Spine

CHAPTER 4

Shoulder Girdle

The shoulder complex is a term that is sometimes used to include all of the structures involved with motion of the shoulder. The *shoulder complex* consists of bones (scapula, clavicle, sternum, humerus, rib-cage), and the joints (sterno-clavicular joint, acromioclavicular joint, glenohumeral joint, and possibly scapulothoracic joint also) **(Figs 4.1A and B)**. As such, the shoulder complex mainly comprises of shoulder girdle and shoulder joint, which are two separate functional structures. However, their anatomical inter-relationships and fine cooperation enables a wide range of arm movements and hand placement while carrying out the important functions, and during various other activities of lifting, pushing, etc.

In fact, the upper extremity is suspended from the axial skeleton (head and trunk) by means of shoulder-girdle which consists of sternum and two clavicles in front, and two scapulae in the back though only two bones, the scapula and clavicle are mainly involved in the movements of shoulder-girdle. The sternoclavicular joints connect the sternum and each clavicle, and the acromioclavicular joints connect the acromion process of each scapula with the corresponding clavicle. Since there is no structural union between the two scapulae in the back, it is an incomplete girdle.

The connection of the upper extremity to the shoulder girdle is made through the shoulder joint (also called as gleno-humeral joint, formed between the head of the humerus and the glenoid cavity of the scapula).

As described above, ***the shoulder girdle mainly has two joints or articulations***:
1. **Acromioclavicular joint**, and
2. **Sternoclavicular joint**.

Acromioclavicular joint is formed between the acromion process of scapula and outer end of the clavicle

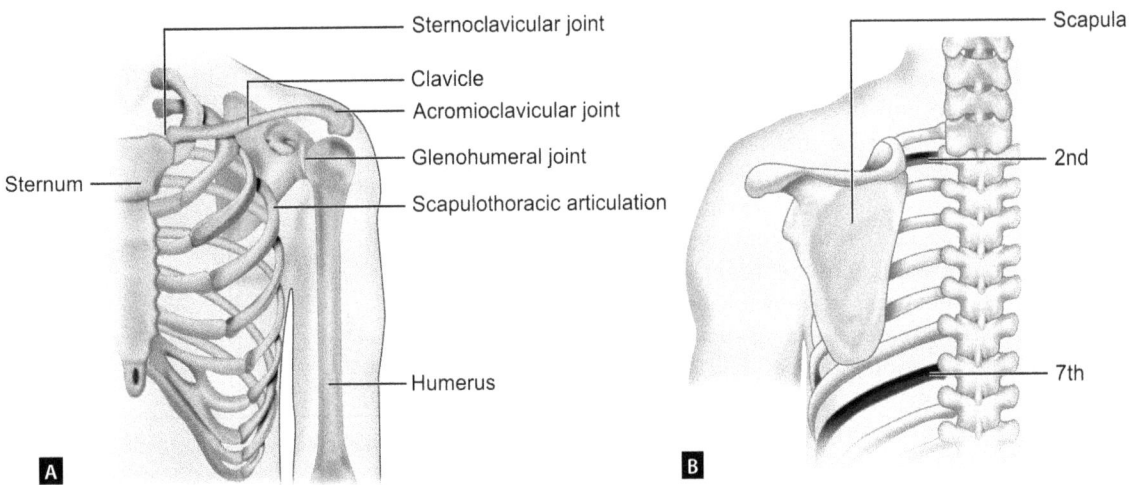

Figs. 4.1A and B: (A) Left shoulder complex (front view); (B) Resting position of left scapula on the thorax (posterior view)

which is a biaxial type of synovial joint. It has a 20–30° total gliding and rotational motion accompanying other shoulder girdle and shoulder joint motions. Its *main ligaments* are: *acromioclavicular ligaments* (located in the front), and *coracoclavicular ligaments* (the main protection against injury to the joint). The joint is strengthened from behind by the trapezius and deltoid muscles.

The sternoclavicular joint is a freely movable (multiaxial) arthrodial, ball and socket type of joint formed between the sternal end of the clavicle, and both sternum and first rib. A fibrocartilage joins the top of the clavicle with the bottom of the cartilage of the first rib near its junction with the sternum. This fibrocartilage acts as a ligament and would help in preventing upward dislocation of the clavicle. The joint capsule is strengthened in front and behind by the thickened bands of fibers called as the *anterior and posterior sternoclavicular ligaments*. There are *two other ligaments* re-inforcing this joint, i.e. *interclavicular*, and *costoclavicular ligaments*.

This sternoclavicular joint is of great importance in the movements of the shoulder girdle and of the arm as a whole as it is the only bony connection between the humerus and the axial skeleton. This joint moves anteriorly about 15° with *protraction* movement, and posteriorly about 15° with *retraction* movement. It allows an upward movement of about 45° with *elevation*, and some 5° downward movement with *depression*. Some *rotation* of the clavicle along its axis during various movements of the shoulder girdle results in slight rotary gliding movement of the sternoclavicular joint.

Scapulothoracic Joint

Certain experts also describe one functional joint of shoulder-girdle by the term *scapulothoracic joint*. Since there is no true joint or point of fixation between scapula and thorax, however, the scapula does move over the ribcage of the thorax. The scapula and thorax are indirectly attached by the clavicle and by several muscles. The moving surfaces of scapula (the medial border) and the ribs (anterolateral surfaces of first nine ribs) form through the muscle serratus anterior the so called *false or functional joint*. It is believed that this joint plays a main function of mobility and stability of the upper extremity. The movements of this joint provide a movable base for the humerus, and therefore increases the range of movements of the arm, and assists the functioning of deltoid muscle, mainly in overhead movements.

In everyday life most of the movements of the scapula are closely integrated with movements of the arm, greatly increasing the possible range of arm movements. The scapula serves as a mobile base from which the arm operates. Whatever action or position is required of the arm to accomplish a given task, the scapula moves to align the glenoid cavity to the best possible position to receive the head of the humerus.

MOVEMENTS

To define the movements of shoulder-girdle customarily involves the description of the movements of the scapulae. It is worth emphasizing here that every movement of the scapula involves motion in both the joints, i.e. the acromioclavicular, and the sternoclavicular. The main shoulder-girdle movements **(Figs. 4.2A to D)** are as follows:

Elevation: An upward movement of scapula, with the vertebral border remaining almost parallel to the spinal column (in other words, the upward movement of scapula without any rotation is called elevation). The range of movement is nearly 60° and it is thought that pure elevation is relatively a rare movement that occurs in shoulder shrugging or picking up a suit case and many other tasks in which the upper limb is not moved forward or sideward away from the body.

Depression: A downward movement of scapula (from an elevated or a normal resting position). According to some,

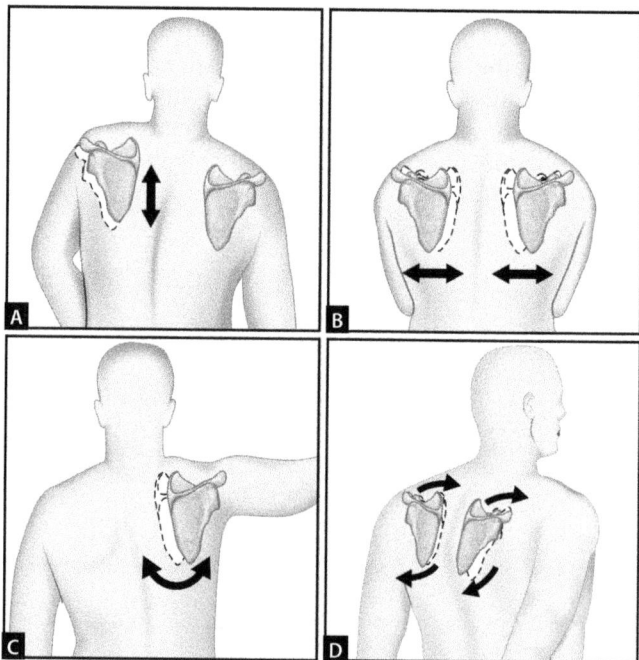

Figs. 4.2A to D: Movements of shoulder-girdle. (A) Elevation/depression; (B) Protraction/retraction; (C) Upward rotation/downward rotation; (D) Scapular tilt

no depression movement occurs below the normal resting position, however, others claim that from a sitting resting position, 5°–10° of depression movement is possible. The depression movement, from the position of maximum elevation can involve elevation of trunk by 4"–6".

Protraction (Abduction of Scapula): A lateral movement of scapula away from the spinal column (since the pure abduction of scapula keeping its vertebral border almost parallel to the spinal column is not possible, this movement is always associated with some degrees of lateral tilt of the scapula. Clavicle also moves more forward at the acromioclavicular joint, with medial border of scapula moving some 2"–3" away from the mid line).

Retraction (Adduction of Scapula): A medial movement of the scapula toward the spinal column, combined with a reduction of lateral tilt of scapula (The medial border of the scapula approaches the midline). The total range of protraction and retraction at the sternoclavicular joint has been measured nearly 25°.

Upward rotation of scapula is a movement of scapula in frontal plane that involves some what upward turning of the glenoid cavity. This movement is always associated with movement of humerus, either forward or sideward. ***Downward rotation of scapula*** is a return downward movement of scapula from the position of upward rotation, so that the glenoid cavity faces more downward. There may be slight downward rotation beyond the normal resting position also. The center of rotation is within the scapula itself. The total range of upward and downward rotation is about 60° and occur at the acromioclavicular joint. The complete range of downward rotation occurs when the dorsum of the hand is placed on the lower back region.

Forward and backward tilt of scapula are the other specialized movements of small extent, which are of very little significance to normal functioning. However, *forward tilt of scapula* occurs when the inferior angle moves backward away from the rib-cage and the superior border of scapula turns slightly forward and downward. This movement is mainly accompanied by the hyperextension of shoulder and rotation of the clavicle. *Backward tilt of scapula* is the opposite movement of forward tilt in which the inferior angle and costal surface of scapula return to the surface of rib-cage and the superior border comes to normal resting position. This movement is also accompanied by hyperflexion of shoulder joint.

Summary of muscular analysis of shoulder-girdle movements is given in **Table 4.1**.

Table 4.1: Summary of muscular analysis of shoulder-girdle movements

Movement	Prime movers	Assistant movers
Elevation	• Upper trapezius • Levator scapulae	• Rhomboids
Depression	• Lower trapezius • Pectoralis minor	• Subclavius
Protraction (Abduction of scapula)	• Serratus anterior • Pectoralis minor	
Retraction (Adduction of scapula)	• Rhomboids (Major and minor) • Middle trapezius	
Upward rotation of scapula	• Upper and lower trapezius • Serratus anterior (lower part)	• Middle trapezius
Downward rotation of scapula	• Rhomboids (Major and minor) • Levator scapulae • Pectoralis minor	

MUSCLES OF THE SHOULDER GIRDLE

There are six main muscles acting on the shoulder-girdle, which are classified as anterior or posterior muscles according to their location on the trunk **(Table 4.2)**. **Figure 4.3** shows actions of scapular muscles. The summary of major muscles acting on shoulder girdle is given in **Table 4.3**.

Serratus Anterior

This is one of the most important muscles of the shoulder-girdle which is named from its serrated or saw toothed anterior edge. This muscle lies on the outer lateral surface of the ribs and is covered by pectoralis major in front, and by the scapula at the back. The muscle may be palpated on the anterior lateral surface of the upper thorax, especially on a thin, muscular subject **(Fig. 4.4)**.

Origin: The outer surface of the upper eight or nine ribs at the side of the chest.

Insertion: The muscle is inserted to the anterior surface of the medial border of the scapula, from its superior to the inferior angle.

Table 4.2: Classification of shoulder girdle muscles according to their location on the trunk

Anterior	Posterior
Serratus anterior	Trapezius
Pectoralis minor	Levator scapulae
Subclavius	Rhomboids

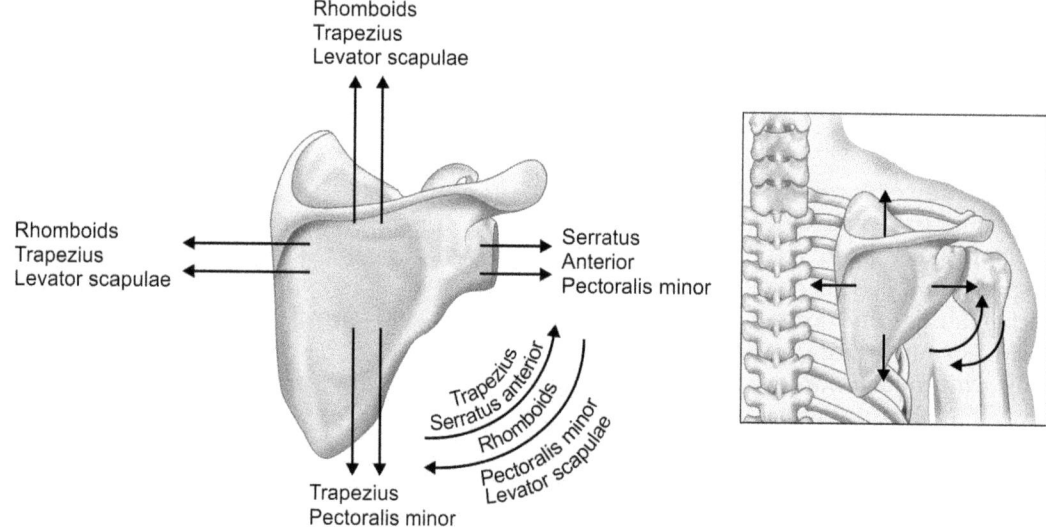

Fig. 4.3: Actions of scapular muscles

Action: Prime mover for abduction and upward rotation of the scapula.

The serratus anterior and trapezius combinedly produce upward rotation of the scapula. Activity of these two muscles is clearly evident during elevation of the arm; the trapezius being more active during abduction of arm, and serratus anterior during active flexion of the arm. Serratus anterior is said to be very important muscle in reaching and pushing activities. The paralysis of this muscle prevents elevation of the arm above 100°, and on attempting forward reaching movement, a typical *winging of scapula* is seen (in which the medial border of the scapula is lifted-up).

Pectoralis Minor

This small muscle is located on the front of the upper chest and is covered by pectoralis major **(Fig. 4.5)**. The muscle may be palpated midway between the clavicle and the nipple when the arm is elevated backward against resistance (resisted shoulder extension), particularly when the pectoralis major is relaxed.

Origin: The outer surfaces of 3rd, 4th and 5th ribs near the costal cartilages.

Insertion: The muscle is inserted to the coracoid process of the scapula.

Action: This muscle participates in several movements such as depression, downward rotation and forward tilt

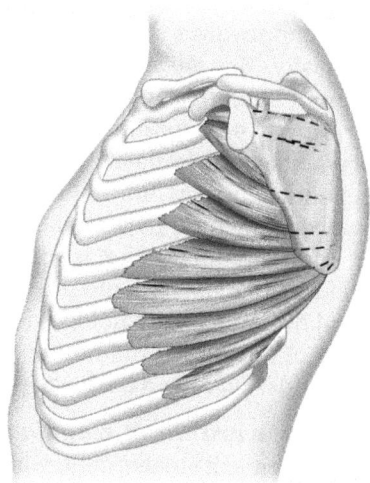

Fig. 4.4: Serratus anterior muscle

of the scapula. It also assists elevation of 3rd–5th ribs. This muscle is very important that influences good posture in stabilizing the scapula by the rhomboids and middle trapezius (the adductors of the scapulae).

Subclavius

As the name of the muscle indicates, it is located beneath the clavicle, which cannot be palpated. This is the smallest of all the muscles of the shoulder-girdle.

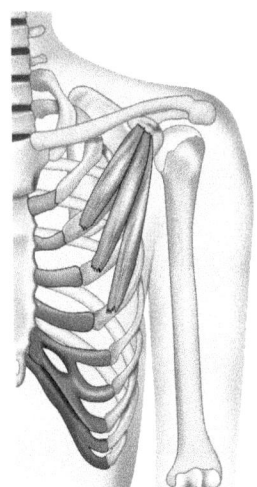

Fig. 4.5: Pectoralis minor muscle

Origin: The upper surface of the first rib, just where it joins its cartilage.

Insertion: To a groove extending along the middle half of the under side of the clavicle.

Action: Its chief function is to protect and stabilize the sterno- clavicular joint. As such, its main action is depression and the medial pull of the clavicle.

Trapezius

The trapezius is a large superficial muscle of the neck and upper back located directly under the skin and is very easy to palpate. The name of the muscle refers to its geometric figure. The muscle has three main parts such as upper, middle and lower trapezius **(Figs. 4.6 and 4.7)**. The reason for this division of the muscle into three parts is because of three different lines of pull (up, in, down) of these parts resulting in different muscle action.

Origin: The muscle originates from base of skull (occipital bone), ligament of the neck, and spinous processes of vertebrae from seventh cervical to twelfth thoracic.

Insertion: To acromion, acromial end of clavicle, and the upper border of the spine of the scapula.

Action:
- The upper trapezius performs elevation and upward rotation of the scapula. It also assists neck extension, lateral flexion and rotation of the neck to the opposite side.
- The middle trapezius performs adduction and upward rotation of the scapula.
- The lower trapezius performs, depression, adduction and upward rotation of the scapula.

As such, the trapezius shows definite activity during elevation, adduction (retraction) movements of the scapula. All three parts of the trapezius muscle work together (synergists) in the retraction (adduction) movement. However, the middle trapezius is the prime mover, and the upper and lower parts can only assist in this movement. The upper and lower trapezius are antagonistic to each other in depression/elevation, and agonist in upward rotation.

Levator Scapulae

As the name indicates this muscle is an elevator of the scapula, and as such, it shares with the upper part of trapezius and rhomboids in the movement of elevation.

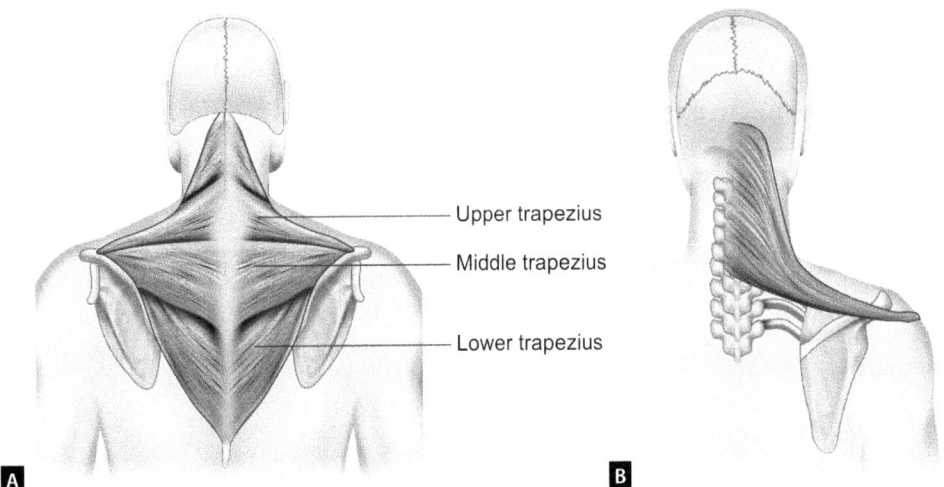

Figs. 4.6A and B: (A) Three parts of trapezius together (both sided); (B) Upper trapezius

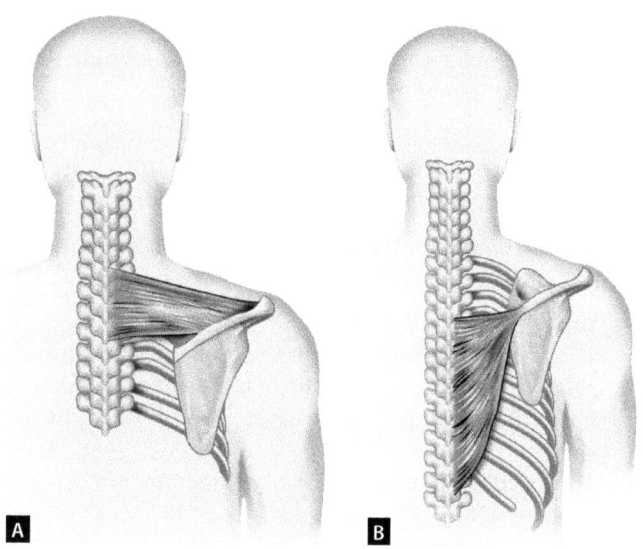

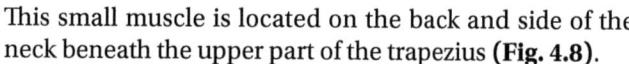

Figs. 4.7A and B: (A) Middle trapezius; (B) Lower trapezius

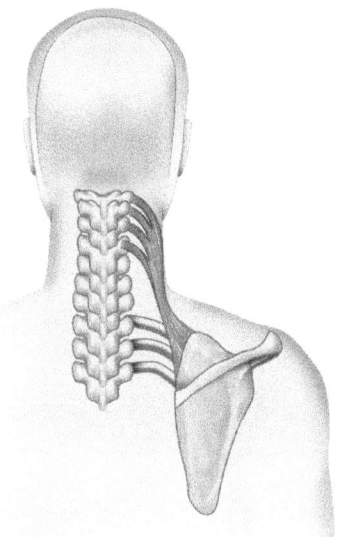

Fig. 4.8: Levator scapulae muscle

This small muscle is located on the back and side of the neck beneath the upper part of the trapezius **(Fig. 4.8)**.

Origin: From the transverse processes of upper four or five cervical vertebrae.

Insertion: To the medial border of the scapula and its spine and near the superior angle of the scapula.

Action: Its main action is elevation and downward rotation of the scapula when the trunk is in erect position. This muscle also assists the lateral flexion and rotation of cervical spine to the same side.

Rhomboids

Both the muscles rhomboid major, and rhomboid minor are functionally regarded as one muscle the rhomboids **(Fig. 4.9)**. This muscle connects the scapula to the vertebral column and lies beneath the trapezius. The name of the muscle is derived from its shape that of an oblique parallelogram.

Origin: From the spinous processes of the vertebrae of seventh cervical to fifth thoracic.

Insertion: To the medial border of the scapula from its spine to the inferior angle.

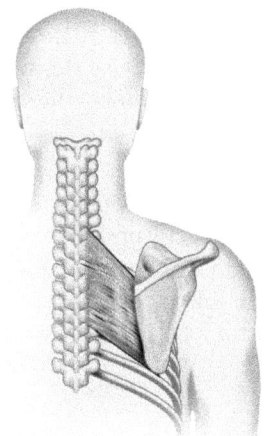

Fig. 4.9: Rhomboid muscles

Action: Adduction (*retraction*) and downward rotation of scapula. It also assists the elevation of the scapula. The cooperative action of rhomboids and middle trapezius helps in the maintenance of good shoulder posture particularly in stabilizing the scapula during abduction of the arm.

Table 4.3: Summary of major muscles acting on shoulder-girdle

Muscle	Location	Origin	Insertion	Main action	Assistive action
Trapezius	Upper back and neck	*Upper*, base of skull, (occipital bone), posterior neck ligament.	Lateral third of clavicle (posteriorly).	Scapular elevation and upward rotation.	Neck-extension, scapular retraction.
		Middle, spinous processes of 7th cervical and upper three thoracic vertebrae.	Spine of scapula.	Scapular retraction.	Upward rotation of scapula.
		Lower, spinous processes of middle and lower thoracic vertebrae.	Base of spine of scapula.	Scapular depression and upward rotation.	Scapular- retraction (adduction).
Levator-scapulae	Side of the neck (posterior)	Transverse processes of first 4 cervical vertebrae	Medial border of scapula (between spine of scapula and superior angle)	Scapular elevation and downward rotation	Lateral flexion and rotation of cervical spine to the same side.
Pectoralis minor	Front of chest (deep to pectoralis major)	Anterior surfaces of 3rd, 4th and 5th ribs	Coracoid process of scapula	Scapular depression, downward rotation and forward tilt of scapula	Scapular protraction (abduction)
Subclavius	Beneath clavicle	Upper surface of 1st rib	Underside of clavicle	Scapular depression, medial pull of clavicle	
Serratus anterior	Lateral thorax	Lateral surface of upper eight ribs	Medial border of scapula (anterior surface)	Scapular protraction (abduction), and upward rotation	
Rhomboids (Major and Minor)	Deep upper back	Spinous processes of 7th cervical to upper five thoracic vertebrae	Medial border of scapula (between spine of scapula and inferior angle)	Scapular retraction (adduction), downward rotation	Scapular elevation

Shoulder

CHAPTER 5

The shoulder or glenohumeral joint is formed by the articulation of the rounded larger head of the humerus with a shallow, pear-shaped glenoid cavity of the scapula **(Fig. 5.1)**. It is a **ball and socket type of the joint**, and the most freely movable *multi-axial joint* in the body permitting wide range of movements in all three planes and around all three axes. Both the head of the humerus and the glenoid fossa are covered with hyaline cartilage, which is thicker on the center of the head of the humerus, and in case of the glenoid fossa it is thicker around the circumference of it. This shallow glenoid cavity is further deepened by a cup of cartilage called *glenoid labrum*, which is firmly attached to the inner surface of the cavity, and the head of the humerus fits into this deepened cup. This also serves as a cushion against the impact of the humeral head in the forceful movements.

The joint is surrounded by an articular capsule *(capsular ligament)* which is attached proximally to the circumference of the glenoid cavity and distally to the anatomic neck of the humerus. This ligament is further reinforced on the front side by a strong ligament called *coracohumeral ligament* which connects the humerus with the coracoid process. The tendons of four muscles, supraspinatus, infraspinatus, subscapularis and teres minor *(rotator cuff)* strengthen the capsule and so the joint stability. The capsule is so loose that it permits the head of the humerus to be drawn out of the socket by 1, but its further pull is limited by the tone of the muscles. The joint is protected by the *acromion* which projects over it, by the *coracoid* in front, and by the *coracohumeral, coracoacromial* and *glenohumeral ligaments*. The only attachment of the shoulder joint to the axial skeleton is with the clavicle at the sternoclavicular joint.

This is highly mobile and least stable joint in the body. It is difficult to determine exactly the range of each movement of shoulder joint because of accompanying

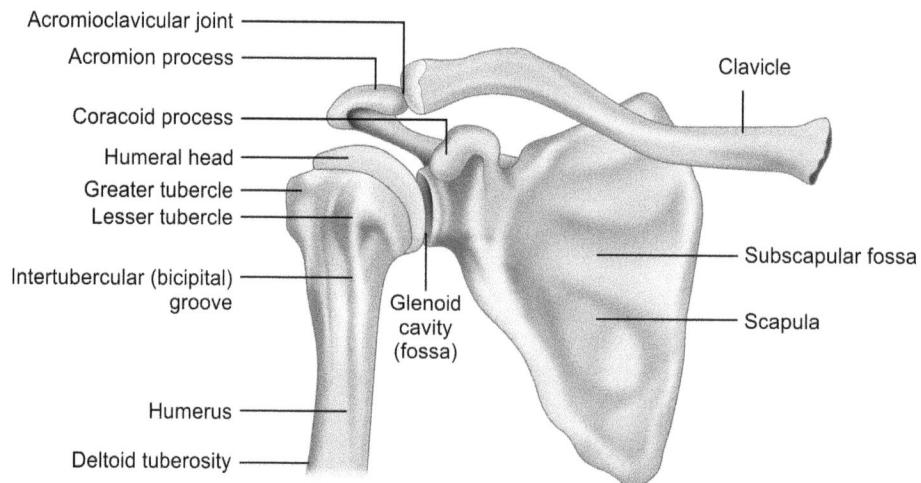

Fig. 5.1: Right shoulder joint (anterior view)

movements of shoulder girdle. When the humerus is flexed above shoulder level, the scapula is elevated, rotated upward and abducted. Similarly, with glenohumeral abduction above shoulder level, the scapula is rotated upward and elevated. Adduction of humerus results in rotation downward and depression, whereas extension of the humerus results in depression, rotation downward and adduction of scapula. The scapula abducts with humeral internal rotation and horizontal adduction. Scapula adduction accompanies external rotation and horizontal abduction of the humerus. **Table 5.1** summarizes these accompanying movements.

Table 5.1: Summary of pairing of shoulder girdle movements accompanied with shoulder (glenohumeral) joint movements

Shoulder (glenohumeral joint)	Shoulder girdle
Abduction	Upward rotation
Adduction	Downward rotation
Flexion	Elevation/upward rotation
Extension	Depression/downward rotation
Internal rotation	Abduction (protraction)
External rotation	Adduction (retraction)
Horizontal abduction	Adduction (retraction)
Horizontal adduction	Abduction (protraction)

MOVEMENTS

Flexion: From the resting anatomic position, a forward upward movement of the arm is called flexion. If it exceeds 180°, then it is called a hyper-flexion movement **(Fig. 5.2A)**.

Extension: The return movement from the flexion is the extension, and further backward continuation of the arm is called hyperextension **(Fig. 5.2B)**.

Abduction: Abduction is a sideward upward movement of the arm **(Fig. 5.3A)**.

Adduction: Adduction is a return movement from the abduction **(Fig. 5.3B)**.

Rotation: When the humerus turns around its long axis so that its anterior aspect moves medially, it is the *inward or medial rotation* **(Fig. 5.4B)** of the humerus, and if it turns laterally then it is the *outward or lateral rotation* **(Fig. 5.4A)**.

Horizontal Extension-Abduction: From the horizontally flexed arm, the horizontally backward movement is called *horizontal extension* and also as *horizontal abduction*

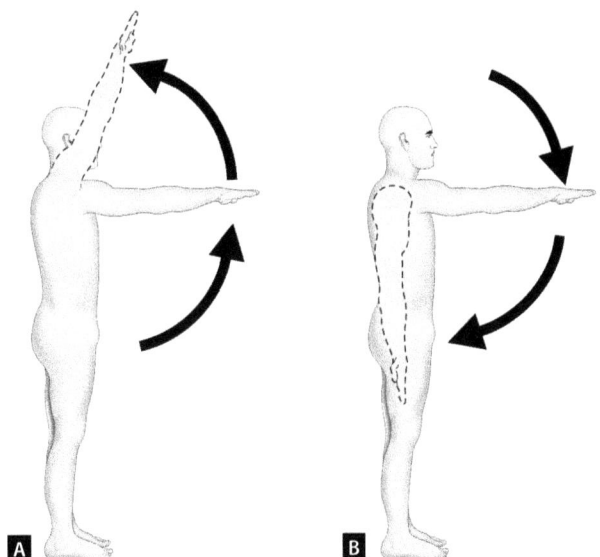

Figs. 5.2A and B: Movements of shoulder. (A) Flexion; (B) Extension

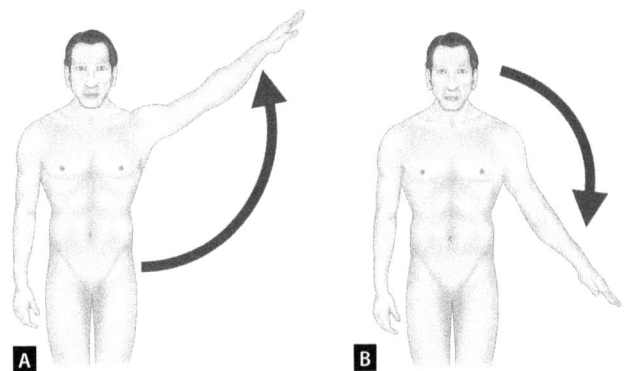

Figs. 5.3A and B: Movements of shoulder. (A) Abduction; (B) Adduction

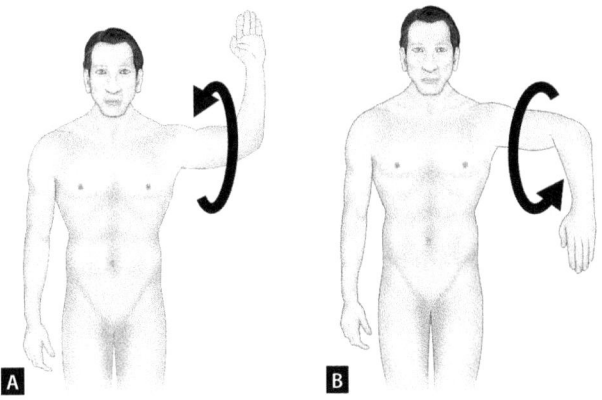

Figs. 5.4A and B: Movements of shoulder. (A) Lateral rotation; (B) Medial rotation

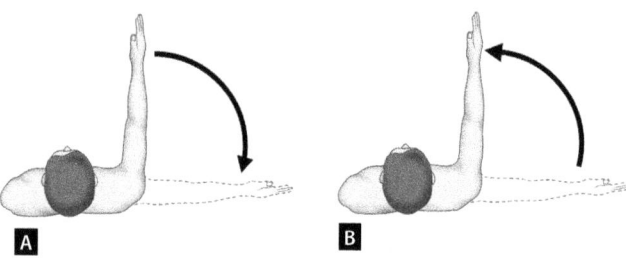

Figs. 5.5A and B: Shoulder (glenohumeral joint) movements. (A) Horizontal abduction; (B) Horizontal adduction

(**Fig. 5.5A**). From this horizontal extension-abduction position, the horizontal forward movement (the return movement) is called *horizontal flexion-adduction* movement (**Fig. 5.5B**).

Circumduction: It is not a pure movement, rather it is a combination of flexion, abduction, extension and adduction movements performed sequentially in either clockwise or anticlockwise direction.

Summary of muscular analysis of shoulder movements is given in **Table 5.2**.

As already explained above, the free movements of the arm at the shoulder or glanohumeral joint are always associated with the movements of the shoulder girdle. The flexion and abduction movements of the shoulder beyond 90° are only possible because the head of humerus rotates externally in the glenoid cavity.

It is also important to remember that during the fair range of abduction movement of head of humerus in the glenoid cavity, the larger humeral head will easily dislocate in the shallow smaller glenoid cavity. The vertical pull of deltoid muscle would also pull the humeral head up against the acromion process. However, the head of humerus is prevented from dislocating in the glenoid cavity by the action of rotator cuff muscles.

Scapulohumeral Rhythm

Normally, the movements of the shoulder joint are precisely co-ordinated series of synchronized movements of scapula and humerus. According to most of the researchers, the early phase of shoulder abduction is individually variable, but after the initial 30° of shoulder abduction throughout upto 170°, a 2:1 ratio occurs for the movements between the main shoulder joint (glenohumeral), and the so called scapulo-thoracic joint. In other words, for every 15° of shoulder abduction, the humerus moves 10°, and the scapula moves 5°. On the other hand, the other workers have found out that the greater proportion of the movement of the humerus is

Table 5.2: Summary of muscular analysis of shoulder movements

Movement	Prime movers	Assistant movers
Flexion	• Anterior deltoid • Pectoralis major (clavicular part)	• Biceps brachii (long head) • Coracobrachialis
Extension	• Teres major • Posterior deltoid • Latissimus dorsi	• Pectoralis major (sternal part) • Triceps brachii
Abduction	• Supraspinatus • Deltoid (middle)	
Adduction	• Pectoralis major (sternal) • Latissimus dorsi • Teres major	• Coracobrachialis
Internal (medial) rotation	• Subscapularis • Teres major	• Pectoralis major • Latissimus dorsi • Anterior deltoid
External (lateral) rotation	• Infraspinatus • Teres minor	• Posterior deltoid
Horizontal flexion-adduction	• Pectoralis major • Anterior deltoid • Coracobrachialis	• Biceps brachii (short head)
Horizontal extension-abduction	• Infraspinatus • Teres Minor • Deltoid (middle-posterior)	• Latissimus dorsi • Teres major

at the beginning and end range of shoulder abduction, and more scapular movement occurs during 80°–140° of shoulder abduction.

MUSCLES OF THE SHOULDER JOINT

The shoulder joint muscles can easily be distinguished from the shoulder girdle muscles, as the shoulder muscles are attached to the humerus. The various movements of shoulder joint are produced by 11 muscles. Two of these muscles, biceps brachii and triceps are primarily the muscles of elbow joint, but since they also cross the shoulder joint, they are known as the two joint muscles. The long and short heads of biceps act as two different muscles at the shoulder joint. Similarly, only the long head of triceps (out of its three heads) acts at the shoulder. Summary of major eleven muscles acting on shoulder (glenohumeral) joint is given in **Table 5.3**. These eleven muscles may be grouped according to their position in relation to the shoulder joint.

Anterior Aspect: Pectoralis major, coraco brachialis, anterior deltoid, subscapularis and biceps (short head).

Posterior Aspect: Posterior deltoid, infraspinatus, teres minor.

Superior Aspect: Middle deltoid, supraspinatus.

Inferior Aspect: Latissimus dorsi, teres major, long head of triceps.

Flexors: Flexion at the shoulder joint is produced mainly by the anterior fibers of *deltoid* and the clavicular portion of *pectoralis major* muscles. Their action is re-inforced by the assistance of *biceps* and sternocostal portion of pectoralis major when the arm is drawn into line with the body from a fully extended position. *Coracobrachialis* mainly works for the horizontal flexion and adduction movements of shoulder joint. This muscle also assists the flexion movement at the shoulder and brings the arm to the neutral position whenever the arm is rotated internally or externally.

The shoulder flexors work with scapular protractors (abductors) and elbow extensors to produce the group movement of joint which results in a forward thrust of the arm.

Deltoid

Deltoid is a triangular muscle which gives shoulder a rounded contour and covers its all three sides. Functionally the muscle is divided into three parts, i.e. anterior, posterior and middle **(Fig. 5.6)**. In structure, the anterior and posterior parts are simple penniforms, and the middle part is multipennate. The middle part is more powerful than the other two parts. Deltoid is used continuously in any lifting movement.

Origin: The muscle takes its origin from the anterior border of the outer third of the clavicle, top of the acromion process, and lower margin of the spine of scapula.

Insertion: It inserts over the lateral aspect of the humerus, at the deltoid tuberosity at the midpoint of the arm.

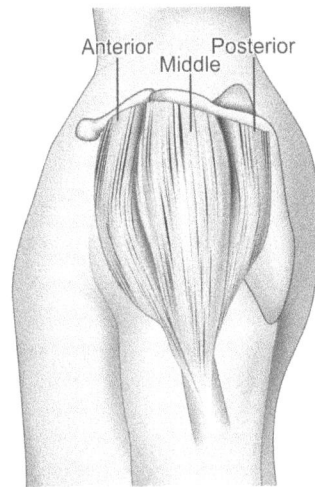

Fig. 5.6: Three parts of deltoid muscle (lateral view of left arm)

Action: The anterior part is a prime mover for the flexion and horizontal flexion movements, and an assistant mover for inward rotation and abduction. The posterior part is a prime mover for extension and horizontal extension, and an assistant mover for the outward rotation and abduction. The middle part is a prime mover for abduction and horizontal extension. The anterior fibers of middle deltoid assist the anterior part in flexion, and the posterior fibers support the posterior deltoid in extension movement.

The deltoid muscle is said to be relatively less effective in abduction when the arm is at an angle of less than 60°, and most effective when it is at between 90° and 180°. The deltoid, apart from being a powerful abductor is also frequently called upon to hold the arm in an elevated position for long periods allowing the hand to work at a height.

Pectoralis Major

This is a large, fan-shapped multipennate muscle lying over the front of the chest **(Fig. 5.7)**.

Origin: The muscle takes its origin from anterior aspect of medial two-thirds of clavicle, anterior surface of sternum, and the cartilages of the first 6 ribs (near their junction with sternum).

Insertion: The muscle inserts at the lateral border of the humerus just below the head, through a tendon 2″ – 3″ wide. The fibers arise directly from the bone and converge to insert through the tendon. Near the insertion the muscle is so twisted that the lowest fibers become the upper most.

Action: The clavicular portion is a prime mover for the flexion and an assistant mover in abduction when the arm is above the horizontal level. Its sternal part is an adductor and extensor. Both the parts help in internal rotation and horizontal flexion-adduction.

Coracobrachialis

This small muscle is named because of its attachments. The muscle is situated on the front and inner side of the arm under cover of the deltoid and pectoralis major **(Fig. 5.8)**.

Origin: The muscle takes origin from the coracoid process of the scapula and *inserts* over the anteromedial surface of the humerus opposite to deltoid attachment.

Action: The muscle is a prime mover for horizontal flexion-adduction at the shoulder joint. The line of pull of the muscle is nearly parallel to the long axis of the humerus, so its chief function is that of stabilization of the shoulder joint. The muscle also helps in flexion of the shoulder.

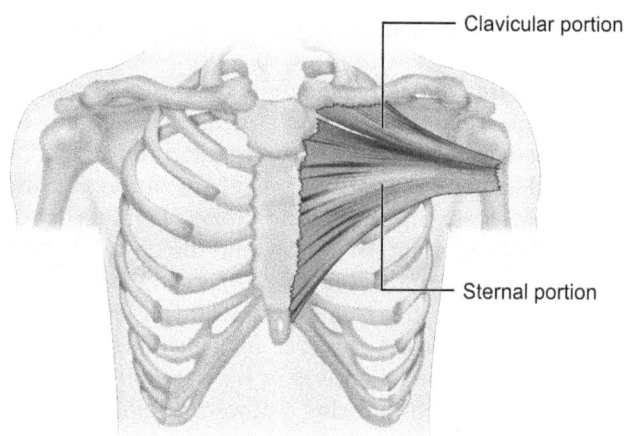

Fig. 5.7: Two parts of pectoralis major

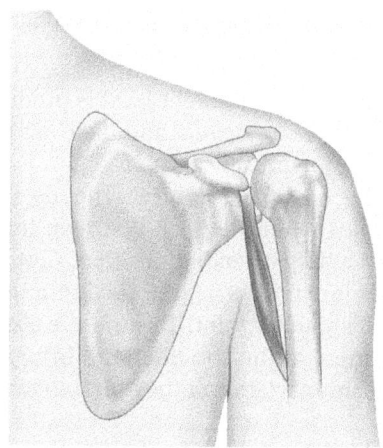

Fig. 5.8: Coracobrachialis muscle

Biceps Brachii

This is primarily a flexor muscle of elbow joint but its long head also helps in flexion at the shoulder joint. It is a very prominent muscle on the front of the upper arm. It is a fusiform type of muscle. (*See* in Chapter 6, **Figure 6.6**).

Origin: Its long head arises from the top of the glenoid fossa, and short head from the coracoid process of the scapula.

Insertion: It inserts at the bicipital tuberosity of radius. It is actually a three joint muscle which may act on shoulder, elbow and forearm.

Action: At shoulder joint, the long head of biceps helps to stabilize the shoulder joint. Contraction of its short head assists in shoulder flexion, particularly against resistance, and also in horizontal flexion-adduction and internal rotation movements.

Extensors: The movement of extension is mainly produced by the *teres major,* and *latissimus dorsi* assisted by the *posterior deltoid* and *sternocostal part of pectoralis major* muscles. Latissimus dorsi and pectoralis major are the powerful extensors from the fully-flexed position of the arm until it is brought to the line with the trunk. The teres major and posterior deltoid are primarily responsible for extending the arm from the plane of the trunk. Triceps also works as an assistant to extension of shoulder.

Latissimus Dorsi

It is a very broad muscle situated on the lower half of the back (**Fig. 5.9**).

Origin: Takes origin from the spinous processes of lower 5–6 thoracic, and all lumbar vertebrae, back of the sacrum, iliac crest and lower three ribs.

Insertion: It inserts over the anterior surface of the humerus below the head by a flat tendon. Like pectoralis major its fibers also converge from their wide origin and the tendon is twisted such as the upper fibers insert at the lower level and vice-versa.

Action: It is a prime mover for extension, adduction and hyperextension at the shoulder joint. This muscle does help in horizontal extension and internal rotation, and is involved powerfully in swimming, rope climbing, row chinning etc. This muscle as stated earlier is more effective in depressing the arm when the arm is raised between 30° and 90° angle.

Teres Major

This is a small round muscle lying along the axillary border of the scapula (**Fig. 5.10**).

Origin: It takes origin from the posterior surface of inferior angle of the scapula.

Insertion: The muscle gets inserted over the anterior surface of the humerus below the head, medial to the tendon of latissimus dorsi.

Action: Its actions are same as of latissimus dorsi- extension, adduction, hyperextension and internal rotation. The muscle is often regarded as a little helper to the latissimus dorsi. This muscle is also a prime mover for inward rotation against resistance.

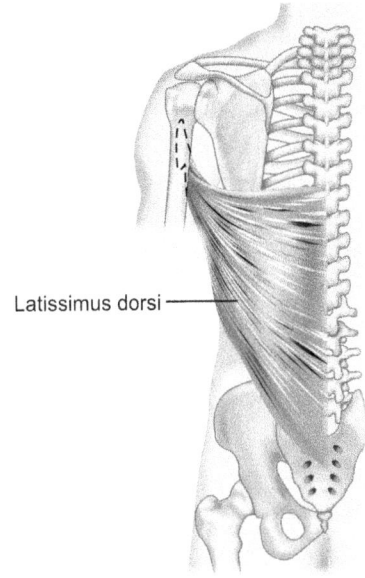

Fig. 5.9: Latissimus dorsi muscle

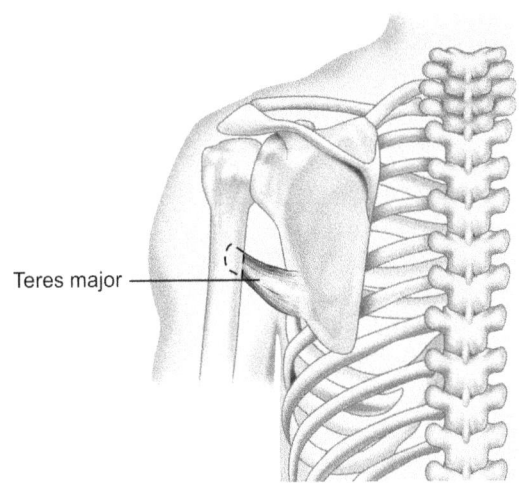

Fig. 5.10: Teres major muscle

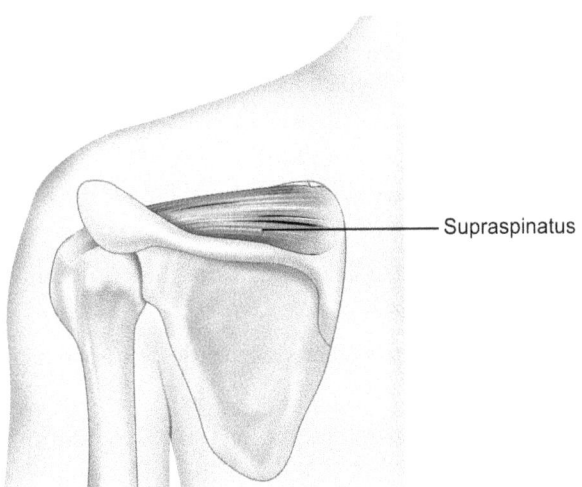

Fig. 5.11: Supraspinatus muscle

Abductors: Shoulder abduction is mainly produced by the *supraspinatus* and *deltoid* muscles. Abduction is initiated by supraspinatus and there after continued by middle deltoid. When the shoulder girdle is free to move, the lateral rotation of the scapula invariably accompanies the movement at the glenohumeral joint. To localize the work of shoulder abductors, the scapula must be fixed, and in this condition abduction is only possible to 85–90°. Therefore, some lateral rotation of humerus is essential to permit the full range of abduction.

Supraspinatus (Fig. 5.11)

Origin: This small, powerful muscle takes origin from the medial two-thirds of the supraspinatus fossa above the spine of the scapula.

Insertion: The muscle gets inserted over top of the greater tubercle of the humerus. It is a penniform type of muscle.

Action: It is a prime mover for abduction. In abduction it creates a first class lever. Its angle of pull is good for initiating abduction, and compensates the weak angle of pull of middle deltoid. This muscle also holds head of humerus against glenoid fossa during abduction. In deltoid paralysis this muscle can carry the arm through a complete range of abduction.

Adductors: In the erect position the movement of adduction is usually performed by *gravity* and controlled by *the eccentric work* of the *shoulder abductors*. However, the shoulder adduction against gravity and resistance calls upon a primary work of the adductors, mainly *pectoralis major* (sternal part), *latissimus dorsi, teres major*, and assited by *coracobrachialis*.

Rotators: Rotation movements at the glenohumeral joint are very essential in all those movements in which the arm is lifted above shoulder level, or when the hand is put into a side pocket or behind the back. Therefore, the rotation movements are of greatest importance in all the functional activities. In shoulder lesions, both inward and outward rotation are often limited, therefore these must be paid special attention.

Inward Rotators: Prime movers for this movement are subscapularis and teres major muscles, assisted by pectoralis major, latissimus dorsi and anterior deltoid

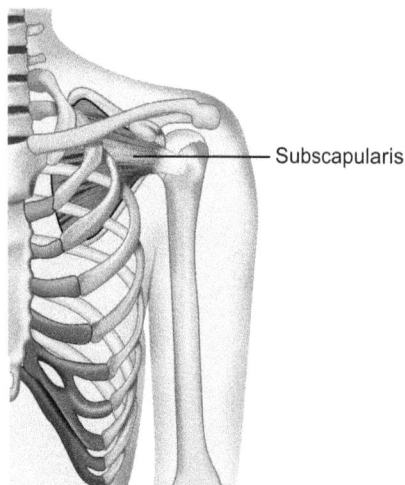

Fig. 5.12: Subscapularis muscle

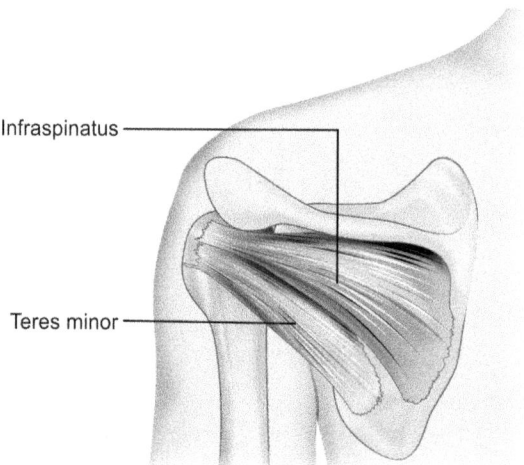

Fig. 5.13: Infraspinatus and teres minor muscles

muscles. All these muscles except subscapularis have already been described above.

Subscapularis

The name of muscle refers to its position on the anterior surface of the scapula (**Fig. 5.12**).

Origin: This muscle originates from the entire anterior surface of the scapula.

Insertion: The muscle inserts over the lesser tubercle of the humerus.

Action: This is a prime mover for inward rotation which is best performed when the arm is by the side, or elevated backward. It is antagonistic to teres minor and infraspinatus muscles. This muscle plays an important role in preventing dislocation of shoulder joint.

Outward Rotators: Infraspinatus and teres minor are the main movers for external or outward rotation, assisted by posterior deltoid.

Infraspinatus and Teres Minor

These muscles are located on the back of scapula. They are actually one muscle like two parts of pectoralis major, or three parts of deltoid (**Fig. 5.13**). Moreover, both these parts have the same function, therefore, according to some there seems no logic in two separate names for these muscles.

Infraspinatus

Origin: The muscle originates from the infraspinatus fossa.

Insertion: The muscle is inserted over the middle of greater tubercle of humerus.

Teres Minor

Teres minor takes *origin* from the dorsal surface of the lateral border of the scapula, and *inserts* to the lower part of greater tubercle of the humerus.

The main *action* of these muscles is external rotation and horizontal extension of the shoulder joint. These muscles also aid in holding the head of humerus in the glenoid fossa. They move the arm (humerus) effectively only when the scapula is stabilized.

Horizontal Flexors-Adductors: The movement of horizontal flexion-adduction is mainly produced by the *pectoralis major, anterior deltoid,* and *coracobrachialis muscles*, assisted by the short head of biceps.

Horizontal Extensors-Abductors: The movement of horizontal extension-abduction is mainly done by *infraspinatus, teres minor* and *middle-posterior deltoid muscles* and assisted by latissimus dorsi, and teres major.

In both of these movements, the scapulae are stabilized by scapular muscles and the rotational element is checked by their opposing forces.

Table 5.3: Summary of major muscles acting on shoulder (glenohumeral) joint

Muscle	Location	Origin	Insertion	Main action	Assisted action
Deltoid	Upper part of arm (anterior, lateral and posterior aspects)	*Anterior* – lateral third of clavicle	Deltoid tuberosity	Shoulder flexion, horizontal flexion	Shoulder abduction, inward rotation
		Posterior – spine of scapula	Deltoid tuberosity	Shoulder extension, horizontal extension	Shoulder abduction, external rotation
		Middle – acromion process of scapula	Deltoid tuberosity	Shoulder abduction, horizontal extension	Flexion of shoulder (anterior fibers) Extension of shoulder (posterior fibres)
Pectoralis major	Chest (anterolateral aspect)	*Clavicular part* – medial third of clavicle	Lateral lip of bicipital groove of humerus	Shoulder flexion	• Shoulder abduction
		Sternal part – front of sternum and costal cartilages of first 6 ribs	Lateral lip of bicipital groove of humerus	Shoulder adduction and extension	• Shoulder internal rotation, horizontal flexion- adduction
Coracobrachialis	Upper arm (medial)	Coracoid process of scapula	Medial border of humerus (middle part)	Horizontal flexion-adduction (stabilization of shoulder)	• Flexion of shoulder
Biceps brachii	Details described in Chapter 6				
Latissimus Dorsi	Lower back (superficial)	Spinous processes of lower 5–6 thoracic and all lumbar vertebrae, posterior surface of sacrum and iliac crest	Bicipital groove of humerus (medial side)	Shoulder extension, adduction	Horizontal extension, internal rotation of shoulder
Teres major	Posterior surface of upper arm	Posterior surface of inferior angle of scapula	Anterior surface of humerus (below its head, and medial to latissimus dorsi tendon)	Shoulder extension, adduction	Horizontal extension, internal rotation of shoulder
Triceps brachii	Details described in Chapter 6				
Supraspinatus	Posterior surface of scapula (below its spine)	Supra spinatus fossa of scapula	Greater tubercle of humerus	Shoulder abduction	
Infraspinatus	Posterior surface of scapula (below its spine)	Infraspinatus fossa of scapula	Greater tubercle of humerus (middle region)	External rotation, horizontal extension-abduction of shoulder	Shoulder stabilization
Teres minor	Posterior surface of scapula (lower part)	Lateral border of scapula (posterior surface)	Greater tubercle of humerus (lower part)	External rotation, horizontal extension-abduction of shoulder	Shoulder stabilization
Subscapularis	Underside of scapula (next to rib cage)	Anterior surface of scapula	Lesser tubercle of humerus	Shoulder internal rotation	Prevention of shoulder dislocation

CHAPTER 6

Elbow and Radioulnar Joints

Almost any movement of the upper extremity will involve the elbow and radioulnar joints. Quite often, these joints are grouped together because of their close anatomical relationship. For this reason, a learner may confuse motions of the elbow with those of the radioulnar joints. However, with close inspection, movements of the elbow joint can be clearly distinguished from those of the forearm or radioulnar joints, hence both of these will be described separately (except the muscles).

ELBOW

The elbow is a **uniaxial hinge type of joint** that allows only flexion and extension movements. In elbow, the lower end of humerus articulates with the upper ends of both radius and ulna **(Fig. 6.1)**. For this reason, the elbow by some, is actually thought of as two interrelated joints, the humeroulnar and the radiohumeral joints. Elbow motions primarily involve movement between the articular surfaces of the humerus and ulna occurring at the *trochlea* of humerus and the *trochlear notch of the ulna*. The *capitulum* (the lower end of the humerus on the lateral aspect) articulates with the circular facet on the head of the *radius* which is relatively a small amount of contact between the two bones. A joint capsule and the synovial membrane enclose both of these joints and also the *superior radioulnar joint*. The capsule is strengthened on all four sides by the bands of the fibers known as *anterior, posterior, ulnar* and *radial collateral ligaments*.

The movements of flexion and extension take place at elbow joint, occurring about the frontal axis in sagittal plane.

Flexion: Movement of the forearm to the shoulder by bending the elbow to decrease its angle **(Fig. 6.2A)**.

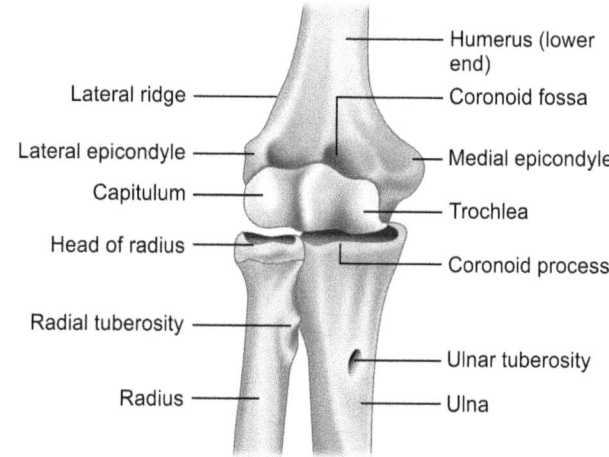

Fig. 6.1: Right elbow joint (anterior view)

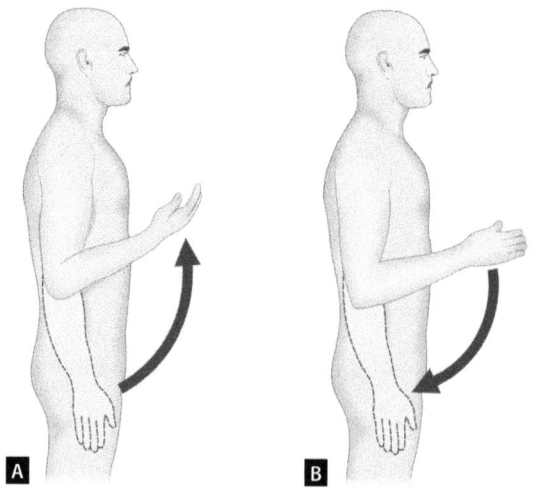

Figs. 6.2A and B: Elbow movements: (A) Flexion, and (B) Extension

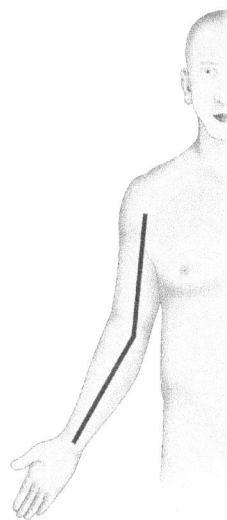

Fig. 6.3: Carrying angle

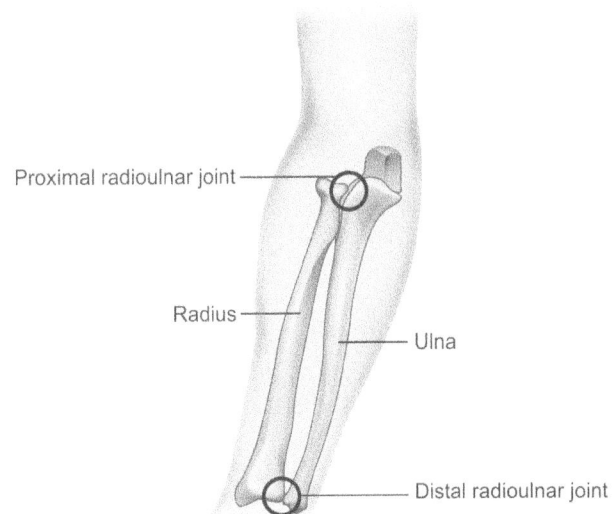

Fig. 6.4: Proximal and distal radioulnar joints

Extension: Movement of the forearm away from the shoulder by straightening the elbow to increase its angle **(Fig. 6.2B)**.

In most of the individuals, approximately 145° of flexion is measured from the 0° position of elbow extension. Flexion is limited by the contact of the soft tissues of the arm and forearm. There is no active hyperextension at the elbow as this movement is limited by locking of the olecranon process of the ulna with the humerus. However, hyperextension is possible in some gymnasts, weight lifters and others who often perform complete forceful extensions.

In full extension, the supinated forearm is about 10°–15° abducted from the axis of the arm and is called **carrying angle (Fig. 6.3)**, but in flexion the arm and forearm both lie in the same plane. The elbow and the radioulnar joints work together in harmony allowing the hand to do useful activities. In human beings, the hand can only perform the useful activities made possible by the co-ordinated positions of the joints of the elbow, forearm and shoulder. If a person loses his ability to rotate the forearm, use of his hand will be so much limited that it will adversely affect his lifestyle. The strength and stability of the elbow is much because of the bony shape which is further supplemented by the ligaments.

RADIOULNAR JOINTS

The movements of supination and pronation take place at the superior and inferior radioulnar joints **(Fig. 6.4)**. **The proximal or superior radioulnar joint is a pivot joint formed between the head of radius and the radial notch of ulna.** The superior radioulnar joint shares the synovial membrane and the capsule of the elbow joint. *Annular ligament* is a strong band of fibers encircling the radial head and holds the head of radius firmly in place. **The distal or inferior radioulnar joint is formed between the ulnar notch of the radius and head of ulna.** The distal radioulnar joint has its own synovial cavity and a thin capsule. The fibrous connection between the shafts of radius and ulna is sometimes referred by some as the middle radioulnar joint. This interosseous membrane helps to prevent the separation of both the bones.

Pronation is the rotational movement of the radioulnar joints in which the radius rotates on its long axis and the distal part of radius is swung in front of ulna such that both the bones resemble the letter *x*. The thumb moves from lateral to medial position **(Fig. 6.5A)**. In **supination**, the radius returns to its original lateral position and the thumb returns from medial to lateral position **(Fig. 6.5B)**. Range of pronation movement varies from 70° to 90°, whereas supination approximately 80° to 90°.

When the arm is extended the movements of pronation and supination are often accompanied by medial and lateral rotation at the shoulder joint respectively. Therefore, to localize the movements at the radioulnar joints, the elbow should be flexed to a right angle and the arm held still. Both the pronation and supination movements are very useful for the functions of the hand.

MUSCLES OF ELBOW AND RADIOULNAR JOINTS

Muscles of the elbow and radioulnar joints may be classified according to their actions **(Table 6.1)** and also

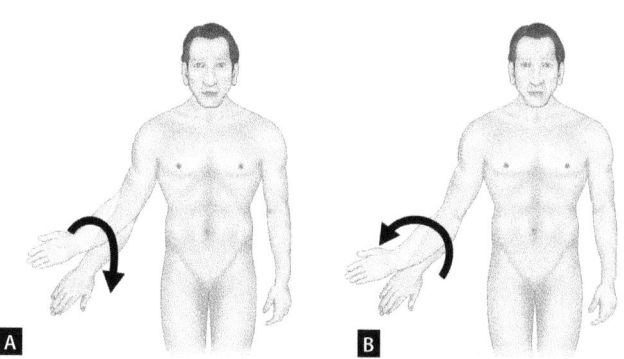

Figs. 6.5A and B: Radioulnar joints movements: (A) Pronation; (B) Supination

Table 6.1: Summary of muscular analysis of elbow and radioulnar joints movements

Movement	Prime movers	Assistant movers
Elbow		
• Flexion	• Biceps Brachii • Brachialis • Brachioradialis	• Pronator teres
• Extension	• Triceps brachii	• Anconeus
Radioulnar Joints		
• Pronation	• Pronator Teres • Pronator quadratus	• Brachioradialis • Anconeus
• Supination	• Supinator • Biceps Brachii	• Brachioradialis

Table 6.2: Summary of muscles of elbow and radioulnar joints according to their location

Anterior	Posterior
Biceps brachii	Triceps brachii
Brachialis	Anconeus
Brachioradialis	Supinator
Pronator teres	
Pronator quadratus	

to their location **(Table 6.2)**. The muscles lying anteriorly are generally the flexors and pronators of the elbow and radioulnar joints and include biceps brachii, brachialis, brachioradialis, pronator teres, and pronator quadratus. The muscles lying posteriorly are the extensors and supinators of these joints, and are mainly the triceps brachii, anconeus and supinator.

The summary of major muscles acting on elbow and radioulnar joints is given in **Table 6.3**.

Elbow Flexors: Prime movers in elbow flexion are *biceps*, *brachialis* and *brachioradialis*. Pronator teres assists the elbow flexion against resistance movement.

Biceps Brachii

We have already studied this muscle while describing the shoulder joint muscles **(Fig. 6.6)**. In case of elbow the biceps is most effective in the outer range upto 90° flexion when it has supinated the forearm. Biceps brachii is a three joint muscle, but its main function is elbow flexion and supination of proximal radioulnar joint, though it does act at the shoulder joint. Biceps is not likely to work as flexor of the elbow when the forearm is pronated, as such, it is a flexor of the elbow only when the forearm is supinated.

In elbow extension, biceps may work as a supinator against resistance, as in driving a screw. In this case its flexion tendency is neutralized by the action of triceps so that biceps may work as supinator.

Brachialis

As the name refers, it is a muscle of the arm situated near the elbow joint **(Fig. 6.7)**. The muscle takes its *origin* from the anterior surface of the lower half of humerus, and *inserts* over the anterior surface of the coronoid process of ulna. Flexion at the elbow is its main action and it is a true flexor of the joint. It is said to be more effective in elbow flexion in the middle and inner range.

Brachioradialis

This is a fusiform type of muscle situated on the lateral border of the forearm **(Fig. 6.8)**. The muscle takes its *origin* from the upper two-thirds of lateral supracondylar ridge of the humerus, and gets *inserted* over the lateral side of the styloid process of the radius.

It is primarily a flexor of the elbow, particularly when a quick and against resistance movement is performed. It also helps to maintain the mid-prone or mid-supine position of the elbow during flexion movement. During elbow flexion it tends to move the radioulnar joint to mid position if it is either supinated or pronated. It is said this muscle hardly works in elbow flexion while the forearm is supinated, however, it works only when heavy resistance is lifted. Similarly, during extended elbow, this muscle does not help in either supination or pronation unless the movement is strongly resisted. As such, the muscle is a flexor of the elbow in its mid position and maintains the neutral position of the forearm.

CHAPTER 6: Elbow and Radioulnar Joints | 97

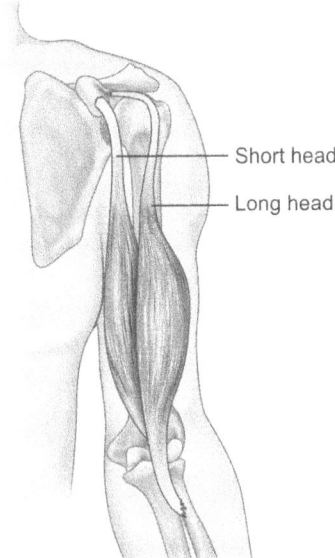

Fig. 6.6: Biceps brachii muscle

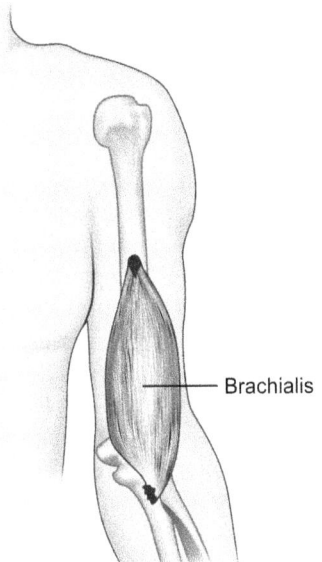

Fig. 6.7: Brachialis muscle

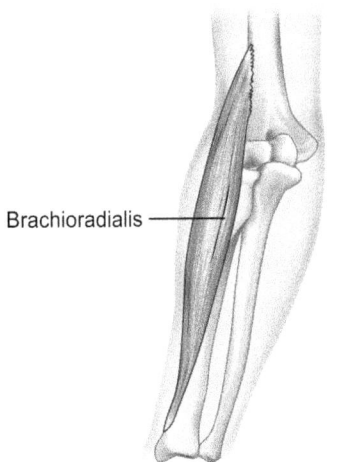

Fig. 6.8: Brachioradialis muscle

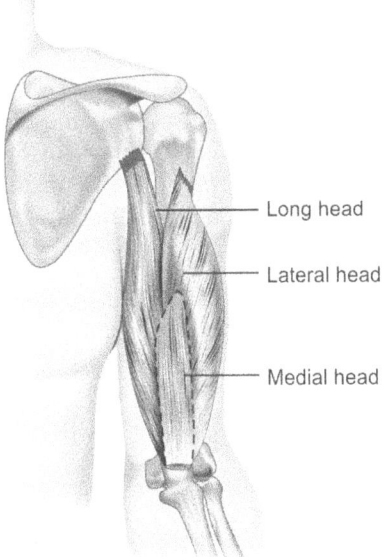

Fig. 6.9: Triceps brachii muscle

Elbow-Extensors: The movement of elbow extension is primarily produced by *triceps brachii* muscle and is assisted by anconeus.

Triceps Brachii

This muscle is located on the posterior aspect of the arm and makes up the entire muscle mass on the back of the arm. This muscle derives its name from its three heads **(Fig. 6.9)**. The *long head* of triceps derives its origin from the inferior rim of glenoid fossa of the scapula. The *lateral* head originates from the upper half of the posterior surface of humerus, whereas the *medial head* takes its origin from the distal two-thirds of humerus on its posterior aspect. The three heads come together to form the muscle belly and are *inserted* to the olecranon process of ulna. At elbow, it is a very strong extensor because of its favorable angle of pull. In a comparison of three heads, the medial head is noted to be the principal extensor of the elbow, usually accompanied by the lateral head. All its three heads often come into action when an against resistance movement of elbow extension is required. Long head of triceps muscle also assists extension movement of shoulder.

Anconeus

This is a small triangular muscle situated on the back of the arm and appears as a continuation of the triceps **(Fig. 6.10)**. It takes *origin* from the posterior surface of lateral epicondyle of the humerus, and gets *inserted* over the lateral aspect of olecranon process of ulna. The muscle helps in elbow extension. Various experiments have proved that this muscle initiates elbow extension, and then maintains the extended position, and therefore it stabilizes the joint. This muscle is also said to be active during pronation of the forearm.

Forearm Pronators: The *pronator teres* and *pronator quadratus* **(Fig. 6.11)** are the prime movers which are assisted by brachioradialis and anconeus muscles.

Pronator Teres

This is a small spindle-shaped muscle lying obliquely in front of the elbow. It is primarily a muscle of the radioulnar joint (pronator), but it does help in elbow flexion. The muscle takes *origin* from the medial epicondyle of the humerus, and the coronoid process of ulna, and gets *inserted* over the lateral surface of the radius near its middle. It is primarily a pronator of the forearm, but does help in elbow flexion when against resistance movement is performed.

Pronator Quadratus

It is a thin rectangular-shaped muscle which lies in front of lower part of the forearm. It takes *origin* from the anterior surface of lower one-fourth of ulna and *attaches* to the anterior surface of lower one-fourth of radius. The main *action* is pronation of forearm. This muscle is said to work greater than pronator teres in pronation.

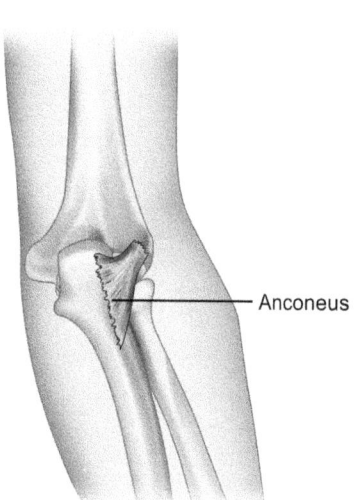

Fig. 6.10: Anconeus muscle

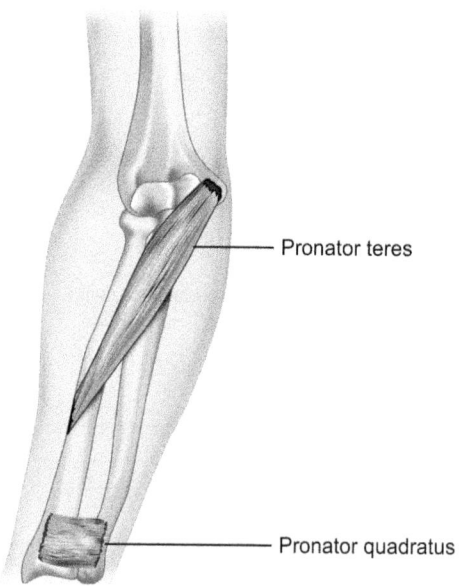

Fig. 6.11: Pronator muscles

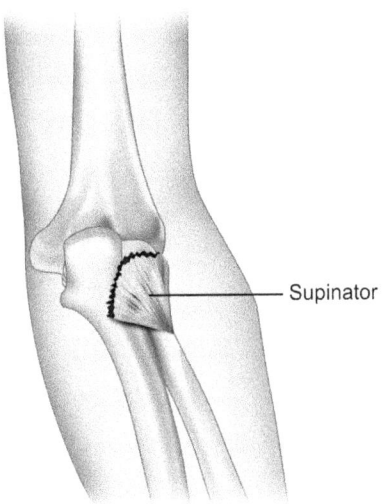

Fig. 6.12: Supinator muscle

Forearm Supinators: Supination is performed primarily by *supinator* and *biceps brachii,* assisted by brachioradialis.

Supinator

It is a broad muscle situated under the brachioradialis and the extensor muscles coming from the lateral epicondyle of the humerus **(Fig. 6.12)**. The muscle takes *origin* from the lateral epicondyle of humerus and adjacent part of ulna (supinator crest of ulna), and gets *inserted* over lateral surface of upper third of radius. Supination of forearm is the main action of this muscle. When the elbow is extended it performs the supination alone in all the slow, fast and unresisted movements. However, biceps comes to play assistance in supination when the elbow is flexed and the movement is resisted.

Table 6.3: Summary of major muscles acting on elbow and radioulnar joints

Muscle	Location	Origin	Insertion	Main action	Supportive action
Biceps brachii	Front of upper arm (superficial)	*Long head:* Supraglenoid tubercle of scapula *Short head:* Coracoid process of scapula	Radial tuberosity of radius	• Elbow flexion • Supination of forearm (radioulnar joints)	• Long head assists • Shoulder flexion • Horizontal adduction • Internal rotation
Brachialis	Front of upper arm (lower two-thirds), deep	Front of distal half of humerus	Coronoid process of ulna	Elbow flexion	
Brachioradialis	Lateral forearm	Lateral supracondylar ridge (distal aspect) of the humerus	Styloid process of the radius (distally)	Elbow flexion	Restores mid-prone or mid-supine postion from pronated or supinated forearm
Pronator teres	Front of upper forearm (obliquely)	Medial epicondyle of humerus and coronoid process of ulna	Lateral aspect of radius (at midpoint)	Pronation of forearm (radioulnar joints)	Elbow flexion
Triceps brachii	Posterior side of arm (humerus)	*Long head:* Infraglenoid tubercle of scapula *Lateral head:* Posterior surface of humerus (upper half) *Medial head:* Posterior surface of humerus (distal two-third)	Olecranon process of ulna	Elbow extension	Long head in shoulder extension
Anconeus	Posterior of elbow	Lateral epicondyle of humerus	Posterior surface of olecranon process of ulna	—	Elbow extension
Pronator quadratus	Front of lower end of forearm (deep)	Distal one-fourth of front of ulna	Distal one-fourth of front of radius	Pronation of forearm (radioulnar joints)	—
Supinator	Anterolateral aspect of upper forearm (deep)	Lateral epicondyle of humerus and adjacent ulna	Anterior surface of upper radius	Supination of forearm (radioulnar joints)	—

CHAPTER 7

Wrist and Hand

The joints of the wrist, hand and fingers are often overlooked in their importance, while compared with larger joints of the body needed for ambulation. The fine motor skills and movement characteristics of wrist and hand are needed for various manipulative and other activities, i.e. drawing, painting, writing etc. There are many sports with skilled activities that require precise functioning of the wrist and hand. Several sports like archery, golf, bowling, baseball and tennis, etc. require the combined use of all of these joints.

The hand is the distal end of the upper extremity and made up of the thumb and finger metacarpals and phalanges. The hand is the key point of function for the upper extremity which we use to accomplish a number of activities ranging from very simple to quite complex tasks. The main purpose of the other joints of the upper extremity is to place the hand in various positions to accomplish these tasks.

The functional anatomy of the wrist and hand is complex because of many muscles, bones and ligaments combined with relatively small joint size. **Wrist joint is formed between lower ends of radius and ulna with proximal row of four carpal bones** (though many experts do not consider ulna, a part of wrist joint). Wrist has a series of articulations (joints), (i) **radiocarpal joint** formed between distal end of radius and carpal bones, mainly three bones—scaphoid, lunate and triquetral, and (ii) **intercarpal** or midcarpal joints formed between the individual carpal bones themselves **(Fig. 7.1)**.

The wrist and hand region has 29 bones including radius and ulna, more than 25 joints, and more than 30 muscles (out of which 18 muscles are intrinsic having both origin and insertion in the hand).

In wrist, there are 8 carpal bones **(Fig. 7.2)** arranged in two rows (4 bones in each row). The junction between the

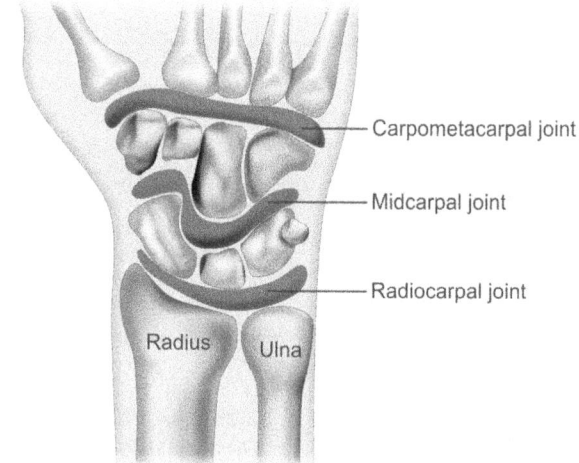

Fig. 7.1: Joints of left wrist (anterior view)

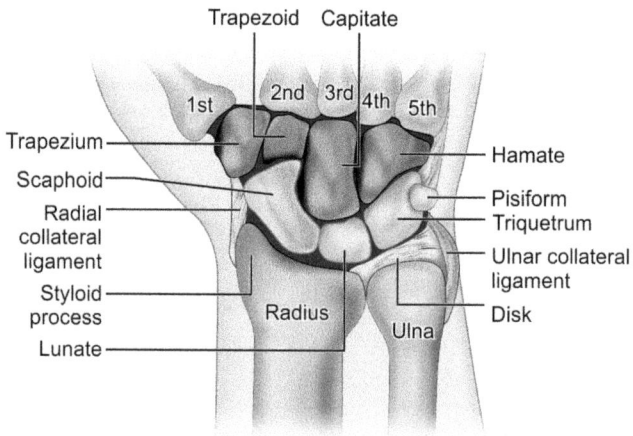

Fig. 7.2: Bones of left wrist (anterior view)

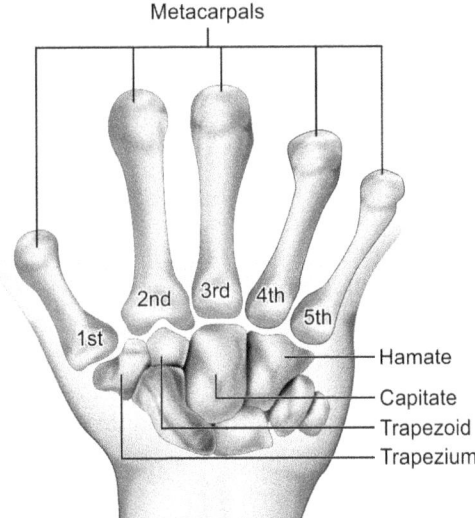

Fig. 7.3: Carpometacarpal (CMC) joints of thumb and fingers

rows forms the *midcarpal joint*. *The proximal row of carpal bones has 4 bones namely—scaphoid, lunate, triquetral* and *pisiform* while moving from radial side to ulnar side. Likewise *the distal row of carpal bones also has 4 bones namely—trapezium, trapezoid, capitate* and *hamate*, while moving from radial side to ulnar side. Five metacarpal bones, numbered one to five from the thumb to the little finger join proximally the distal row of carpal bones and form the carpometacarpal joints of thumb and fingers **(Fig. 7.3)**. There are 14 phalanges (digits), 3 phalanges for each finger except the thumb which has only 2 phalanges. The phalanges are named as proximal, middle and distal phalange respectively as moving from the metacarpals side. **Figure 7.4** shows the bones and joints of fingers and thumb.

LIGAMENTS OF WRIST AND HAND

There are many ligaments in the wrist and hand that provide support and static stability to the joints. Wrist is strengthened by following ligaments; collateral (ulnar and radial collateral ligaments), palmar (radiocarpal, ulnocarpal ligaments) and dorsal ligament (dorsal radiocarpal ligament).

- *Ulnar collateral ligament* is a rounded cord stretching from ulnar styloid process to triquetral and pisiform bones.
- *Radial collateral ligament* passes from radial styloid process to scaphoid and then to trapezium.
- *Dorsal* or *posterior radiocarpal ligament* runs from lower aspect of the radius to scaphoid, lunate and triquetral.

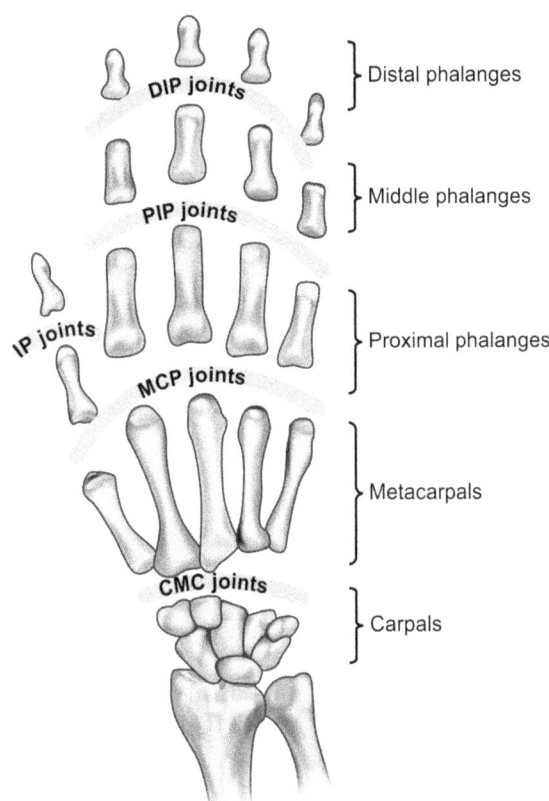

Fig. 7.4: Bones and joints of fingers and thumb

- On palmar surface, the *anterior radiocarpal ligament* and *ulnocarpal ligament* attach from the lower ends of radius and ulna to proximal carpal bones.

A number of other important ligaments are also present in the hand region:

- *Flexor retinaculum ligament* extends over the anterior surface of the wrist in a mediolateral (horizontal) direction.
- *Palmar carpal ligament* attaches to the styloid process of the radius and ulna and crosses over the flexor muscles.
- *The transverse carpal ligament* attaches medially to the pisiform and hamate and laterally to the scaphoid and trapezium bones.
- *The extensor retinaculum ligament* extends on the posterior side of the wrist in a horizontal direction, from the styloid process of ulna medially to the triquetrum, pisiform and lateral side of radius bones.
- In the hand, particularly in the fingers there are pairs of collateral and palmar ligaments for each of metacarpophalangeal (MCP), proximal interphalangeal (PIP), and distal interphalangeal (DIP) joints.

MOVEMENTS OF WRIST AND HAND

The wrist joint is a **condyloid** type of joint permitting movements of flexion (70°-90°) **(Fig. 7.5A)**, extension (65°-85°) **(Fig. 7.5C)**, abduction or radial flexion (15°-25°) **(Fig. 7.5E)**, and adduction or ulnar flexion (25°-40°) **(Fig. 7.5D)**. Wrist movements occur primarily at radiocarpal joint formed between the distal end of radius and proximal row of carpal bones mainly consisting of scaphoid, lunate and triquetral. The radiocarpal joint is also classified as a biaxial joint which does not permit rotation at the wrist.

The midcarpal or intercarpal joints are classified as *plane joints* which are nonaxial joints that allow gliding motions, which collectively contribute to radiocarpal joint motion.

The available range of motion (ROM) at the wrist is a combination of radiocarpal and midcarpal movements. Flexion of wrist occurs more at midcarpal joints, and extension of wrist is greater at radiocarpal joint, and the combined movement is about 85° in each flexion and extension. Abduction occurs mostly at midcarpal joint (range is about 15°). Adduction involves most movement at radiocarpal joint and has a range of 45° or so.

Carpometacarpal (CMC) joints occur between the distal row of carpal bones and proximal end of the metacarpal bones.

Flexion: Movement of palm of the hand and/or the phalanges toward the anterior or volar aspect of forearm.

Extension: Movement of back of the hand and/or phalanges toward the posterior or dorsal aspect of forearm (sometimes also referred as hyperextension).

Abduction (Radial Flexion): Movement of thumb side of hand toward the lateral aspect or radial side of forearm, whereas abduction at fingers is the movement of fingers away from the middle finger.

Adduction (Ulnar Flexion): Movement of little finger side of the hand toward the medial aspect or ulnar side of forearm. In case of fingers, adduction is a movement of the fingers back together toward the middle finger.

Opposition: Movement of the thumb (comprising of combined flexion, abduction with some rotation) across the palmar aspect to oppose any or all of the phalanges of the fingers. Likewise, opposition of little finger also takes place which is a movement of the little finger across the palm to oppose the phalanges of the thumb.

Each finger has three joints. The metacarpophalangeal joints (MCP) are classified as biaxial condyloid type of joints. The proximal interphalangeal joint (PIP) is classified as uniaxial hinge type that can move from full extension to nearly 90°-120° of flexion. The distal interphalangeal joint (DIP) is also classified as uniaxial hinge type permitting only flexion upto 80°-90° from full extension.

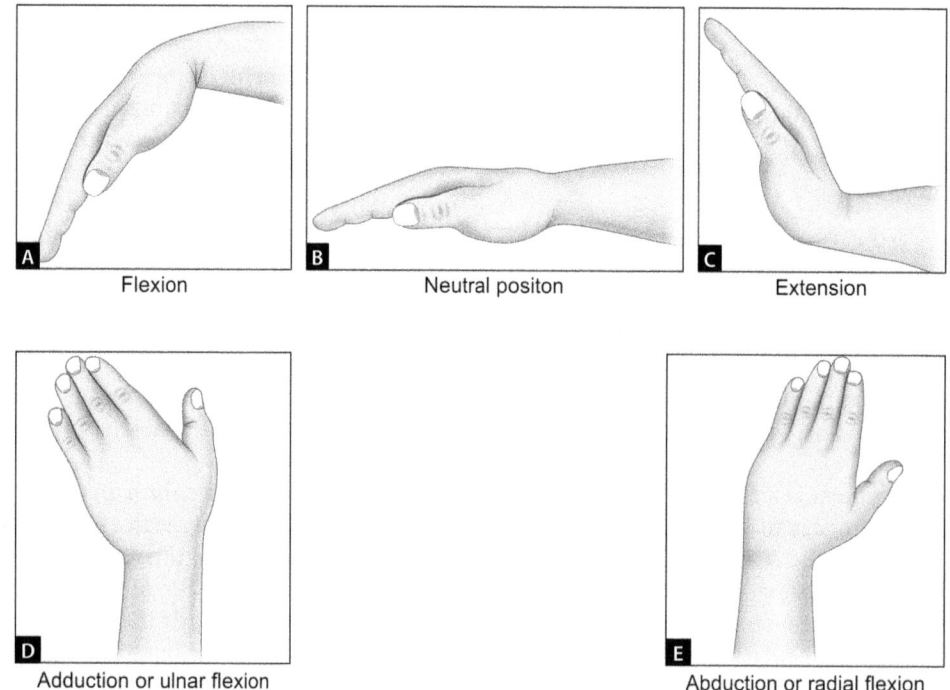

Figs. 7.5A to E: Movements of wrist

- The *thumb* has only two joints, and both of which are ginglymus type. The metacarpophalangeal (MCP) joint moves from full extension into 40°–90° of flexion, and interphalangeal joint (IP) can flex upto 80°–90°. The carpometacarpal joint (CMC) of thumb is a unique saddle type of joint permitting 50°–70° of abduction, flexion upto 15°–45°, and extension upto 0°–20°.

Movements of carpometacarpal (CMC) joint of thumb are shown in **Figure 7.6**.

- Fingers can only flex and extend permitting extension 0°–40°, flexion 0°–85°, except at metacarpophalangeal joints (MCP) where abduction and adduction movements are controlled by the intrinsic hand muscles. In the hand, the middle finger is regarded as the reference point to differentiate abduction and adduction of the fingers. *Abduction* occurs when the second, fourth and fifth fingers move away from the middle (third) finger, and when the middle finger moves in either direction. *Adduction* is the return movement from abduction and occurs with the second, fourth and fifth fingers. There is no adduction of the middle finger. **Figure 7.7** shows the movements of metacarpophalangeal (MCP) joints of fingers.

- Thumb is abducted when it moves away from the palm, and adduction movement takes place if thumb moves toward the palmar aspect of 2nd metacarpal. Summary of musclar analysis of wrist and hand movements is given in **Table 7.1**.

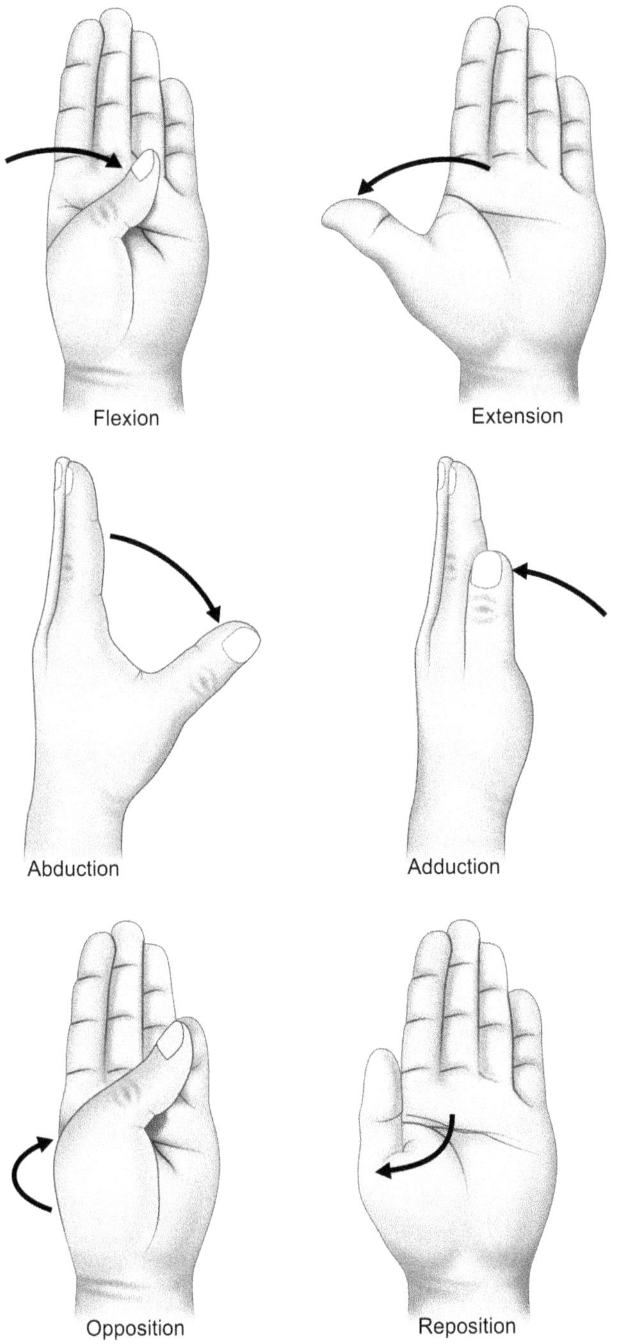

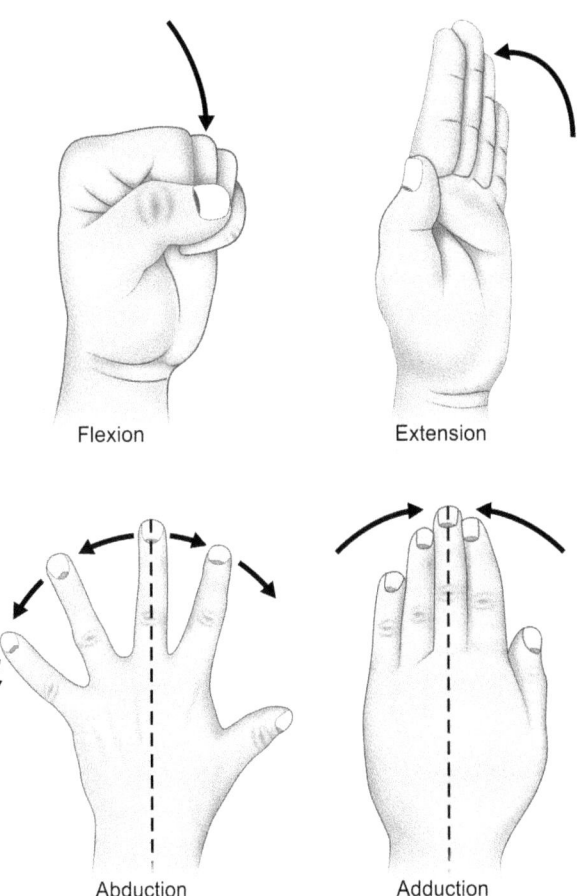

Fig. 7.6: Movements of carpometacarpal (CMC) joint of thumb

Fig. 7.7: Movements of metacarpophalangeal (MCP) joints of fingers

Table 7.1: Summary of muscular analysis of wrist and hand movements

Movement	Prime movers	Assistant movers
Wrist		
Flexion	• Flexor carpi radialis • Flexor carpi ulnaris • Palmaris longus	• Flexor digitorum superficialis • Flexor digitorum profundus • Flexor pollicis longus
Extension	• Extensor carpi radialis longus • Extensor carpi radialis brevis • Extensor carpi ulnaris	• Extensor digitorum • Extensor indicis • Extensor digiti minimi • Extensor pollicis longus and brevis
Abduction (radial flexion)	• Flexor carpi radialis • Extensor carpi radialis longus and brevis	• Abductor pollicis longus • Extensor pollicis longus and brevis
Adduction (ulnar flexion)	• Flexor carpi ulnaris • Extensor carpi ulnaris	• Flexor digitorum superficialis • Flexor digitorum profundus
Hand or CMC		
Thumb (Carpo metacarpal joint)		
Flexion	• Flexor pollicis brevis • Flexor pollicis longus	
Extension	• Extensor pollicis brevis • Extensor pollicis longus	
Abduction	• Abductor pollicis brevis • Abductor pollicis longus	Flexor pollicis brevis
Adduction	• Adductor pollicis	
Opposition	• Opponens pollicis	
Thumb (metacarpophalangeal or MCP joints)		
Flexion	• Flexor pollicis brevis • Flexor pollicis longus • Adductor pollicis	
Extension	• Extensor pollicis brevis • Extensor pollicis longus	
Thumb (interphalangeal joints)		
Flexion	• Flexor Pollicis Longus	
Extension	• Extensor Pollicis Longus	
Fingers (metacarpophalangeal joints)		
Flexion	• Lumbricals • Flexor digitorum superficialis • Flexor digitorum profundus • Flexor digiti minimi brevis (only 5th finger)	
Extension	• Extensor digitorum • Extensor indicis • Extensor digiti minimi	
Abduction	• Dorsal interossei • Abductor digiti minimi	
Adduction	• Palmar interossei	
Opposition (fifth finger)	• Opponens digiti minimi	
Fingers (interphalangeal joints)		
Flexion		
PIP	• Flexor digitorum superficialis • Flexor digitorum profundus	
DIP	• Flexor digitorum profundus	
Extension		
PIP	• Lumbricals, extensor digitorum, extensor digiti minimi	
DIP	• Extensor indicis	

MUSCLES OF WRIST AND HAND

Wrist

A large number of muscles are used in these movements. Anatomically and structurally, the human wrist and hand have highly developed complex mechanisms that are capable of a variety of movements. There are 6 **extrinsic muscles** that move the wrist, but do not cross the hand to move fingers and thumb. These muscles are listed as *prime movers/flexors and extensors of wrist*. There are another *9 extrinsic muscles* that function primarily to move the phalanges, but are also involved in the movements of wrist joint because of their origin on forearm and crossing the wrist.

Flexors (wrist): All of these muscles generally have their *origin* on anteromedial aspect of forearm and medial epicondyle of humerus.

Prime Movers: Flexor Carpi Radialis, Flexor Carpi Ulnaris, Palmaris Longus. All of the prime movers *originate* from medial epicondyle of humerus. *Insertion* is on the anterior aspect of wrist and hand.

Assistant Movers: Flexor digitorum superficialis, flexor digitorum profundus, and *flexor pollicis longus.*

Extensors (wrist): Generally these muscles take *origin* on posterolateral aspect of proximal forearm and lateral epicondyle of humerus. *Insertion* is on the posterial aspect of wrist and hand.

Prime Movers: Extensor carpi radialis longus, extensor carpi radialis brevis, extensor carpi ulnaris. All of these muscles *originate* from lateral aspect of humerus.

Assistant Movers: Extensor digitorum, extensor indicis, extensor digiti minimi and extensor pollicis longus and brevis.

Wrist Abductors (Radial Flexors)
Prime Movers: Flexor carpi radialis, extensor carpi radialis longus and brevis.
Assistant Movers: Abductor pollicis longus, extensor pollicis longus, extensor pollicis brevis.

Wrist Adductors (Ulnar Flexors)
Prime Movers: Flexor carpi ulnaris, extensor carpi ulnaris.

Hand

Fingers Flexors: Flexor digitorum superficialis, flexor digitorum profundus.

Fingers Extensors: Extensor digitorum, extensor indicis, extensor digiti minimi.

Thumb Flexors: Flexor pollicis longus, flexor pollicis brevis.

Thumb Extensors: Extensor pollicis longus, extensor pollicis brevis.

Thumb Abductors: Abductor pollicis longus, abductor pollicis brevis.

Thumb Adductor: Adductor pollicis.

Thumb Opposition: Opponens pollicis.

Actions performed by major extrinsic muscles of wrist and hand: These extrinsic muscles take their origin (proximal attachment) from the forearm or above the wrist, and apart from wrist movements also perform movements of fingers and thumb. The details of these muscles being exhaustive are avoided here and only their actions are described.

Table 7.2 shows the summary of extrinsic and intrinsic muscles acting on wrist and hand describing their location, origin, insertion, main and supporting actions.

- *Flexor Carpi Radialis (Fig. 7.8):* Flexion of wrist, radial flexion or abduction of wrist.
  Assistant: Weak flexion of elbow.
- *Palmaris Longus (Fig. 7.9):* Flexion of wrist.
  Assistant: Weak flexion of elbow.
- *Flexor Carpi Ulnaris (Fig. 7.10):* Flexion of wrist, adduction or ulnar flexion of wrist (along with extensor Carpi ulnaris).
  Assistant: Weak flexion of elbow.
- *Extensor Carpi Ulnaris (Fig. 7.11):* Extension of wrist, adduction or ulnar flexion of wrist (along with flexor carpi ulnaris).
  Assistant: Weak extension of elbow.
- *Extensor Carpi Radialis Longus (Fig. 7.12):* Extension of wrist, abduction or radial flexion of wrist. Weak extension of elbow.
- *Extensor Carpi Radialis Brevis (Fig. 7.13):* Same actions as described above but somewhat weaker than that.
- *Flexor Digitorum Superficialis (Fig. 7.14):* Flexion of fingers at MCP (metacarpophalangeal), PIP (proximal interphalangeal) joints.
  Assistant: Flexion of wrist, weak flexion of elbow.
- *Flexor Digitorum Profundus (Fig. 7.15):* Flexion of four fingers at MCP, PIP and DIP joints.
  Assistant: Flexion of wrist.
- *Flexor Pollicis Longus (Fig. 7.16):* Flexion of thumb at carpometacarpal, metacarpophalangeal (MCP), and IP (interphalangeal) joints.
  Assistant: Flexion of wrist.

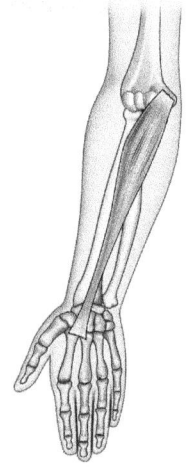

Fig. 7.8: Flexor carpi radialis

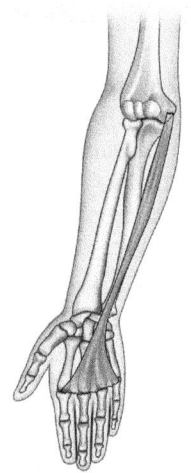

Fig. 7.9: Palmaris longus

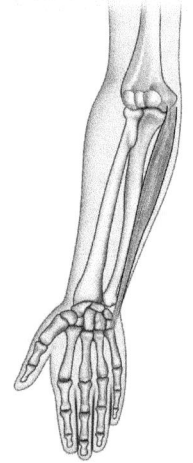

Fig. 7.10: Flexor carpi ulnaris

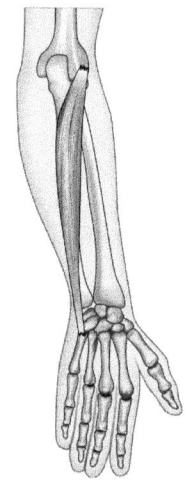

Fig. 7.11: Extensor carpi ulnaris

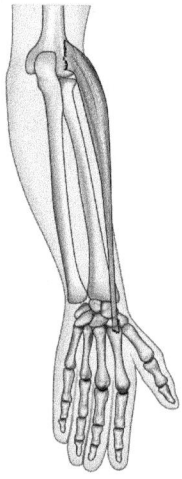

Fig. 7.12: Extensor carpi radialis longus

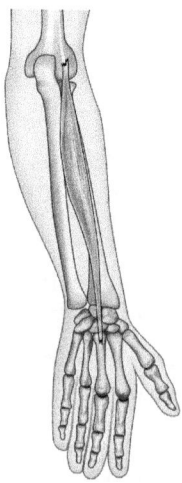

Fig. 7.13: Extensor carpi radialis brevis

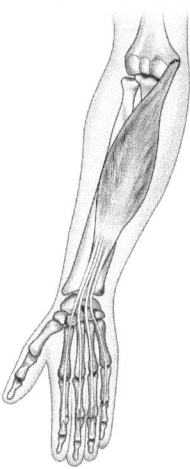

Fig. 7.14: Flexor digitorum superficialis

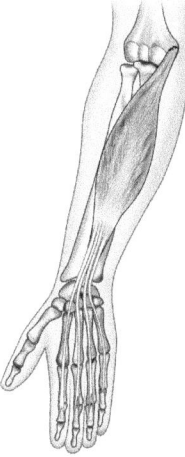

Fig. 7.15: Flexor digitorum profundus

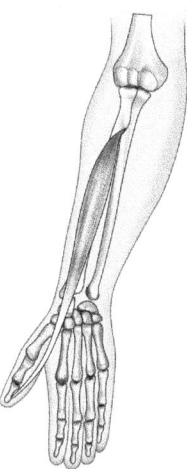

Fig. 7.16: Flexor pollicis longus

- *Extensor Digitorum (Fig. 7.17):* Extension of 2nd, 3rd, 4th and 5th fingers at metacarpophalangeal joints, and weak extensor of wrist and elbow.
- *Extensor Indicis (Fig. 7.18):* Extension of index finger at MCP joint, and weak wrist extension.
- *Extensor Digiti Minimi (Fig. 7.19):* Extension of little finger at MCP joint, and weak wrist extension.
- *Extensor Pollicis Longus (Fig. 7.20):* Extension of thumb, and weak extension of wrist.
- *Extensor Pollicis Brevis (Fig. 7.21):* Extension of thumb at metacarpophalangeal (MCP) joint, and weak wrist extension.
- *Abductor Pollicis Longus (Fig. 7.22):* Abduction of thumb at carpometacarpal joint, and weak abduction of wrist (radial flexion).

Intrinsic muscles of hand have their origins and insertions on the bones of hand. These intrinsic muscles produce movements of thumb and fingers. These muscles are responsible for fine motor control and precision movements of the hand. These muscles may be grouped according to their location as well as according to the parts of the hand they act upon **(Figs. 7.23A and B)**.

- **On radial side of hand there are 4 muscles of thumb:**
    1. Opponens pollicis
    2. Abductor pollicis brevis
    3. Flexor pollicis brevis
    4. Adductor pollicis.

These muscles make *thenar eminence* (muscular pad on the palmar surface of the 1st metacarpal).

- **On ulnar side of hand there are 3 muscles of little finger:**
    1. Opponens digiti minimi
    2. Abductor digiti minimi
    3. Flexor digiti minimi brevis.

These muscles make *hypothenar eminence* which is the muscular pad forming the ulnar border on the palmar surface of the hand.

- There are 11 intermediate hand muscles out of which there are: (i) **Lumbricals**-four **(Fig. 7.24C)**. These muscles flex the index, middle, ring and little fingers at MCP joints, and extend the PIP and DIP joints. (ii) **Dorsal Interossei**-four **(Fig. 7.24A)**. These muscles flex and abduct the index, middle and ring fingers at proximal phalanges, and also assist extension of middle and distal phalanges of these fingers. (iii) **Palmar Interossei**-three. These muscles are the adductors of 2nd, 4th and 5th fingers phalanges **(Fig. 7.24B)**.
- *Four intrinsic muscles act on the carpometacarpal joint of thumb;* out of which the main muscles are opponens pollicis (causes opposition of thumb metacarpal), and

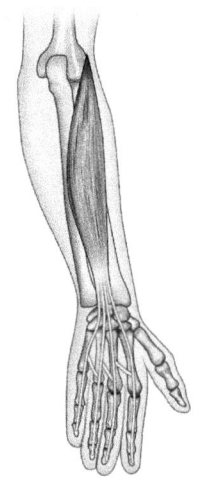

Fig. 7.17: Extensor digitorum

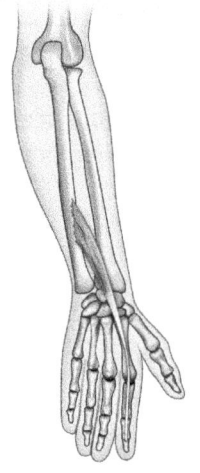

Fig. 7.18: Extensor indicis

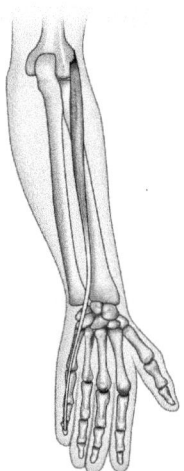

Fig. 7.19: Extensor digiti minimi

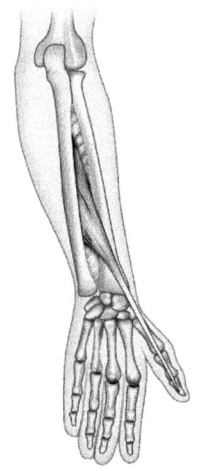

Fig. 7.20: Extensor pollicis longus

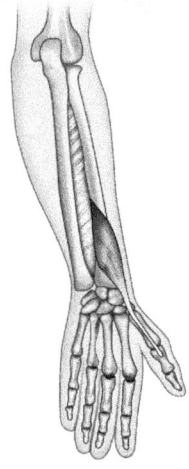

Fig. 7.21: Extensor pollicis brevis

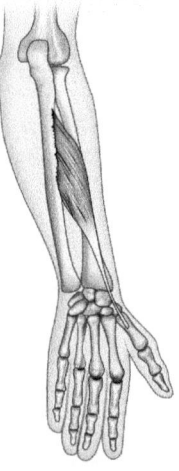

Fig. 7.22: Abductor pollicis longus

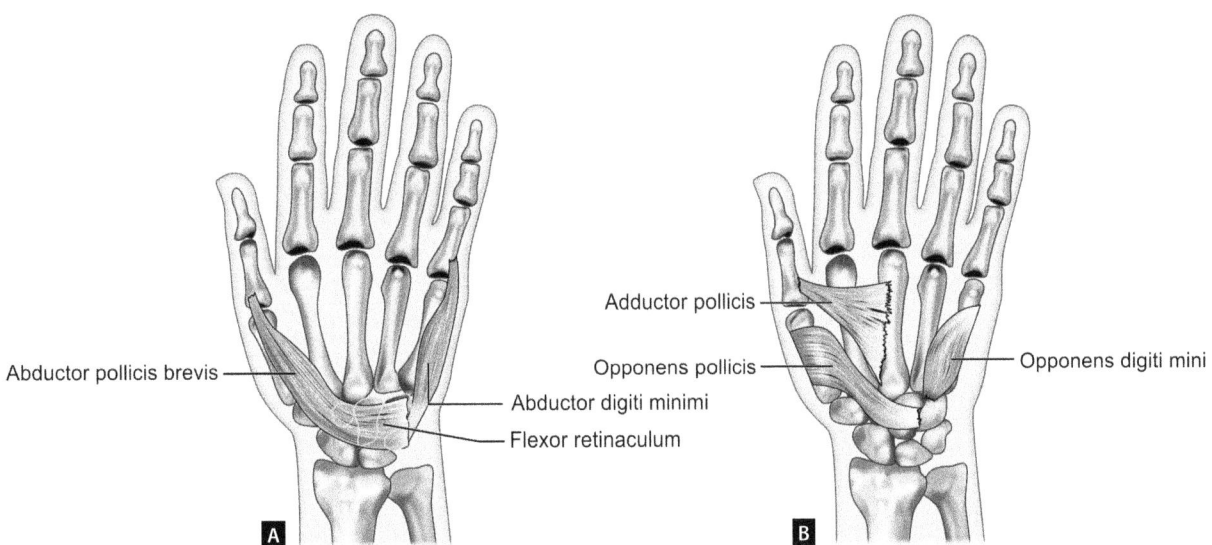

Figs. 7.23A and B: Intrinsic muscles of hand: (A) Abductor pollicis brevis and abductor digiti minimi muscles; (B) Opponens pollicis, adductor pollicis and opponens digiti minimi muscles

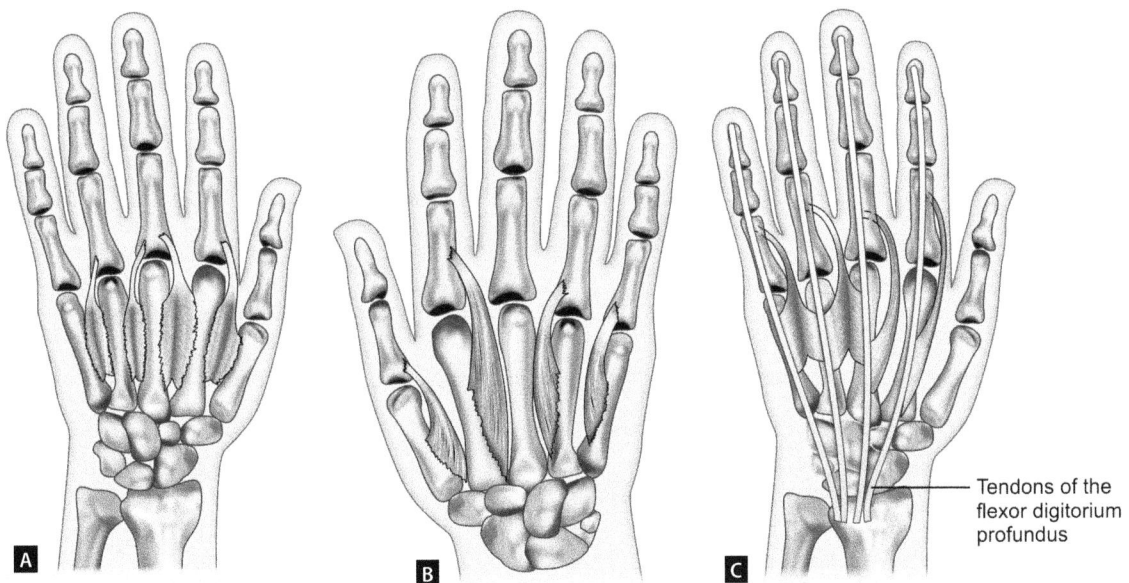

Figs. 7.24A to C: (A) Dorsal interossei muscles (on the deep posterior side of the hand); (B) Palmar interossei muscles; (C) Lumbrical muscles

abductor pollicis brevis (abducts thumb's metacarpal) which is assisted by flexor pollicis brevis.
- *Flexion of proximal phalynx of thumb* is done by flexor pollicis brevis and adductor pollicis.
- *Three muscles act on little finger:* (i) opponens digiti minimi (causes opposition of little finger's metacarpal), (ii) abductor digiti minimi (causes abduction of little finger's metacarpal), and (iii) flexor digiti minimi brevis (flexes the metacarpal of 5th finger).

Carpal bones form a concave transverse arch on their palmar aspect, which is maintained by *flexor retinaculum* (a ligament) which attaches medially to pisiform and hamate. The flexor retinaculum attaches laterally to scaphoid tubercle and to the groove of

trapezium. The space formed underneath the flexor retinaculum is called '**carpel tunnel**', which allows the passage of following:
- Tendon of flexor pollicis longus,
- Tendon of flexor digitorum profundus,
- Tendon of flexor digitorum superficialis, and
- Median nerve.
- On the posterior (or dorsal) aspect of the wrist, the *extensor retinaculum* (a ligament) stretches from the radius to the hamate and pisiform bones, and extends inferiorly to form six longitudinal compartments for the passage of following extensor tendons:
  - Extensor carpi ulnaris
  - Extensor digiti minimi
  - Extensor digitorum, and extensor indicis
  - Extensor pollicis longus
  - Extensor carpi radialis longus, and
  - Extensor pollicis brevis.

Table 7.2: Summary of extrinsic and intrinsic muscles acting on wrist and hand

Muscle	Location	Origin	Insertion	Main action	Supporting action
Flexor carpi radialis	Anterior forearm (superficial)	Medial epicondyle of humerus	Base of 2nd and 3rd metacarpals	• Flexion of wrist • Abduction (radial flexion) of wrist	Elbow flexion
Flexor carpi ulnaris	Anterior forearm (superficial)	Medial epicondyle of humerus	Base of 5th metacarpal and pisiform	• Flexion of wrist • Adduction (ulnar flexion of wrist)	Elbow flexion
Palmaris longus	Anterior forearm	Medial epicondyle of humerus	Palmar fascia of 2nd, 3rd, 4th and 5th metacarpals	Flexion of wrist	Elbow flexion
Extensor carpi radialis longus	Posterior forearm	• Lower third of lateral supracondylar ridge of humerus • Lateral epicondyle of humerus	Base of 2nd metacarpal	• Extension of wrist • Abduction (radial flexion) of wrist)	Elbow extension
Extensor carpi radialis brevis	Posterior forearm	Lateral epicondyle of humerus	Base of 3rd metacarpal	• Extension of wrist • Abduction (radial flexion of wrist)	Elbow extension
Extensor carpi ulnaris	Posterior forearm	Lateral epicondyle of humerus	Base of 5th metacarpal	• Extension of wrist • Adduction (ulnar flexion of wrist)	Extension of elbow
Flexor pollicis longus	Anterolateral aspect of forearm	Anterior surface of middle radius	Distal phalanx of thumb (palmar surface)	Flexion of thumb of CMC, MCP and IP joints	Flexion of wrist
Flexor pollicis brevis	Anterolateral aspect of forearm	Trapezium and flexor retinaculum (palmar aspect)	Proximal phalanx of thumb (palmar aspect)	Flexion of CMC and MCP joints of thumb	
Extensor pollicis longus	Posterolateral aspect of lower forearm	Lower middle of ulna (posterior surface)	Distal phalanx of thumb (posterior surface)	Extension of thumb (all three joints)	Extension of wrist
Extensor pollicis brevis	Posterolateral aspect of lower forearm	Posterior aspect of distal radius	Proximal phalanx of thumb (posterior surface)	Extension of CMC and MCP joints of thumb	Extension of wrist
Adductor pollicis	Anterolateral aspect of middle palm (thumb side)	Capitate, 2nd and 3rd metacarpals (palmar surface)	Proximal phalanx of thumb (palmar surface)	Adduction of thumb at CMC joint	
Abductor pollicis longus	Posterolateral aspect of forearm	Posterior aspect of radius and mid shaft of ulna	Base of 5th metacarpal (dorsal aspect)	Abduction of thumb at carpometacar-pal joint	

Contd...

Contd...

Muscle	Location	Origin	Insertion	Main action	Supporting action
Abductor pollicis brevis	Lateral palm (thumb side)	Scaphoid, trapezium and flexor retinaculum (palmar aspect)	Proximal phalanx of thumb (palmar aspect)	Abduction of thumb at CMC joint	
Opponens pollicis	Anterolateral aspect of proximal palm (thumb side)	Proximal phalanx of thumb (palmar aspect)	First metacarpal (palmar aspect)	Opposition of thumb at CMC joint	
Flexor digitorum superficialis	Anterior forearm (medial)	Medial epicondyle of humerus, coronoid process and radius	Each tendon splits and attaches to the sides of middle phalanx of each of four fingers	Flexion of fingers at MCP and PIP joints	Flexion of wrist Flexion of elbow
Flexor digitorum profundus	Anterior forearm (deep)	Upper three-fourths of ulna (anterior and medial side)	Distal phalanx of four fingers	Flexion of fingers of all three joints (MCP, PIP and DIP joints)	Flexion of wrist
Flexor digiti minimi	Anterior wrist	Hamate and flexor retinaculum (anterior aspect)	Proximal phalanx of 5th finger (palmar surface)	Flexion of little finger at CMC and MCP joints	
Extensor digitorum	Posterior forearm	Lateral epicondyle of humerus	Base of middle and distal phalanges of four fingers	Extension of all three joints of all four fingers	• Extension of wrist • Extension of elbow
Extensor indicis	Posterior forearm	Distal ulna (posterior aspect)	Middle and distal phalanges of second finger (dorsal side)	Extension of all three joints of index finger	Extension of wrist
Extensor digit minimi	Posterior forearm	Lateral epicondyle of humerus	Middle and distal phalanges of fifth finger (dorsal side)	Extension of all three joints of little finger	Extension of wrist
Abductor digiti minimi	Anterolateral aspect of lower palm (5th metacarpal region)	Pisiform and tendon of flexor carpi ulnaris (palmar aspect)	Proximal phalanx of fifth finger (palmar surface)	Abduction of little finger at MCP joint	
Opponens digiti minimi	Lower palm (lateral surface)	Hamate and flexor retinaculum (palmar surface)	Fifth metacarpal (palmar surface)	Opposition of little finger at CMC joint	
Dorsal interossei (04)	Back side of metacarpals (deep)	Adjacent metacarpals (posterior side)	Proximal phalanx of each finger (posterior side)	Flexion and abduction of MCP joints (proximal phalanx) of index, middle and ring fingers	Extension of middle and distal phalanges of index, middle and ring fingers
Palmar interossei (04)	Palm (deep)	Metacarpals of thumb and fingers (except middle finger)	Proximal phalanx of respective thumb/fingers (palmar side)	Adduction of index, ring and little fingers at MCP joint	
Lumbricals (04)	Palm (middle partdeep)	Tendon of flexor digitorum profundus muscle	Tendon of extensor digitorum muscle	• Flexion of MCP joints • Extension of PIP and DIP joints of all four fingers	

8

CHAPTER

Pelvic Girdle

The relationship between the hip joint and pelvic girdle is very much like the shoulder joint and shoulder girdle. In shoulder joint, the scapula rotates to place the glenoid cavity in a favorable position for the movements of the humerus. Similarly, the pelvic girdle tilts and rotates to put the acetabulum in a favorable position for the movements of femur. The only difference is that either of the shoulder girdles can move independently, whereas the pelvic girdle only moves as one unit. Secondly, the movements of pelvic girdle are dependent on the movements of the lumbosacral, other lumbar joints and hip joints. Sport skills such as running or kicking a football are good examples of co-ordinated movements of hip and pelvic girdle. Pelvic rotation helps to increase the length of stride in running; in kicking, it can result in a greater range of motion, which results to a greater distance or more speed to the kick.

The pelvic girdle performs several **functions**. Perhaps the most important function from the point of view of movement and posture is that it supports the weight of the body through the vertebral column and passes that weight/force onto the hip bones. Conversely, it receives the ground forces generated when the foot contacts the ground, and transmits them upward toward the vertebral column.

During walking, the pelvic girdle moves as a unit in all three planes to allow for relatively smooth function. In addition, it supports and protects the pelvic viscera, provides attachment for muscles, and makes up the bony portion of the birth canal in females.

The pelvis is a rigid bony basin that serves as a massive connecting link between the trunk and the lower extremities. Each pelvic bone (or each half of the pelvic girdle) consists of three bones; the *ilium* (above at the side of the hip), the *pubis* (below and forward), and the *ischium* (below and to the back side). These three bones

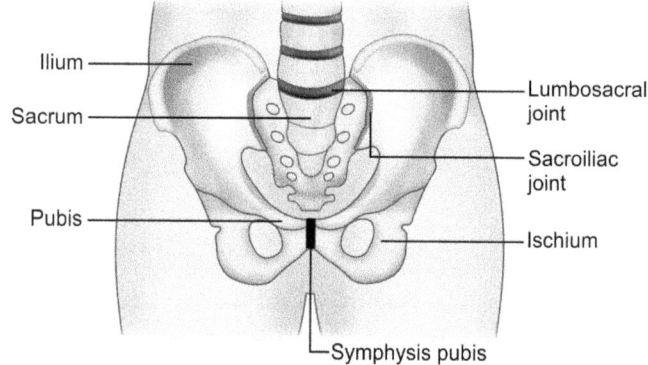

Fig. 8.1: Joints of pelvic girdle (anterior view)

are separate in early life, however these bones become fused into a single solid bony structure by about the time of puberty. The two pelvic bones together form the pelvic girdle (**Fig. 8.1**).

The pelvic basin is closed and attached posteriorly and firmly by the sacrum at the **sacroiliac joint**. This joint formed between the sacrum and ilium is regarded as a synovial, non-axial joint having very irregular articular surfaces. This irregularity of the articulating surfaces helps to interlock the two surfaces together such that practically no voluntary movement is possible at this joint. Any movement that does occur is involuntary, and is regarded by some experts as 'slight giving' movement which occurs as a shock absorption device. Others claim that no motion occurs at this joint normally, except in women during pregnancy when the ligaments relax to permit a slight spreading of the bones.

The sacroiliac joint is held firmly by three of the strongest ligaments in the human body; the anterior, posterior and interosseous sacroiliac ligaments, and by the

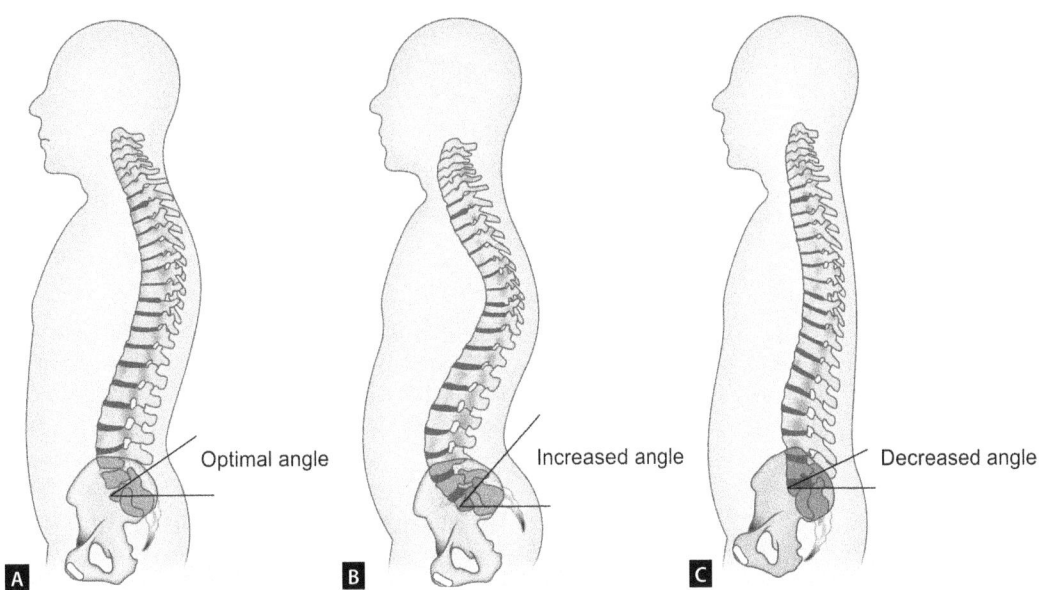

Figs. 8.2A to C: (A) Optimal lumbosacral angle; (B) Increased lordosis; (C) Decreased lordosis

lower portion of the erector spinae muscle. Because of this firm attachment and from the point of view of function, the sacrum might well be considered a part of the pelvic girdle rather than of the spine. *The main function of the sacroiliac joint is to transmit the weight of the upper body through the vertebral column to the hip bones.* As such, this joint is designed more for stability rather than mobility.

The pelvis is tied together anteriorly at the *pubic symphysis*, which is a joint located in the midline of the body. The right and left pubic bones (pubes) are separated by a heavy disc of fibrocartilage. This joint is also a tight joint and is heavily re-inforced by two ligaments, the *superior pubic ligament* above, and the *inferior pubic ligament* below.

Because the pelvic girdle, including the sacrum, acts as a unit, the **lumbosacral joint** becomes the important articulation when the pelvis moves in relation to the spine. This lumbosacral joint is made up of fifth lumbar vertebra and the first sacral vertebra. The joint between these two vertebrae is similar to all other nearby intervertebral joints, and supported by various ligaments i.e. *anterior* and *posterior longitudinal ligaments, supraspinal, interspinal, ligamentum flavum, iliolumbar*, and *lumbosacral* ligaments which are described in details in the chapter on spinal column.

Lumbosacral Angle

The lumbosacral angle is of great significance while assessing the posture and mechanics of low back region, including the pelvic region. This angle is determined by drawing one line parallel to the ground and another line along the base of the sacrum. This angle increases with anterior tilting of the pelvis and also with increased lumbar lordosis **(Fig. 8.2B)**. The angle decreases with posterior tilting of the pelvis, and with decreased lumbar lordosis **(Fig. 8.2C)**. The optimal lumbosacral angle is approximately 30° **(Fig. 8.2A)**.

MOVEMENTS OF PELVIC GIRDLE

The movements of pelvis occur in all three planes. The movements of pelvis or changes in the position of the pelvis are actually brought about by virtue of the movements of lumbar spine (particularly of the lumbosacral joint), and the hip joints **(Table 8.1)**. Movements in the lumbar spine and hip joints permit the pelvis to tilt forward, backward, and sideward, and to rotate horizontally. While analyzing the movements of pelvis in standing in the upright position, the pelvis should be level in the sagittal plane, and the anterior superior iliac spine (ASIS) and pubic symphysis should be in the same vertical plane.

Anterior or Forward Tilt

Anterior or forward movement, or tilting of the pelvis in the sagittal plane about a frontal-horizontal axis so that the ASIS move forward, the pubic symphysis turns downward, and the posterior surface of the sacrum turns upward **(Fig. 8.3C)**.

Table 8.1: Movements of pelvic-girdle in association and co-operation with that of lumbosacral spine and hip joints

Pelvic Girdle	Lumbosacral Spine	Hip
Anterior (forward) tilt	Hyperextension	Flexion
Posterior (backward) tilt	Flexion	Extension
Lateral (sideward) tilt to left	Lateral or sideward flexion to right	• Adduction—right hip • Abduction—left hip
Pelvic rotation to left (without turning the head or moving the feet)	Rotation to right	• External rotation to right • Internal rotation to left

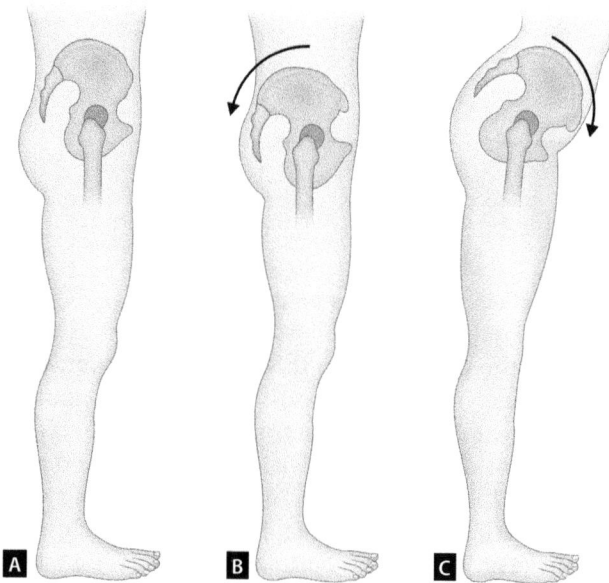

Figs. 8.3A to C: Anteroposterior tilt of the pelvis: (A) Mid position; (B) Backward or posterior tilt; (C) Forward or anterior tilt

Posterior or Backward Tilt

Posterior or backward movement, or tilting of the pelvis in sagittal plane about a frontal-horizontal axis so that the ASIS move backward, the pubic symphysis turns forward-upward and the posterior surface of the sacrum turns somewhat downward (**Fig. 8.3B**).

Lateral or Sideward Tilt

The lateral or sideward tilting of the pelvis in the frontal plane about a sagittal-horizontal axis so that one iliac crest is lowered and the other is raised. The lateral tilt is named in terms of the side that moves downward. Thus, in a lateral tilt of the pelvis to the left, the left iliac crest is lowered, and since the pelvis moves as a unit in the tilt, the right iliac crest is raised (**Figs. 8.4A and B**).

Pelvic Rotation

A rotation of the pelvis in the horizontal plane about a vertical axis when one side of the pelvis moves forward in relation to the other side. The movement of pelvic rotation is named in terms of the direction toward which the front or anterior of the pelvis turns. Thus, in a pelvic rotation to the right, the right sided pelvis moves to the front in relation to the left sided pelvis which moves backward (**Fig. 8.5**).

Relationship of the Pelvis to the Trunk and Lower Extremities

The pelvis is located strategically from the architecture point of view. While linking the trunk with the extremities, it must co-operate with the movements of both trunk and lower extremities, and the same time also contribute to the stability of the total structure. When the human body is in the erect standing position, the pelvis receives the weight of the head, trunk, and upper extremities, divides it equally, and transmits it to both the lower extremities. However, when an individual stands only on one foot, the pelvis automatically adapts itself to this position and transmits the entire weight of the upper part of the body to one of the two lower extremities, and this requires a fine adjustment in such a way that the balance of the total structure is maintained.

The movements of the pelvis depend on the movements of lower spine and hips. As such, sometimes, the movements of the pelvis result secondary to the movements of the trunk (spine), or that of the thighs. However, sometimes the pelvic movements are initiated in the pelvis itself, (as when an individual is standing erect on both the lower extremities and tilting the pelvis forward or backward on the hips) and the spine and thighs precisely co-operate with the pelvis.

For example, for the body to remain upright during anterior or forward tilting of the pelvis, the movement in the opposite direction must occur in the joints above (lumbosacral spine) and below (hip joints) the pelvis, and as such, the lumbosacral spine goes into hyperextension and the hip joints flex. Similarly, during the lateral or sideward tilting of the pelvis to one side, the lower spinal column, bends laterally to opposite side, the same sided hip abducts, and opposite sided hip adducts. Similarly, the pelvic rotation to one side will result in associated rotation

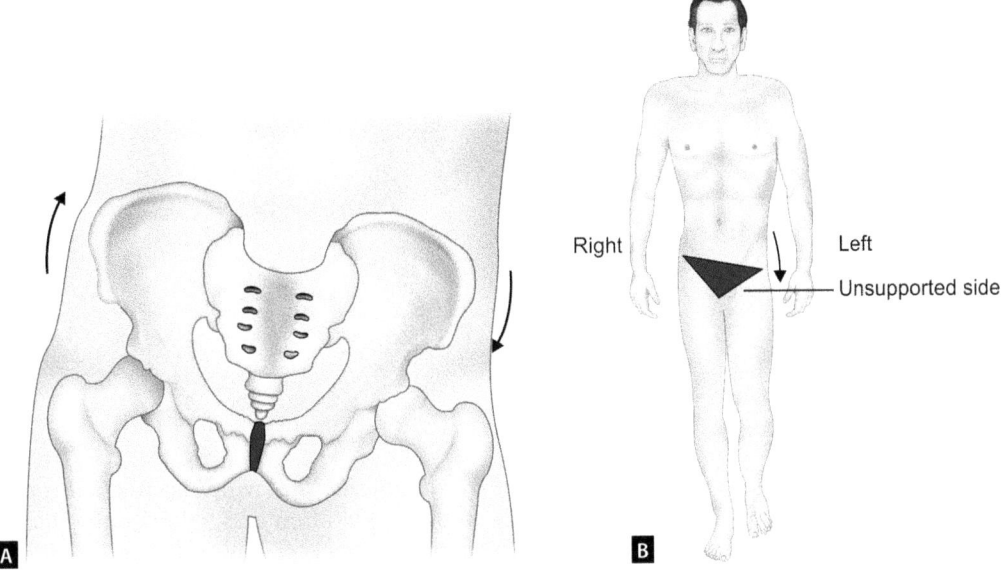

Figs. 8.4A and B: Lateral tilt of pelvis (left): anterior views

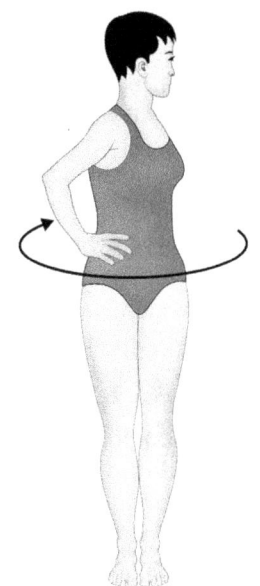

Fig. 8.5: Pelvic rotation to the right

of lumbosacral spine and external rotation of the hip to the opposite side, and the internal hip rotation to the same side.

MUSCLES OF THE PELVIS

All the muscles that are attached to the pelvic bones or to the sacrum serve either to initiate or to control pelvic movements. Summary of various aspects of the pelvis providing origin attachment to the muscles of pelvic girdle is given in **Table 8.2**. All these muscles attached to the pelvic bones are primarily the muscles of hip joint or of the lumbar spine and are described in details in the respective chapters 9 and 12.

The muscles providing the force for primary movements of the pelvis when an individual stands with both feet facing forward is shown in **Table 8.3**.

Table 8.2: Summary of various aspects of the pelvis providing origin attachment to the muscles of pelvic girdle

Anterior pelvis (Hip flexors are generally attached)	*Posterior pelvis* (Hip extensors originate)
• Sartorius	• Gluteus maximus
• Tensor fascia lata	*Posteroinferior Pelvis*
• Rectus femoris	• Hamstrings
Lateral pelvis (Hip abductors originate)	*Medial pelvis* (Hip adductors originate)
• Gluteus medius	• Adductor magnus
• Gluteus minimus	• Adductor longus
	• Adductor brevis
	• Pectineus
	• Gracilis

Table 8.3: Summary of muscular analysis of movements of pelvis

Forward or anterior tilt	*Backward or posterior tilt*
• Hip flexors	• Hip extensors
• Lumbosacral spinal extensors	• Lumbosacral spinal flexors
Lateral or sideward tilt to right	*Rotation to right*
• Left sided lateral lumbosacral flexors	• Left sided lumbosacral rotators
• Right sided hip abductors	• Left hip outward rotators
• Left sided hip adductors	• Right hip inward rotators

9
CHAPTER

Hip

The lower extremity includes the pelvis, hip, thigh, knee, leg, ankle and foot. Bones of the pelvis are the two hip bones, the sacrum, and the coccyx. The anterior views of bones of lower extremities, and of the pelvis with right hip joint are shown in **Fig. 9.1 and 9.2** respectively.

The hip joint, the most proximal of the lower extremity joints, is a typical example of **ball and socket type of joint formed between the rounded head of femur and the deep cup shaped acetabulum of the pelvis**. The acetabulum, formed by the junction of three pelvic bones (ilium, ischium, and pubis), is also described as 'horseshoe-shaped' because there is a gap (the acetabular notch) at the lower part of the cup. It has a usual joint capsule (called *acetabular labrum*) which is attached above to the margin of the acetabulum and below to the distal margin of the anatomic neck of femur. Since the acetabular labrum is considerably thicker at the circumference than at the center, it adds to the depth of the acetabulum. This joint capsule is further strengthened and re-inforced by the three *ligaments* (one for each of the pelvic bones), which help to keep the head of the femur in the acetabulum and protect the joint from all sides. On the front side, the *iliofemoral ligament*, also called the *Y ligament* because of its supposed resemblance to an inverted Y, is an extraordinarily strong band of fibers, which because of its position, serves to prevent hip hyperextension, and also both inward and outward rotation; medially, the *pubofemoral ligament*, which consists of narrow band of fibers, prevents excessive hip abduction and extension; and *ischiofemoral ligament*, a strong triangular ligament on the posterior side, which prevents the excessive hip adduction and internal rotation, are the three important ligaments concerned with the stability of hip joint. The *transverse acetabular ligament* is a strong, flat band of fibers continuous with the acetabular labrum, bridges the acetabular notch, completing the acetabular ring. The *teres femoris* or the round ligament connects the transverse acetabular notch with the pit on the femoral head (fovea capitis). Its function is to 'tie' the head of femur to the lower part of the acetabulum and thus provide the joint with reinforcement from within.

MOVEMENTS

The range of motion possible in the hip joint is comparatively less than that of shoulder joint. This is because of the weight bearing function of the hip joint

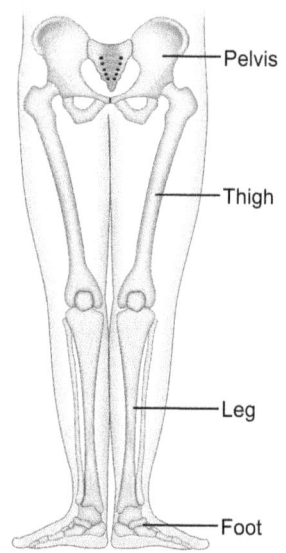

Fig. 9.1: Bones of lower extremities (anterior view)

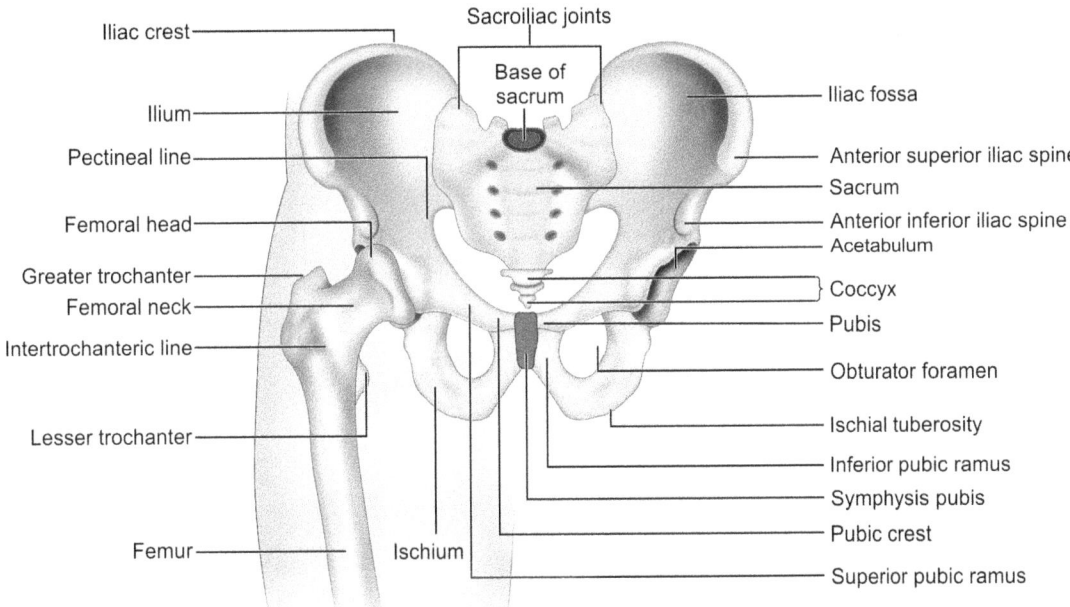

Fig. 9.2: Pelvis with right hip joint (anterior view)

demanding more stability, and since the stability of hip is more important than its mobility, the same is met with by the bony shape, strong ligaments, muscles, and gravity holding the head into the socket. However, it still has a fair range of all the movements possible at a synovial joint, i.e. flexion, extension, abduction, adduction, rotation and circumduction **(Figs. 9.3A to F)**. As such, the functions of hip in the weight bearing and locomotion (which are significantly enhanced by its wide range of movements) provide the ability to run, crossover cut, side step cut, jump, and many other directional changes as required in various physical education and sports activities.

Flexion: This is forward movement of femur in the sagittal plane **(Fig. 9.3A)**. Range of movement is approximately around 150° which is the most free among all the movements of hip. Flexion of hip with a straight knee is limited by the tension of the hamstrings muscles.

Extension: This is a return movement from flexion (or a downward-backward movement from the flexed position in a sagittal plane) upto the vertical position of the limb **(Fig. 9.3B)**. The backward movement of femur known as hyper-extension **(Fig. 9.3B1)** is very limited in range. This movement is limited in range because of the tension of iliofemoral ligament and iliopsoas muscles lying on the anterior aspect of hip.

Abduction: This is a sideward movement of femur in the frontal plane **(Fig. 9.3C)** so that the thigh moves away from the midline of the body. A greater range of abduction is possible when the femur is rotated externally.

Adduction: It is return movement from the abduction **Fig. 9.3D)**.

Circumduction: It is a combination of the flexion, abduction, extension and adduction movements in sequence in either direction.

Rotation: It is the turning or rotation of the femur around its long axis. When the knee is turned inward, it is an *medial or internal rotation* **(Fig. 9.3E)**, and when it is turned outward, it is called *lateral or external rotation* **(Fig. 9.3F)**.

The summary of muscular analysis of hip movements is shown in **(Table 9.1)**.

MUSCLES

A number of muscles act at the hip joint, some of the muscles act on one joint (i.e. hip) which provide most of the control of the hip and most of other muscles are two joint muscles, which because of their length act upon the hip as well as knee and provide the range of motion at hip. The muscles of the hip can be grouped according to their location **(Table 9.2)** and somewhat by their function also.

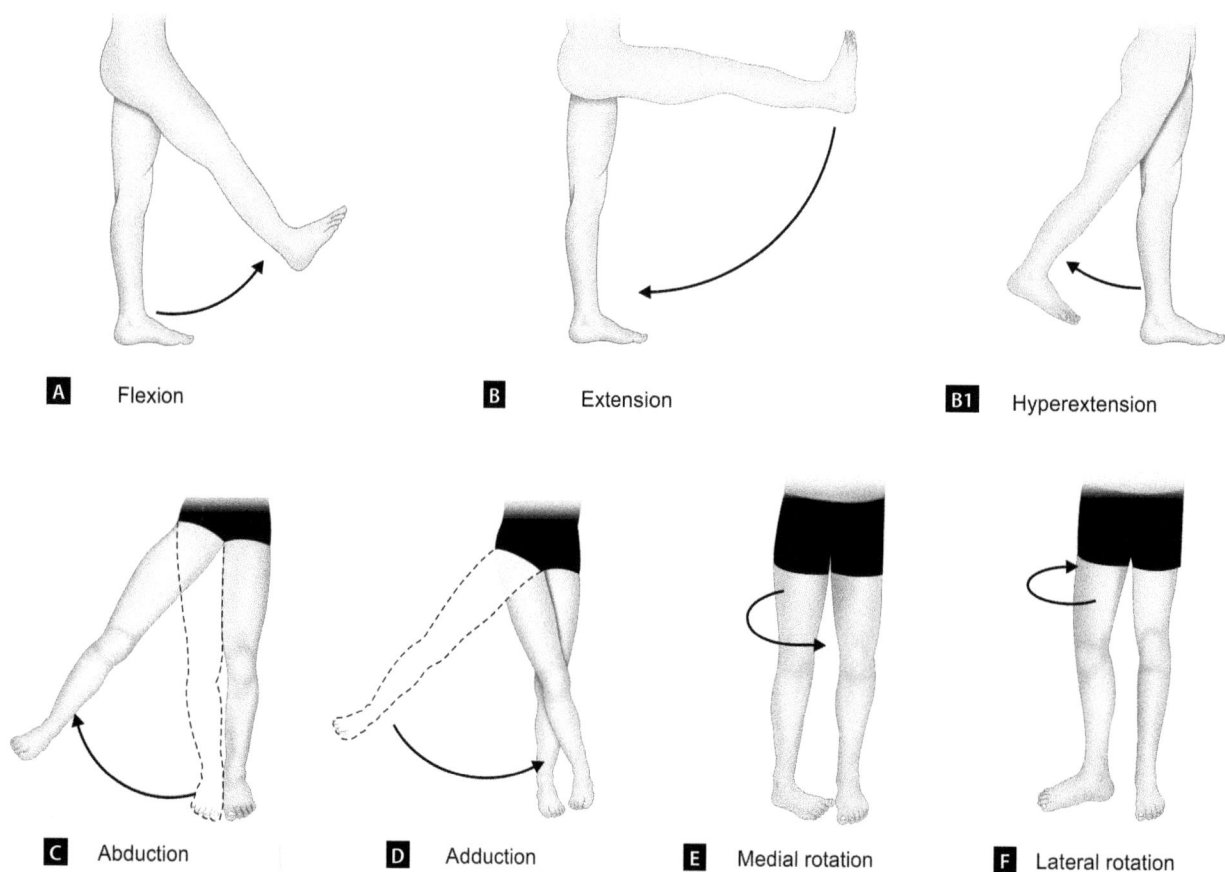

Figs. 9.3A to F: Movements of hip: (A) Flexion; (B) Extension; (B1) Hyperextension; (C) Abduction; (D) Adduction; (E) Medial rotation; (F) Lateral rotation

Table 9.1: Summary of muscular analysis of hip movements

Movement	Prime Movers	Assistant Movers
Flexion	• Iliopsoas • Rectus femoris • Pectineus	• Sartorius • Tensor fascia lata
Extension	• Gluteus maximus • Hamstrings	
Abduction	• Gluteus medius	• Gluteus minimus • Tensor fascia Lata • Sartorius • Gluteus maximus (upper part)
Adduction	• Adductor longus • Adductor magnus • Adductor brevis	• Pectineus • Gracilis
Lateral or External Rotation	*Six outward Rotators* 1. Piriformis 2. Obturator externus 3. Obturator internus 4. Quadratus femoris 5. Gamellus superior 6. Gamellus inferior	• Sartorius
Medial or Internal Rotation	Gluteus minimus	• Gluteus medius (anterior part) • Tensor fascia lata • All hip adductors (3)

Table 9.2: The hip muscles as according to their location

Anterior	Posterior

• Iliopsoas
• Rectus femoris (part of quadriceps) ⎤ Two joint
• Sartorius ⎦ muscles
	• Gluteus maximus
	• Deep external rotators (6)
	• Hamstrings
	  – Biceps femoris ⎤
	  – Semitendinosus ⎬ Two joint muscles
	  – Semimembranosus ⎦

Medial • Adductor longus • Adductor brevis • Adductor magnus • Pectineus • Gracilis (two joint muscle)	*Lateral* • Gluteus medius • Gluteus minimus • Tensor fascia lata (two joint muscle)

For example, the anterior muscles generally tend to be flexors of the hip, posterior muscles tend to be extensors, lateral muscles to be abductors and medial muscles tend to be adductors of the hip. The summary of major muscles acting on hip is given in **Table 9.3.**

Flexors: *Iliopsoas, rectus femoris* and *pectineus* are the prime movers, assisted by sartorius and tensor fascia latae (TFL).

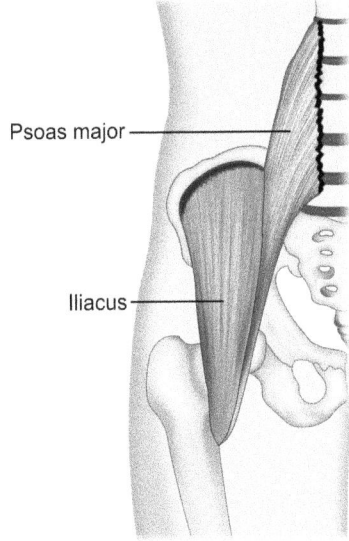

Fig. 9.4: Iliopsoas muscles

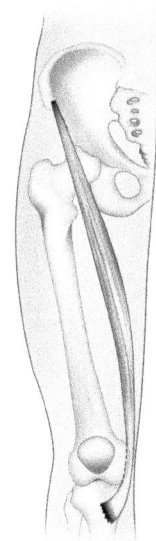

Fig. 9.5: Sartorius

Iliopsoas

Though it is the two muscles, i.e. *iliacus* and *psoas major,* but since they have a common distal attachment and act as one muscle, it is quite usual practice to treat them as one muscle (**Fig. 9.4**). The **psoas major** takes origin from the sides of the bodies of the last thoracic and all lumbar vertebrae, the **iliacus** arises from the anterior surface of ilium and base of sacrum. Distally both are *attached* to the lesser trochanter of femur and shaft just below. *Action:* The main movement is flexion of thigh on trunk, or flexion of trunk on thighs as in sitting from lying position. Probably these muscles also stabilize the hip joint in standing position.

Sartorius

It is the longest muscle in the body which is also known as **tailor muscle (Fig. 9.5)**. This slender muscle is obliquely directed downward and inwards across the front of the thigh. The muscle *originates* from the anterior superior iliac spine and upper half of the notch below it, and gets *inserted* over the anteromedial surface of the tibia just below condyle. *Action:* At the hip, it helps in flexion, abduction and external rotation. This is a two joint muscle and as such, it also acts at knee as flexor and internal rotator.

Pectineus

It is a short, thick muscle situated just below the groin (**Fig. 9.6**). This muscle takes *origin* from the one inch wide

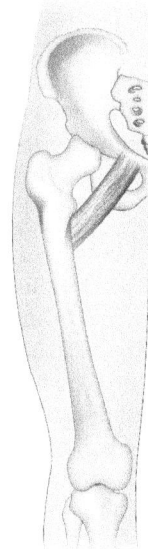

Fig. 9.6: Pectineus

space on front of the pubis just above the crest, and gets *inserted* over the pectineal line extending from the lesser trochanter to the linea-aspera on the femur. *Action:* It works as flexor and adductor of the hip. The adduction action is seen when the hip is in the flexed position.

Rectus Femoris

It is one of the muscles composing quadriceps femoris group of muscles which is an extensor of the knee (**Fig. 9.7**). It takes its *origin* from the anterior inferior iliac spine

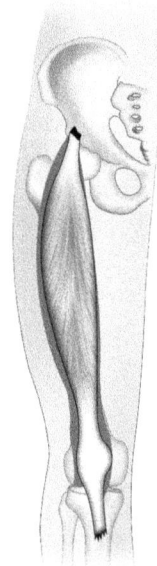

Fig. 9.7: Rectus femoris

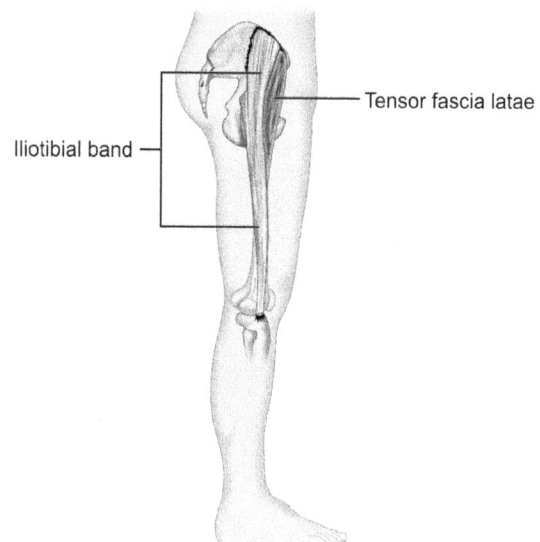

Fig. 9.8: Tensor fascia latae muscle

and *inserts* at the base of patella, and through the patellar ligament to the tibial tuberosity. It is a two joint muscle and works as flexor at the hip and as an extensor at the knee. This muscle is also known as *kicking muscle* (kicking in football). This muscle works in all the movements of combined flexion at the hip and extension at the knee.

Tensor Fascia Latae (TFL)

This small muscle is located on the anterolateral aspect of the thigh. Its name is derived from its action of tightening the fascia of the thigh **(Fig. 9.8)**. This muscle has no distal attachment over the bone. The muscle takes *origin* from the iliac crest near the anterior superior iliac spine and gets *inserted* over the iliotibial band (ITB) on the lateral aspect of the thigh, one-third way down. *Action:* The muscle assists in hip flexion, abduction and internal rotation (in flexed position). Also works in knee extension and knee stability in weight bearing position.

Extensors: *Gluteus maximus* and *hamstrings* are the prime movers for the extension at the hip. Gluteus maximus is regarded as a true antagonist of hip flexors. Hamstrings work whenever strong resistance is offered. Hamstrings work better at the hip when the knee is straight as it is a two joint muscle.

Gluteus Maximus

This is a large fleshy muscle at the back of the hip (buttock) **(Fig. 9.9)**. The muscle takes *origin* from the posterior

Fig. 9.9: Gluteus maximus muscle (posterior aspect of upper hip)

one-fourth of the crest of ilium, adjacent part of sacrum and side of coccyx. It gets *inserted* over the posterior surface of femur on a gluteal line extending from greater trochanter to the linear aspera. *Action:* This muscle is a strong extensor of the hip and an outward rotator in extended hip. Since upper one-third of this muscle lies above the axis of hip joint, it helps in hip abduction when the weight of the body is falling on one limb, as in jogging or running. Its lower two-third fibers are said to help in movement of hip adduction against resistance.

Normally the muscle is not called upon to extend the hip when it is flexed upto 45° unless there is strong resistance. It works effectively and strongly when hip is flexed beyond 45°.

Hamstrings

These muscles are primarily flexors of the knee, and also the hip extensors. It is actually three muscles collectively called hamstrings **(Figs. 9.10A and B)**. The *biceps femoris* (also known as outer hamstrings) lies on the lateral side of the posterior aspect of thigh and knee. The two muscles, i.e. *semitendinosus and semimembranosus* lie on the medial side of thigh posteriorly. These muscles help the gluteus maximus in extension of hip when the knee is almost straight. Hip extension with knee bent is not aided by hamstrings, and only produced by the gluteus maximus muscle.

Biceps Femoris

This muscle takes *origin* through two heads, the long head which contributes only in hip extension takes *attachment* from the ischial tuberosity. The short head taking origin from the linea aspera, does not work in hip extension since it arises from below the hip joint. The muscle as a whole gets *inserted* over the lateral condyle of tibia and head of fibula. Its main *actions* are extension at the hip, and flexion at the knee. It is said to assist external rotation at hip and also at knee.

Semitendinosus

It is one of the medial hamstring muscles taking *origin* from the ischial tuberosity along with the long head of biceps femoris and *inserts* at upper part of anteromedial surface of tibial shaft (along with sartorius). It works as an extensor at the hip.

Semimembranosus

It also takes *origin* from the ischial tuberosity and gets *inserted* over posterior surface of medial condyle of tibia, and works as an extensor of hip and flexor of knee.

Of these two muscles, the semimembranosus is comparatively medial to semitendinosus. Both of the medial hamstrings also contribute to the internal rotation at the hip and knee. All the hamstrings are used in ordinary walking as hip extensors and allow gluteus maximum to relax in this movement.

Abductors: Gluteus medius is primarily a strong hip abductor whereas the *gluteus minimus, tensor fascia latae* (TFL), and upper fibers of gluteus maximus and sartorius all assist abduction at the hip joint. TFL is an abductor when the hip is extended.

Gluteus Medius

This is a short thick muscle situated on the side of the hip giving it a rounded contour **(Fig. 9.11)**. The muscle takes its *origin* from the outer surface of ilium just below the crest and gets *inserted* over the lateral surface of greater trochanter on an oblique ridge. *Action:* Gluteus medius

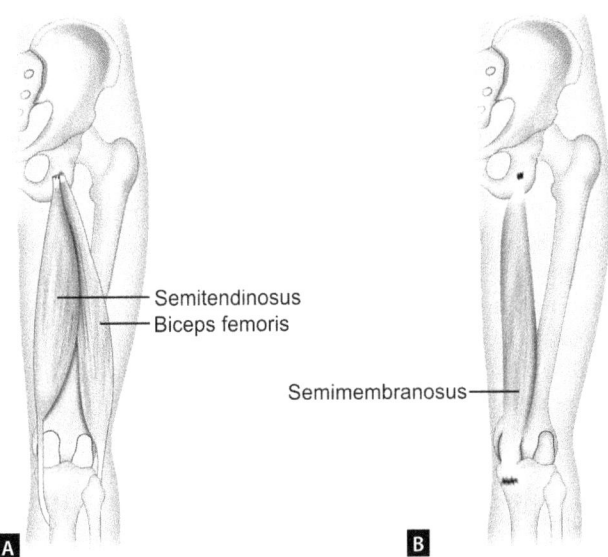

Figs. 9.10A and B: Hamstrings muscles

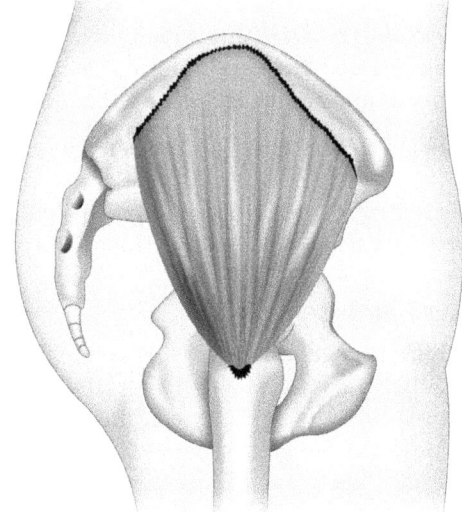

Fig. 9.11: Gluteus medius muscle (lateral view of pelvis)

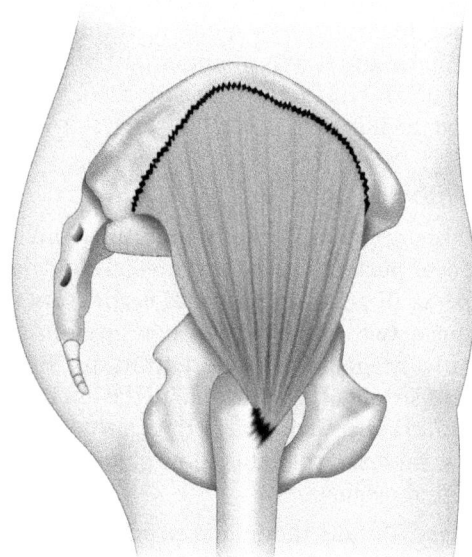

Fig. 9.12: Gluteus minimus muscle (lateral view of pelvis)

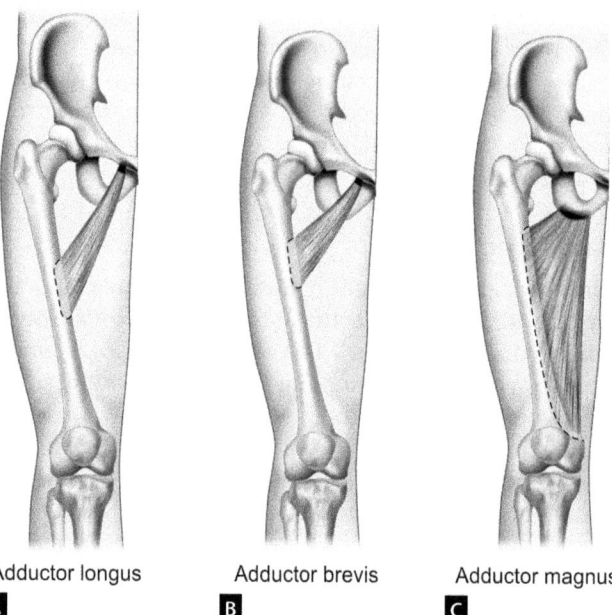

Figs. 9.13A to C: Adductor muscles of right hip

is a powerful abductor of the hip joint. Its anterior fibers help in internal rotation of the hip. This is a very important muscle used in walking and standing in good posture. While standing on one leg, the muscle works statically to stabilize the hip.

Gluteus Minimus

It is a smaller companion of gluteus medius which lies just beneath it **(Fig. 9.12)**. This muscle takes *origin* from the outer surface of ilium below the origin of gluteus medius, and takes its *insertion* over the anterior surface of greater trochanter. *Action:* Gluteus minimus is primarily an internal rotator and a weak abductor at the hip.

Adductors (Figs. 9.13A to C): *Adductor longus, magnus* and *brevis* are the prime movers for the adduction of the hip, assisted by pectineus, and gracilis. They are all collectively very strong and powerful muscles.

Adductor Longus

This is a thick, triangular muscle lying on the inner side of the thigh. This muscle *arises* by a short tendon from the front of the pubis just below the crest and gets *inserted* on the middle third of linea aspera of femur. *Action:* This muscle is an adductor, and an assistant mover in hip flexion and internal rotation. It helps in hip flexion upto 70° only and beyond this, the muscle becomes an extensor of the hip since the line of pull of the muscle changes in relation to axis of the joint.

Adductor Brevis

This is also a short, triangular muscle above and behind the adductor longus. The muscle takes *origin* from the front of the pubis just below the origin of adductor longus, and takes *insertion* at lesser trochanter and upper one-fourth of linea aspera. *Action:* The muscle causes adduction of hip, and also assists in hip flexion and internal rotation.

Adductor Magnus

This is one of the largest muscles of the body and the strongest of the hip adductors situated on the medial side of the thigh. The muscle takes *origin* from the front of the pubis and the ischial tuberosity, and *inserts* at the adductor tubercle on the medial condyle of femur, and whole length of linea aspera. *Action:* This is a powerful adductor of the hip. Its upper fibers assit in internal rotation and flexion at the hip, whereas the lower fibers assist in extension and internal rotation. Since this muscle is used in gripping the sides of the horse in riding, it is often regarded as *rider's muscle.*

Gracilis

This is a slender muscle along the inner side of the thigh (**Fig. 9.14**). This muscle takes *origin* from the front of lower half of pubic symphysis and upper half of pubic arch, and *inserts* over the medial surface of tibia just below the condyle. *Action:* The muscle adducts the hip, assists internal rotation and hip flexion when the knee is straight. Also serves as flexor of the knee and an internal rotator.

Six Lateral or External Rotators of Hip: The six muscles, i.e. *piriformis* (most superior), *obturator externus* and *internus, quadratus femoris, gamellus superior* and *inferior* are collectively termed as lateral or external rotators of the hip, which are situated deeply at the back of the hip joint (**Figs. 9.15A and B**). These muscles take *origin* from the posterior surface of pelvis near obturator foramen and sacrum. Their *insertion* is at greater trochanter of femur. *Action:* These muscles are very important external rotators of the hip, and also help in holding the head of femur in the acetabulum when the hip is flexed about 90° or more. These muscles can also produce horizontal extension of hip.

Inward Rotators: *Gluteus minimus* is the prime mover in internal rotation at the hip, assisted by gluteus medius (anterior fibers), tensor fascia latae (only in hip flexion), and all the adductors of the hip. All these muscles are already described above.

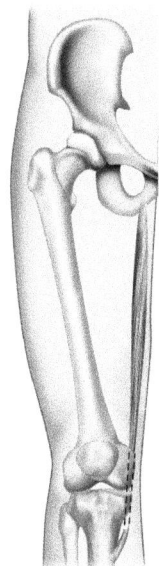

Fig. 9.14: Gracilis Muscle

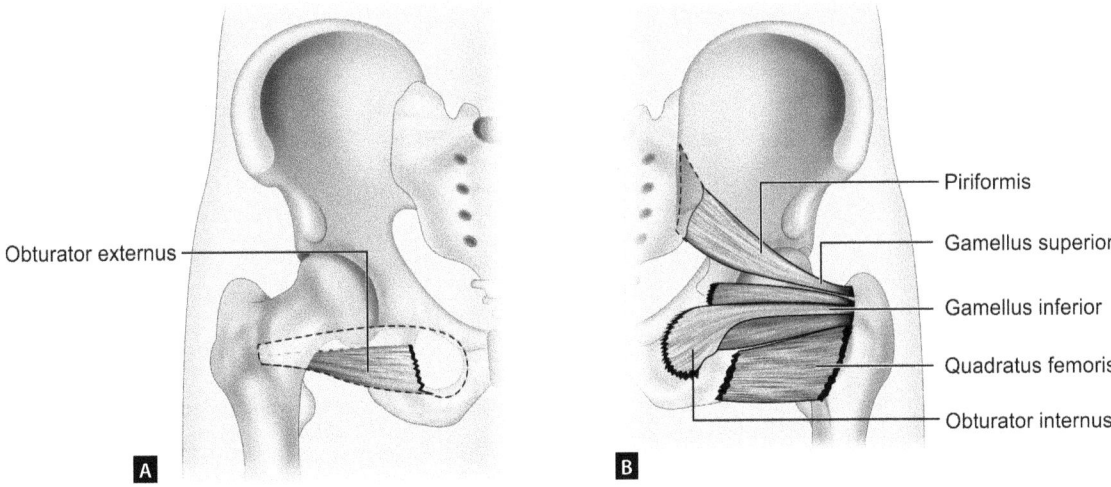

Figs. 9.15A and B: Hip lateral or external rotator muscles

Table 9.3: Summary of major muscles acting on hip

Muscle	Location	Origin	Insertion	Main action	Supportive action
Iliopsoas	Anterior wall of pelvic cavity	Iliac fossa, anterior and lateral surfaces of 12th thoracic and all five lumbar vertebrae	Lesser trochanter of femur	Hip flexion	Stabilization of Hip in standing
Rectus femoris	Anterior thigh (superficial)	Anterior inferior iliac spine of pelvis	Tibial tuberosity	• Hip-flexion • Knee-extension	
Sartorius	Anteromedial thigh (superficial)	Anterior superior iliac spine of pelvis	Upper medial aspect of tibia	Combined movement of hip flexion, abduction and external rotation	Knee flexion and internal rotation
Gluteus maximus	Buttock (superficial)	Posterior sacrum and ilium	Posterior and upper femur, iliotibial band	Hip extension	Hip external rotation (when hip is extended)

Hamstrings	Posterior thigh (superficial)	• *Biceps Femoris*
		  – *Long Head*: Ischial tuberosity
		  – *Short Head*: Lateral lip of linea aspera
		• *Semitendinosus*: Ischial tuberosity
		• *Semimembranosus*: Ischial tuberosity
			Fibular head
			
			• Upper tibia (anteromedial surface)
			
			• Medial condyle of tibia (posteromedial surface)
				Hip extension and knee flexion
				
				Hip extension and knee flexion
				Hip extension and knee flexion
					External rotation of hip and knee
					
					Internal rotation of hip and knee
					Internal rotation of hip and knee

Gluteus medius	Lateral thigh (upper)	Outer surface of ilium	Lateral surface of greater trochanter of femur	Hip abduction	Inward rotation of hip (anterior part)
Gluteus minimus	Lateral thigh (upper part), deep	Lateral aspect of ilium	Anterior surface of greater trochanter of femur	Hip internal rotation	Hip abduction
Adductor longus	Medial thigh	Front of pubis	Middle third of linea aspera of femur	Hip adduction	Hip flexion and internal rotation
Adductor magnus	Medial thigh	Front of pubis and ischium	Adductor tubercle on the medial condyle of femur, and entire linea aspera	Hip adduction	*Upper part*: • Hip flexion • Internal rotation *Lower part*: • Hip extension • Internal rotation
Adductor brevis	Medial thigh	Front of pubis (below the origin of adductor longus)	Lesser trochanter and upper part of linea aspera of femur	Hip adduction	Hip flexion and internal rotation
Pectineus	Medial thigh	Superior ramus of pubis	Pectineal line of femur	Flexion and adduction of hip	
Gracilis	Medial thigh	Pubis	Upper tibia (anteromedial aspect)	Hip adduction	• Hip internal rotation and flexion • Knee flexion

Contd...

Contd...

Muscle	Location	Origin	Insertion	Main action	Supportive action
Tensor fascia lata	Upper and lateral thigh	Anterior superior iliac spine	Lateral condyle of tibia	Combined hip flexion and abduction	Hip internal rotation (when hip is flexed)
Lateral or external rotators: • Piriformis • Obturator externus • Obturator internus • Quadratus femoris • Gamellus superior • Gamellus inferior	Posterolateral buttock (5) and pubis (1)	Posterior surface of pelvis and sacrum	Greater trochanter area of femur	Hip external rotation	Hip horizontal extension

10

CHAPTER

Knee

Knee joint is the largest and the most complex joint in the body. It serves the dual functions of weight bearing and locomotion. Knee joint is generally referred to be a modified, synovial **hinge type of joint** as it permits rotational movement (not as a free movement but an accessory motion accompanying knee flexion and extension).

The main articulation of the knee is formed between the condyles of femur and tibia i.e. **tibiofemoral joint (Fig. 10.1)**, which is actually the knee joint proper, however the other joint is also formed between femur and patella, i.e. **patello-femoral joint (Fig. 10.3)** within the single joint cavity. The lower end of femur terminates into two rocker like (convex) condyles which rest on the two slightly concave areas on the top of tibia. The lateral condyle of femur is broader and prominent than the medial one. They are also not quite parallel. The medial condyle is projected more downward than the lateral one. This is to compensate for the obliquity of the femoral shaft which in normal femur slants inward from above downward.

The anterior view of knee (with patella removed) is shown in **Figure 10.2**.

- The articulating surface of tibia is quite enlarged and almost horizontal. The medial articular surface is oval whereas the lateral is smaller and nearly round. Over the articulating surfaces of the tibia, *medial* and *lateral semilunar cartilages* or *menisci* rest over the medial and lateral condyles respectively. They deepen the articular facets on tibia and enhance stability apart from serving as shock absorbers. They are thicker at the periphery but taper down on inner sides. The *medial meniscus* is attached at its periphery to the tibial collateral ligament, but the *lateral meniscus* has no such attachment to the fibular collateral ligament. They can slip about slightly, and are held in place by ligaments.
- In front, there is slightly concave surface between the condyles of the femur for articulation with patella.

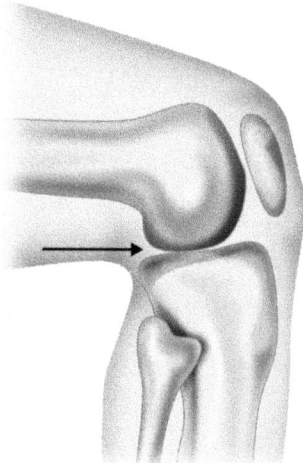

Fig. 10.1: Knee (tibiofemoral) joint in flexion (lateral view)

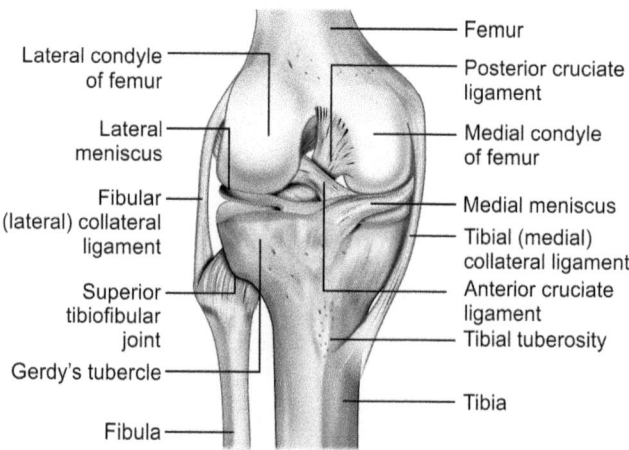

Fig. 10.2: Knee: anterior view with patella removed

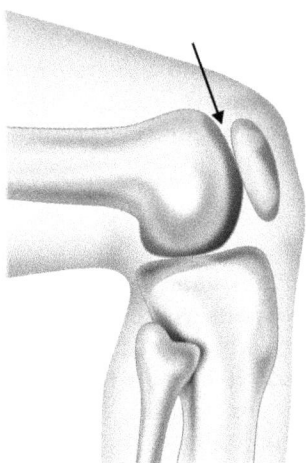

Fig. 10.3: Patellofemoral joint (in flexed knee position)

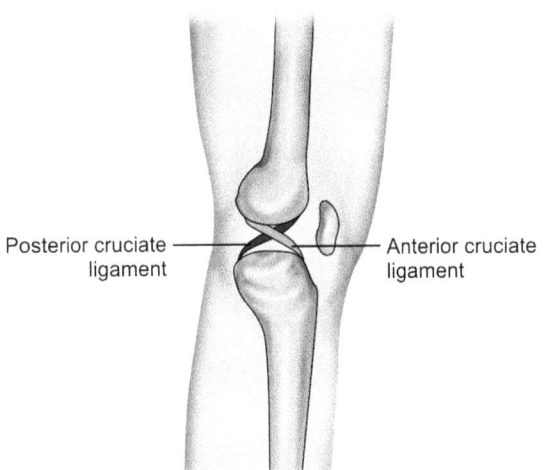

Fig. 10.4: Cruciate ligaments of knee

Patella or *knee cap* is a sesamoid bone implanted in the quadriceps tendon and *patellar ligament* in front of the knee joint. This provides a better angle of pull and therefore greater strength for the quadriceps muscles.
- All the articular surfaces of femur, tibia and patella are covered with hyaline cartilage. The *joint capsule* of the knee is very extensive. Its *synovial membrane* is also very extensive and lines the whole joint capsule excluding the menisci.
- The tibia, the medial bone in the leg bears much of the body's weight than the lateral bone fibula. The fibula serves as the attachment for some very important knee joint structures, though it does not articulate with the femur or patella and is not part of the knee joint.
- The knee joint is well supplied with synovial fluid from the synovial cavity, which lies under the patella and between the surfaces of the tibia and the femur. There are more than 10 *bursae* located in the knee, some of which are connected to the synovial cavity. These bursae are located where they can absorb shock or prevent friction.
- *Stability* of the knee mainly depends upon the powerful muscles of the knee, combined with strong ligaments, with no bony stability. The ligaments provide the static stability to the knee joint, whereas the quadriceps and hamstrings muscles produce dynamic stability. All these make the knee a most strong functioning unit enabling it to bear all the strains and stresses.

The main ligaments of knee are:
- On the medial side of the knee is found a *medial collateral ligament (MCL)* or *tibial ligament* which is attached to the medial condyles of femur and tibia, which on the way is also attached to the joint capsule and medial meniscus. On the outer side of the knee, a *fibular or lateral collateral ligament (LCL)* joins the head of fibula and femur.
- The *anterior and posterior cruciate ligaments* **(Fig. 10.4)** are named because they cross within the knee between tibia and femur. The *anterior cruciate ligament (ACL)* extends upward and backward from the anterior intercondylar area of the tibia to the back part of the medial surface of the lateral condyle of the femur. The *posterior cruciate ligament (PCL)* is a stronger and shorter ligament passing upward and forward from the posterior intercondylar fossa of tibia to the front part of medial condyle of the femur. *These cruciate ligaments maintain the inner stability of knee joint.* They limit extension and prevent rotation in extended position. They also check the anteroposterior stability of the knee.
- *Patellar Ligament:* It is a strong flat ligament which connects the lower margin of patella with tibial tuberosity.
- *Transverse Ligament:* It is a short cord like ligament which connects the anterior margin of lateral meniscus to the anterior margin of medial meniscus. The *iliotibial tract* also serves as a stabilizing ligament at the knee joint.

Various ligaments of knee (posterior view) are shown in **Figure 10.5**.

MOVEMENTS

Since the knee is a modified *hinge joint*, rotation in flexed knee also takes place apart from flexion-extension movements. **Flexion** or bending of the leg occurs at the knee so that the posterior surface of the leg comes in

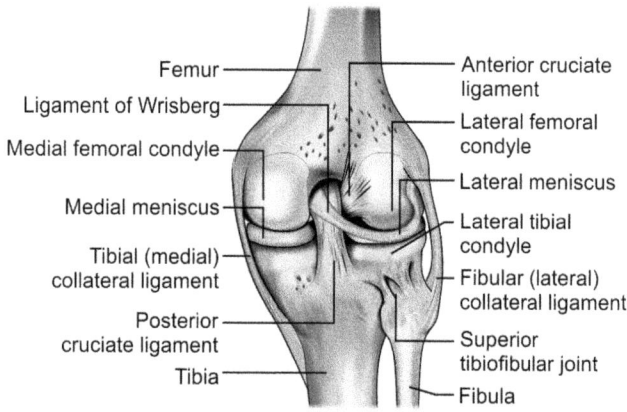

Fig. 10.5: Ligaments of knee: posterior view

contact to the posterior surface of thigh (**Fig. 10.6A**). **Extension** is the reverse movement from the flexed knee, or it is the straightening of leg at the knee (**Fig. 10.6B**).

The knee can usually extend to 180° or a straight line, although it is not uncommon for some knees to hyperextend upto 10° or more. When the knee is in full extension, it can move from there to about 140° of flexion. When the knee is straight, no side to side movement i.e. abduction-adduction movement is possible. However, with the knee flexed to approximately 30° or more in a non-weight bearing position, approximately 30° of internal rotation and 45° of external rotation can occur. This becomes possible only because the collateral ligaments are relaxed.

Due to the shape of the medial femoral condyle, the knee must '*screw home*' to fully extend. As the knee approaches full extension, the tibia must externally rotate approximately 10° to achieve proper alignment of the tibial

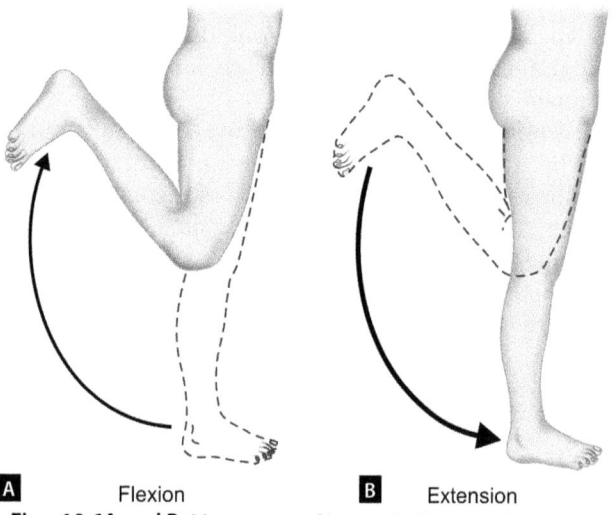

Figs. 10.6A and B: Movements of knee : (A) Flexion, (B) Extension

and femoral condyles. Similarly, during the initial flexion from a fully extended position, the knee 'unlocks' by the tibia rotating internally, to a degree, from its externally rotated position to achieve flexion.

The summary of muscular analysis of knee movements is given in **Table 10.1**.

MUSCLES

There are about *12 muscles* acting on knee joint. The muscles acting on the knee joint, as according to their location on the thigh and knee are regarded as anterior and posterior muscles and are shown in **Table 10.2**. Usually, the anterior muscles cause extension at the knee, and the posterior muscles mainly contribute to the flexion movement. The summary of major muscles acting on knee is given in **Table 10.3**.

Extensors: The quadriceps femoris is a very powerful extensor muscle of the body.

Quadriceps Femoris

This group of muscles is said to be a mirror of the knee joint, as much of the stability and functions of the knee depend

Table 10.1: Summary of muscular analysis of knee movements

Movement	Prime movers	Assistant movers
Flexion	Hamstings muscle group • Biceps femoris • Semitendinosus • Semimembranosus	• Sartorius • Gracilis • Popliteus • Gastrocnemius • Plantaris
Extension	Quadriceps femoris muscle group • Rectus femoris • Vastus medialis • Vastus lateralis • Vastus intermedius	
Outward rotation (Knee in flexion)	• Biceps femoris	
Inward rotation (Knee in flexion)	• Semitendinosus • Semimembranosus • Popliteus	• Sartorius • Gracilis

Table 10.2: Knee muscles as according to their location

Anterior	Posterior

• Quadriceps femoris group
  – Rectus femoris
  – Vastus medialis
  – Vastus lateralis
  – Vastus intermedius
	• Hamstrings group
	  – Biceps femoris
	  – Semitendinosus
	  – Semimembranosus
	• Sartorius
	• Gracilis
	• Popliteus
	• Gastrocnemius
	• Plantaris

upon the strength of this muscle. As the name refers, this is a muscle which takes origin from four heads, i.e. rectus femoris, vastus lateralis, vastus medialis, and vastus intermedius. Rectus femoris **(Fig. 10.7A)** has already been described as a hip flexor in Chapter 9.

Vastus Lateralis

This is a muscle located on the outer aspect of thigh, giving it a rounded shape and contour. This muscle takes origin from the lateral surface of femur below the greater trochanter and upper half of the linea aspera of femur.

Vastus Medialis

This muscle is located on the medial side of the thigh, and takes origin from the whole length of linea aspera of femur and the medial supracondylar line.

Vastus Intermedius

This muscle is a companion of vastus medialis and lateralis, and lies between them under cover of rectus femoris. It takes its origin from the anterior and lateral aspects of upper two-thirds of shaft of femur.

Insertion: All these three vasti **(Fig. 10.7B)**, along with the tendon of rectus femoris unite to form the tendon of quadriceps femoris, which is attached to the base of patella, and indirectly to the tibial tuberosity through the patellar ligament. *Actions:* Rectus femoris is a two joint muscle which acts as a flexor of hip and an extensor of the knee. All three vasti converge towards the patella, and steady the knee joint in all the weight bearing positions. Because they are the one joint muscles only, they are the powerful extensors of knee irrespective of the position of the hip. In the flexed hip position, the extension of knee is mainly produced by three vasti only, and the rectus femoris is now relaxed at the knee. Previously it was thought that the vastus medialis was responsible for terminal extension of the knee and this concept has now been disproved. Quadriceps is a very coarse muscle which rapidly goes wasted if not used for long. This muscle is very frequently used in all the kicking, running, jumping, bicycling, and walking activities.

Flexors of the Knee: All the three hamstring muscles **(Figs. 10.8A and B)**, i.e. *biceps femoris, semitendinosus,* and *semimembranosus* are the prime movers in the knee flexion, assisted by sartorius, gracilis, popliteus, gastrocnemius and plantaris muscles. Hamstrings, gracilis and sartorius muscles have already been described as the hip muscles in Chapter 9.

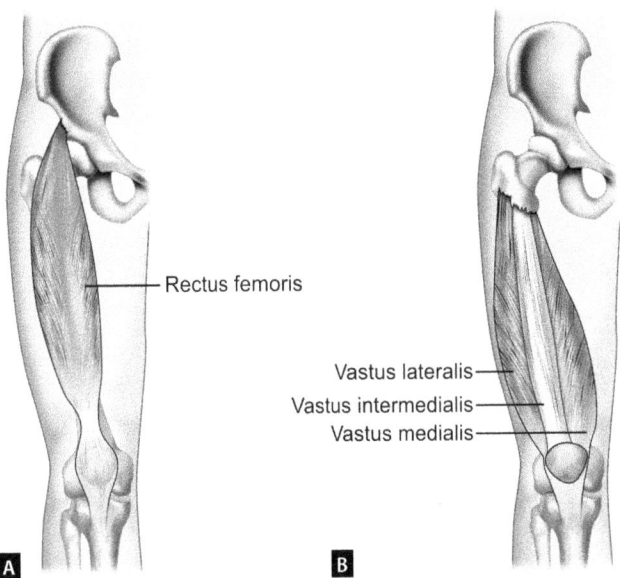

Figs. 10.7A and B: Quadriceps femoris muscle group: anterior views

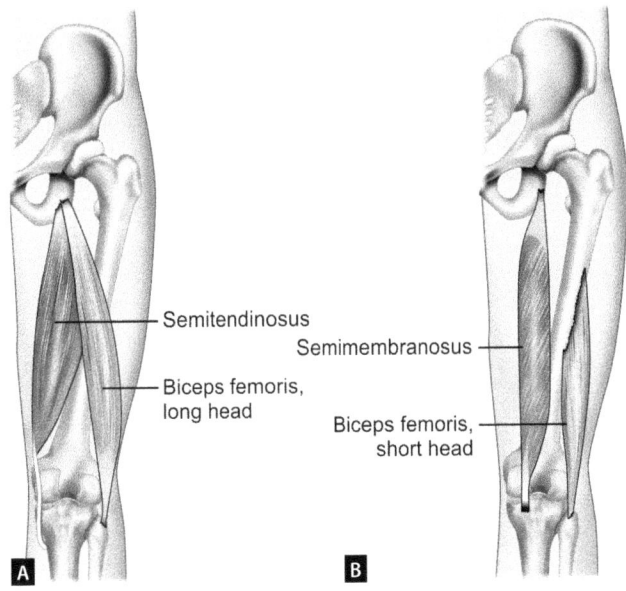

Figs. 10.8A and B: Hamstring muscles

Popliteus

This thin triangular muscle takes origin from the lateral surface of lateral condyle of femur and gets inserted over the posteromedial surface of upper tibia **(Fig. 10.9)**. Action: The muscle is an assistant mover in knee flexion, and also helps in internal rotation of the tibia. This is the

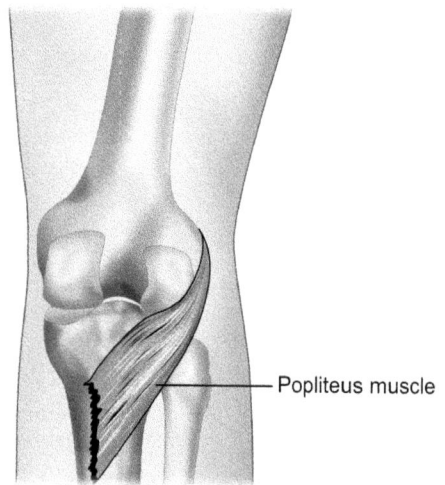

Fig. 10.9: Popliteus muscle (posterior view of knee)

only flexor of the knee which is a single joint muscle. The muscle unlocks the knee in the beginning of flexion and also has the stabilizing role. Along with the cruciate ligaments it prevents the forward dislocation of the knee in flexed position.

Gastrocnemius and plantaris are primarily the muscles of the ankle and are described in Chapter 11. However, since the gastrocnemius lies posteriorly to the knee joint, it also helps in the knee flexion. Its role over the knee is more to protect the joint in the movements of violent extensions and therefore it serves more like a posterior ligament.

Inward rotation at the knee is mainly produced by the semitendinosus, semimembranosus, and the popliteus muscles, assisted by sartorius and gracilis. **Outward rotation** is mainly produced by biceps femoris muscle, which is an outer part of hamstrings.

Table 10.3: Summary of major muscles acting on knee

Muscle	Location	Origin	Insertion	Main action	Supporting action
Quadriceps femoris	Anterior thigh	Rectus femoris	Already described in Chapter 9	• Hip flexion • Knee extension	
		Vastus medialis: Linea aspera of femur	Tibial tuberosity via patellar tendon	• Knee extension	
		Vastus lateralis: Linea aspera and upper lateral femur	Tibial tuberosity via patellar tendon	• Knee extension	
		Vastus intermedius: Anterior femur (upper two-thirds)	Tibial tuberosity via patellar tendon	• Knee extension	
Hamstrings	Posterior thigh	Biceps femoris: Long head: Ischial tuberosity Short head: Lateral lip of linea aspera	Fibular head	• Knee flexion • Hip extension	External rotation of hip and knee
		Semitendinosus: Ischial tuberosity	Upper tibia (anteromedial surface)	• Hip extension	Internal rotation of hip and knee
		Semimembranosus: Ischial tuberosity	Medial condyle of tibia (posteromedial surface)	• Hip extension	Internal rotation of hip and knee
Sartorius	Described in details in Chapter 9			Hip flexion, abduction and external rotation	Knee flexion and internal rotation
Gracilis	Described in details in Chapter 9			Hip adduction	• Hip flexion, internal rotation • Knee flexion
Popliteus	Posterior surface of knee	Lateral condyle of femur (lateral side)	Medial condyle of tibia (posterior side)		• Initiates knee flexion • Internal rotation of leg (flexed knee)
Gastrocnemius	Described in details in Chapter 11			Plantar flexion of ankle	Knee flexion
Plantaris	Described in details in Chapter 11				• Knee flexion • Ankle plantar flexion

11

CHAPTER

Ankle and Foot

The complexity of the foot is evidenced by the 26 bones, 19 large muscles, many small muscles (intrinsics), and more than 100 ligaments that make up its structure.

Support and propulsion are two important functions of the foot. For proper functioning of the foot, adequate development of muscles of the foot, and practice of good foot mechanics is very essential for every one. The foot is united with the leg at the ankle joint, or in other words, the weight of the body is transmitted to the foot through the ankle joint.

Ankle joint is a uniaxial hinge type of joint which is formed between the lower ends of tibia and fibula (malleoli), and the talus (one of the seven tarsal bones of the foot) **(Fig. 11.1)**. The fibular and tibial malleoli, extending down on the lateral and medial sides of the talus respectively serve to deepen the concave area into which the talus fills, and also prevent the sideward movement at the ankle. The lateral malleolus extends about ½' distally to the ankle joint. According to certain experts, the true ankle joint (talotibial or talocrural joint) is considered to be made up of the distal tibia sitting on the talus, with the medial malleolus of the tibia fitting down around the medial aspect of the talus and the lateral malleolus of the fibula fitting down around the lateral aspect. This type of joint is often described using a carpentry term as a 'tenon and mortise joint'. A mortise is a notch cut in a piece of wood to receive a projecting piece (tenon) shaped to fit. Therefore, the malleoli of the tibia and fibula would be the mortise, and the talus would be the tenon.

BONES AND JOINTS OF FOOT

Apart from the support and weight-bearing function of the foot, the foot is quite flexible enough to easily adapt itself to the uneven and rough surfaces while walking.

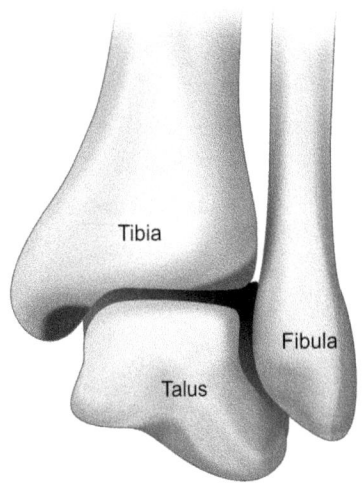

Fig. 11.1: Ankle joint

In foot, there are *26 bones* **(Figs. 11.2A to C)** in all i.e. *7 tarsals, 5 metatarsals* and *14 phalanges*. Each of the toes has got 3 phalanges, whereas the great toe has only two phalanges. The metatarsals are numbered from within outwards.

Out of 7 tarsal bones, the *calcaneum* or the heel bone is the base bone of the foot. It is the most posterior bone in the tarsals. Over the calcaneum rests the *talus* which is the top most bone of the foot. This bone participates in the formation of ankle joint.

The inferior surface of talus articulates with the superior surface of the calcaneum at *subtalar* or *talo-calcaneal joint* **(Fig. 11.3)**. This joint mainly allows gliding motion.

The *navicular* or *scaphoid* is a bone on the medial side which is anterior to the talus, and a *talonavicular joint* is formed between the two bones. On the lateral aspect, the calcaneous articulates with *cuboid* bone (which is little further to the navicular) to form a *calcaneocuboid joint*.

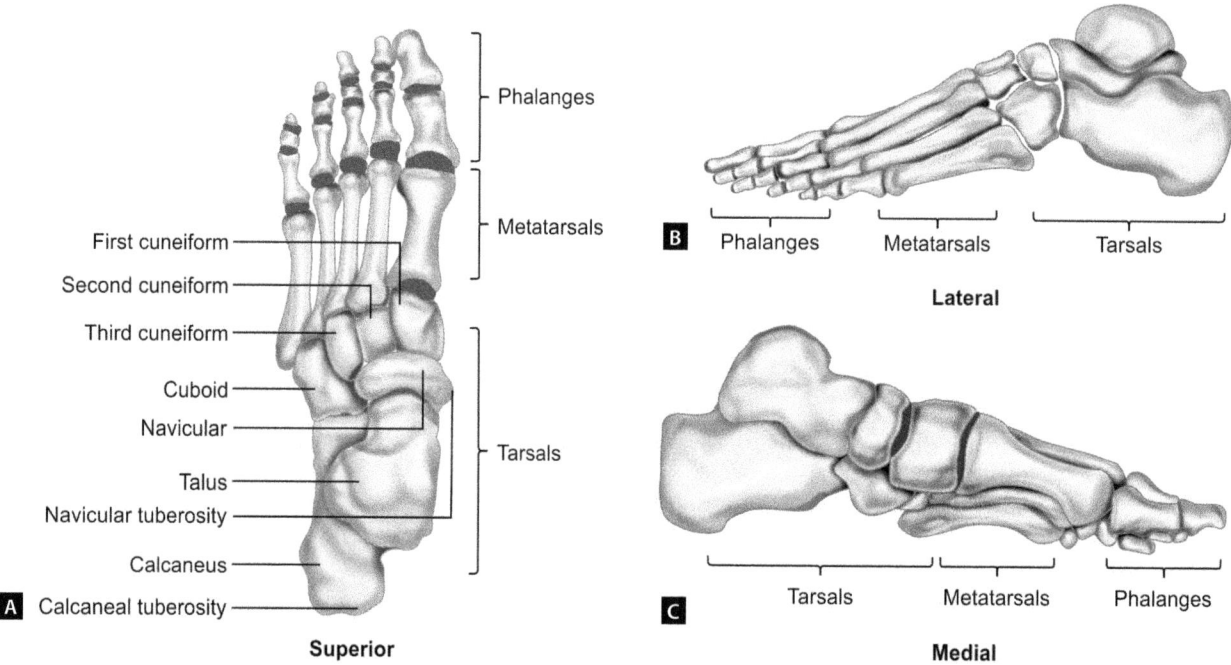

Figs. 11.2A to C: Bones of left foot: (A) Superior; (B) Lateral and; (C) Medial views

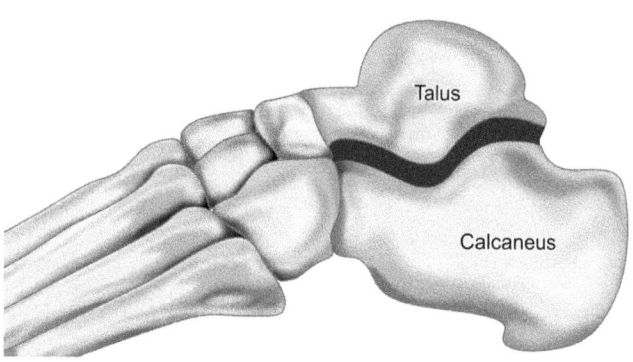

Fig. 11.3: Subtalar joint

These two joints, calcaneocuboid and the talonavicular form the transverse tarsal or midtarsal joints **(Fig. 11.4)**.

The combined motions of subtalar and midtarsal joints allow the foot to assume almost any position in space. This is quite useful in allowing the foot to adapt to irregular surfaces while walking on uneven ground, or climbing about on rocks at mountains.

In front of the navicular, *three cuneiform bones* lie side by side. The first metatarsal articulates with the first cuneiform, second metatarsal with second or middle cuneiform, and the third metatarsal with the third cuneiform. The IV and V metatarsals articulate with cuboid bone.

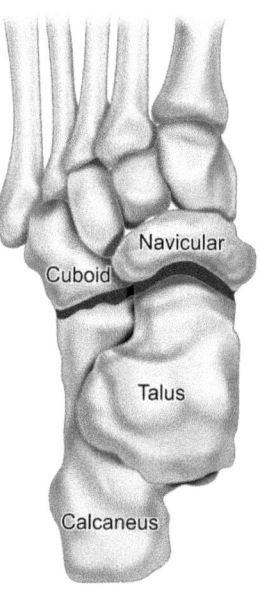

Fig. 11.4: Transverse tarsal joints

Distal end of each metatarsal articulates with the proximal phalanx to form a *metatarsophalangeal (MTP) joint*, and *interphalangeal (IP) joints* are formed between phalanges. The great toe has only one interphalangal joint whereas all other toes have two inter-phalangeal joints (i.e. proximal interphalangeal (PIP), and distal interphalangeal (DIP) joints **(Fig. 11.5)**.

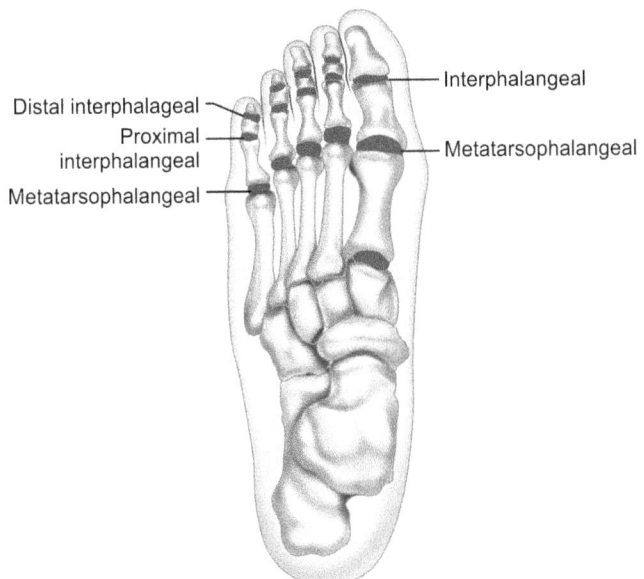

Fig. 11.5: Joints of phalanges of foot

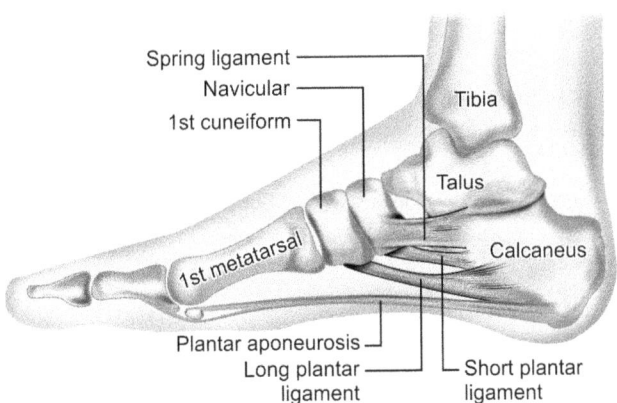

Fig. 11.6: Support structures of foot (medial view)

- The joints are also formed between the heads and bases of the metatarsal bones. These joints are the side by side articulations with the adjacent metatarsals, permitting only slight gliding movements.
- The strength and elasticity of the foot in general depends upon its bony structure, upon the ligaments which bind the bones together, and also upon the muscles which maintain the balance of the foot. The medial view of foot shows its various support structures in **Figure 11.6**.

MOVEMENTS

Because the ankle is a hinge joint and therefore only two movements, flexion and extension take place around a horizontal axis. This axis is actually slightly oblique because the lateral malleolus is little posterior to the medial malleolus.

In the ankle, flexion (or *dorsiflexion*) is the movement of the foot in the upward direction so that the dorsal surface of the foot approaches the front of the tibia (anterior surface) **(Fig. 11.7A)**. The range of dorsiflexion movement is about 20°. Extension (or the *plantarflexion*) is the reverse movement. It is actually a downward movement of the foot in which the dorsal surface of the foot moves away from the anterior surface of tibia **(Fig. 11.7B)**. The range of movement of plantarflexion is approximately 30–50°.

Inversion and *eversion* are two movements that take place technically in the tarsal joints (subtalar and midtarsal joints) of foot, though these movements are commonly thought of occurring at the ankle joint. Inversion is the movement of the foot in which the sole of the foot turns medially and the weight of the body is on the outer edge of the foot **(Fig. 11.7C)**. In eversion, the sole turns outward and the body weight falls on the inner aspect of the foot **(Fig. 11.7D)**. The normal range of motion of inversion is approximately 20 to 30°, whereas the motion of eversion is about 5 to 15°.

In *abduction-adduction* **(Fig. 11.7E)**, the front part of the foot bends laterally and medially, or in other words abduction is the outward movement of the toes away from the mid line of the foot, i.e. the 2nd toe, and adduction is the return movement from the abduction or the toes are drawn inwards towards the 2nd toe.

- All five MTP joints allow flexion, extension, hyperextension, abduction and adduction movements. The first MTP joint (of great toe) is much more mobile permitting approximately 45° of flexion and extension and 90° of hyperextension. The rest four lateral MTP joints allow about 40° of flexion and extension, and only about 45° of hyperextension. This hyperextension movement is very important during the toe-off phase (end of step) of walking.
- Sometimes the terms *pronation* and *supination* are also used to describe the movements between the forefoot and hind foot and occur at the mid-tarsal or transverse tarsal joints. The pronation is the combination of eversion and abduction movement, and the supination is the combination of the inversion and the adduction movements of the foot.

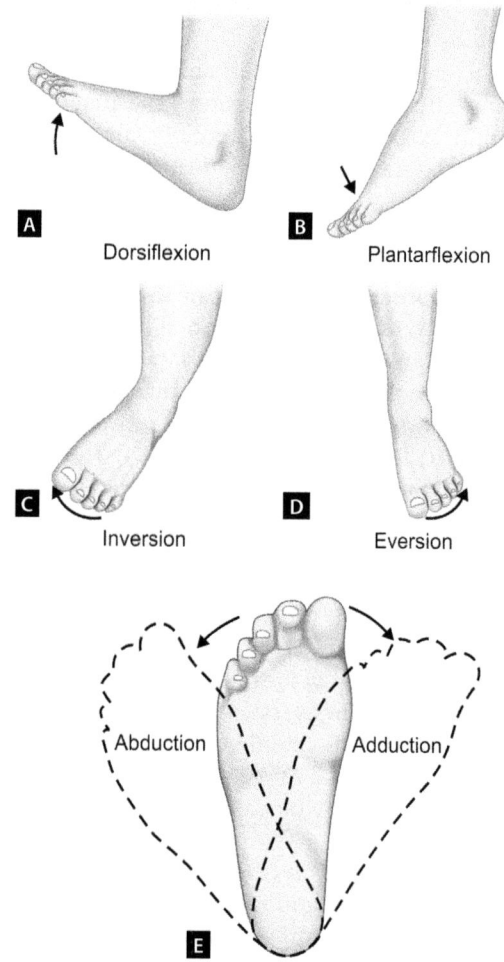

Figs. 11.7A to E: Movements of ankle and foot

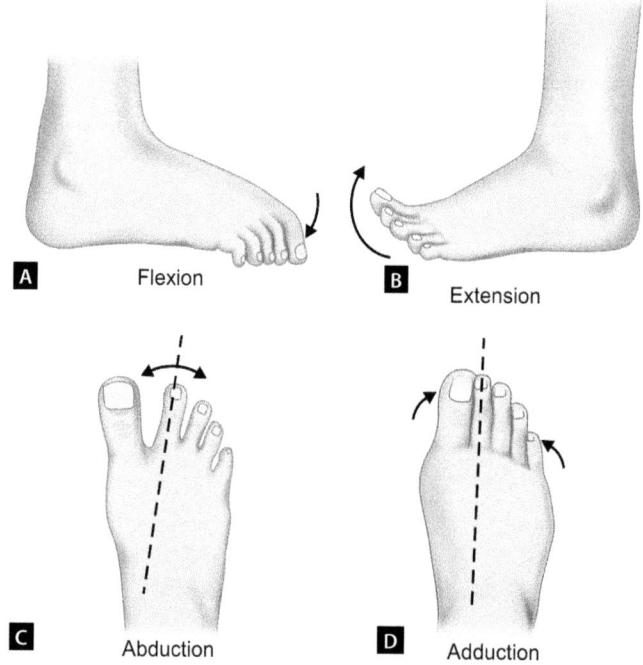

Figs. 11.8A to D: Movements of toes

- *Flexion* and *extension* movements of the toes refer to the curling and straightening movements. In flexion, the toes move towards the floor **(Fig. 11.8A)**, and in extension they move in upward direction **(Fig. 11.8B)**. The sideward movements of four toes (other than 2nd toe) away and toward the second toe (mid-line) refer to their abduction and adduction respectively **(Figs. 11.8C and D)**.

The summary of muscular analysis of ankle and foot movements is given in **Table 11.1**.

Ligaments of the Ankle and Foot

The ankle joint, a synovial joint, has a joint capsule. This joint capsule is reinforced by collateral ligaments on the sides. These collateral ligaments are actually groups of several ligaments. The collateral ligament on the medial side is a triangular-shaped *deltoid ligament* or medial ligament **(Fig. 11.9A)** which is attached proximally to the medial

Table 11.1: Summary of muscular analysis of ankle and foot movements

Movement	Prime movers	Assistant movers
Dorsiflexion	• Tibialis anterior • Extensor digitorum longus • Peroneus tertius	• Extensor hallucis longus
Plantarflexion	• Gastrocnemius • Soleus	• Tibialis posterior • Peroneus longus • Peroneus brevis • Plantaris • Long flexors of toes and hallux
Inversion	• Tibialis anterior • Tibialis posterior	• Flexor hallucis longus • Flexor digitorum longus • Extensor hallucis longus
Eversion	• Peroneus longus • Peroneus brevis • Peroneus tertius	Extensor digitorum longus
Flexion of toes	• Flexor hallucis longus (big toe) • Flexor digitorum longus (4 lateral toes)	
Extension of toes	• Extensor hallucis longus (big toe) • Extensor digitorum longus (4 lateral toes)	

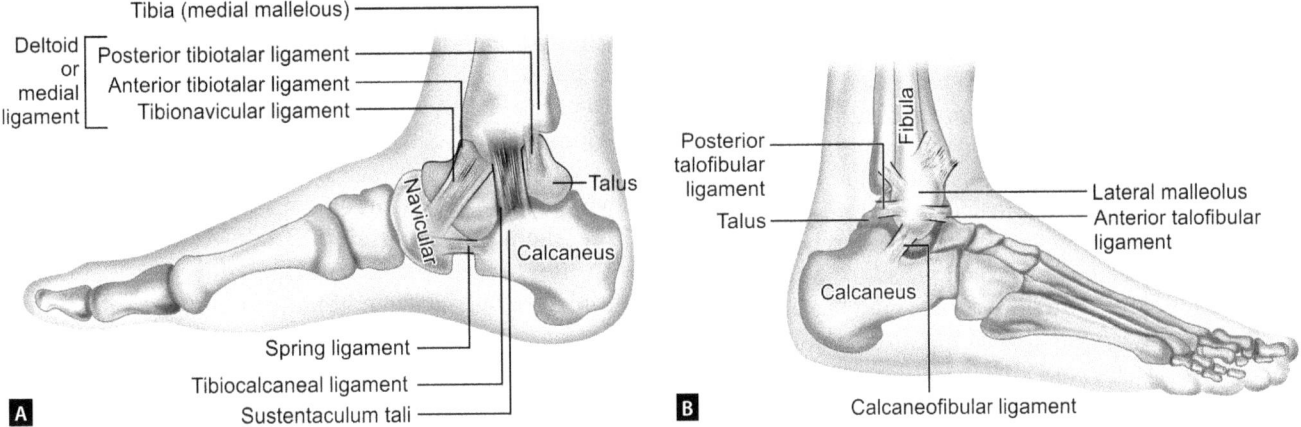

Figs. 11.9A and B: Ligaments of ankle. (A) Medial; (B) Lateral

malleolus of tibia and distally to the talus, calcaneous and navicular bones. This is a very strong ligament which not only protects the medial side of the joint, but also prevents the forward and backward displacement to the tibia.

On the lateral side of the ankle joint is a group of three ligaments commonly referred to as the lateral ligaments **(Fig. 11.9B)**. These three ligaments are *anterior* and *posterior talo-fibular ligaments*, and *calcaneofibular ligament*. The lateral ligament comprising of these three ligaments connect lateral malleolus to the talus, and calcaneous. These ligaments prevent the forward and backward displacement of fibula.

The lateral ligaments are weaker than medial ligaments. Otherwise also, the lateral side of the ankle appears less protected than the medial side, thus explaining higher incidences of injuries to the lateral ankle ligaments.

There are numerous other ligaments in the foot that attach various tarsals to each other, to the metatarsals, and to the phalanges etc. These ligaments are named for the bones to which they are attached. The individual names and further description of these ligaments is not given here.

Functional Aspects of Foot

Functionally, the foot can be divided into three parts; the *forefoot*, the *midfoot*, and the *hindfoot* **(Fig. 11.10)**. The *hindfoot* is made up of the talus and calcaneous tarsal bones. In the gait cycle, the hindfoot is the first part of the foot that makes contact with the ground, and thus influencing the function and movements of other two parts of foot (the forefoot, and the midfoot).

The *midfoot* is made up of the five tarsal bones, i.e., navicular, the cuboid, and three cuneiforms. The mechanics of midfoot provide stability and mobility to the foot as it transmits movements from the hindfoot to the forefoot.

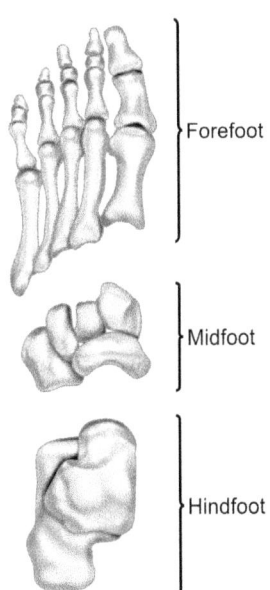

Fig. 11.10: Functional areas of foot

The *forefoot* is made up of the five metatarsals, and all of 14 phalanges of the toes. This part of the foot adapts to the level of the ground. It is also the last part of the foot to be in contact with the ground during stance phase.

The ankle joint and foot perform three main functions:
1. to act as a shock-absorber as the heel strikes the ground at the beginning of the stance phase,
2. to adapt to the level (or unevenness) of the ground, and
3. to provide a stable base of support from which to propel the body forward.

Arches of the Foot

Since the foot is the usual point of impact with the ground, it must be able—(i) to absorb a great deal of shock, (ii) to adapt or adjust to changes in ground/surfaces and (iii) to propel the body forward during movement/walking, etc.

To allow these things to occur, the whole foot is designed to be an elastic structure, and the bones of the foot are arranged in arches. We tend to stand on a triangle with weight bearing borne from the base of the calcaneous to the heads of first and fifth metatarsals **(Fig. 11.11)**.

Between these three weight bearing points/surfaces, the foot has two types of arches in it;
1. Longitudinal arch, and
2. Transverse arch.

(These arches are at right angles to each other).

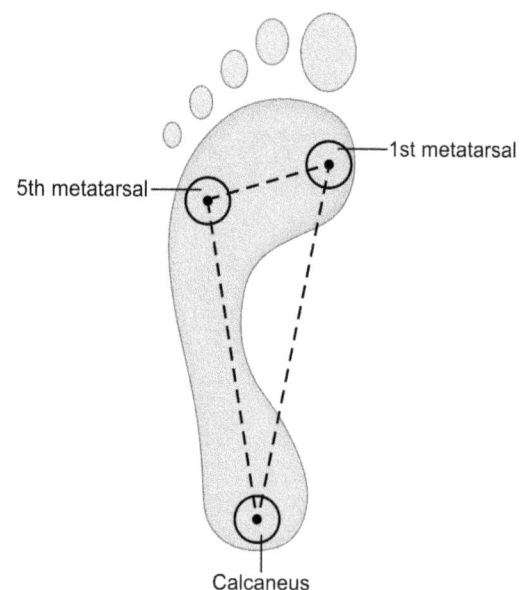

Fig. 11.11: Main weight-bearing surfaces of right foot (plantar view)

Longitudinal arch extends from the heel to the heads of the five metatarsals. This arch has two components and are described separately. The *medial longitudinal arch* **(Fig. 11.12A)**, (the medial or inner component of the longitudinal arch) makes up the medial border of the foot, and runs from the calcaneous posteriorly to the talus, navicular and three cuneiforms anteriorly, to the first three (medial) metatarsals. The talus is at the top of this arch. The medial longitudinal arch depresses somewhat during weight bearing, and then recoils when the weight is removed. Normally, this arch never flattens or touches the ground. The medial longitudinal arch, with its greater flexibility and curving, is better suited for the function of shock absorption, which is very important in all forms of locomotion.

It is also important to be reminded that the talus is the keystone bone of the arches present in the foot. The talus, apart from being the connecting link between foot and the leg, it has, by chance, no muscles attached to it, and by receiving and transmitting the weight of the entire body (except of the foot itself), a function requiring great deal of strength and firm support should be properly met with.

The *lateral longitudinal arch* **(Fig. 11.12B)** (the lateral or outer component of the longitudinal arch) runs from the calcaneous to the cuboid and finally to the anteriorly lying 4th and 5th metatarsals. This arch has a nearly flat contour, lacks mobility and normally rest on the ground during weight bearing. As such, this arch is better suited for the function of support.

The *transverse arch* is the side-to-side concavity on the underside of the foot formed by the anterior tarsal bones (from the three cuneiforms to the cuboid) and the metatarsals **(Fig. 11.13)**. The anterior boundary of this arch is formed by all five metatarsal heads. (The second cuneiform bone is the keystone of this arch).

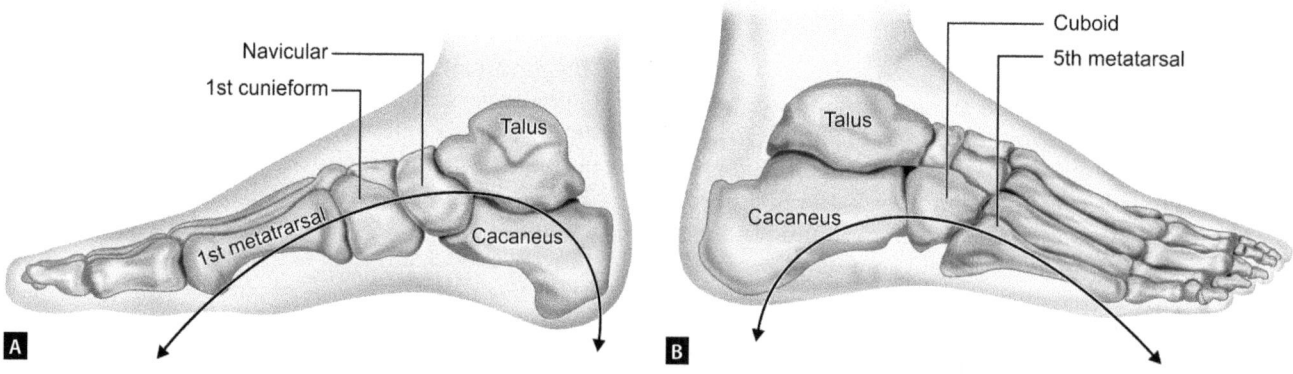

Figs. 11.12A and B: Longitudinal arches of foot: (A) Medial; (B) Lateral

CHAPTER 11: Ankle and Foot | 137

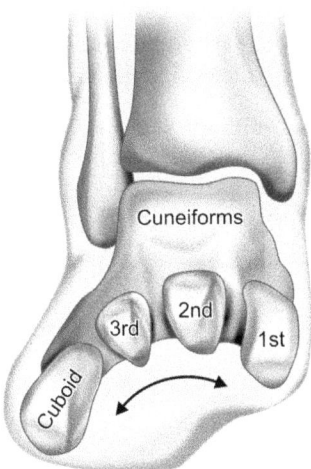

Fig. 11.13: Transverse arch of the foot (frontal view)

These three arches are maintained by.
1. The shape of the bones and their relation to each other,
2. The plantar ligaments and aponeurosis, and
3. The muscles (mainly the invertors, evertors, and the intrinsic muscles).
- The intrinsic muscles of the foot are probably more important as they provide more support than the extrinsic muscles. However, it has been estimated that the total muscular support to the arches is only about 15 to 20% of the total stress to the arches.
- Certain experts are of the opinion that the height of the medial longitudinal arch is not a definite indicative of the strength of the arch. And, therefore, a low arch is not necessarily a weak arch, provided that it is not associated with a pronated (abducted and everted) foot.

MUSCLES OF THE ANKLE AND FOOT

There are extrinsic and intrinsic muscles in the ankle and foot. *The extrinsic muscles originate on the leg whereas the intrinsic muscles originate on the tarsal bones and are located entirely within the foot.* There are about 12 extrinsic muscles that are involved in the movements described around the ankle and the foot.

The *extrinsic muscles* may be classified according to the function, and their location **(Table 11.2)** in relation to the leg. The muscles, generally on the anterior aspect of the leg are dorsiflexors (tibialis anterior, extensor digitorum longus, extensor hallucis longus and peroneus tertius). The muscles on the posterior aspect of the leg are the plantarflexors (gastrocnemius, soleus, tibialis posterior, flexor digitorum longus, flexor hallucis longus, and

Table 11.2: Summary of extrinsic muscles of ankle and foot according to their location

Anterior aspect of leg	Posterior aspect of leg
• Tibialis anterior • Extensor hallucis longus • Extensor digitorum longus • Peroneus tertius	*Superficial* • Gastrocnemius • Soleus • Plantaris
Lateral aspect of leg	*Deep*
• Peroneus longus • Peroneus brevis	• Tibialis posterior • Flexor hallucis longus • Flexor digitorum longus

plantaris muscles). On the lateral aspect the main muscles located are evertors (peroneus longus and peroneus brevis). All of these muscles participate in the ankle and foot movements and are described below. The summary of major muscles acting on ankle and foot is given in **Table 11.3**.

Dorsiflexors: The movement of dorsiflexion is primarily produced by *tibialis anterior, extensor digitorum longus* and the *peroneus tertius* muscles, assisted by extensor hallucis longus muscle.

Tibialis Anterior

It is a slender muscle situated on the anterolateral aspect of the tibia **(Fig. 11.14)**. This muscle takes *origin* from the upper two third of the lateral outer surface of the tibia and is *inserted* over the medial cuneiform and the base of Ist metatarsal bone. This muscle is responsible for the movement of dorsiflexion at the ankle, and inversion at the foot.

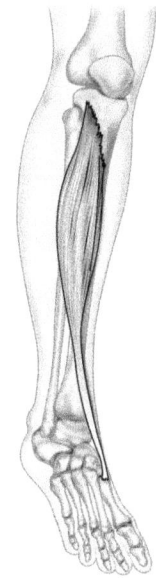

Fig. 11.14: Tibialis anterior

This muscle holds the medial part of the foot up, and its paralysis leads to the foot-drop deformity which makes the walking difficult. The muscle does not work as the invertor when the foot is plantarflexed.

Extensor Digitorum Longus

This muscle is also situated on the upper part of the leg just lateral to the tibialis anterior **(Fig. 11.15)**. The muscle takes *origin* from the lateral condyle of the tibia and upper three fourth of anterior surface of fibula. In front of the ankle joint its tendon divides into four tendons, one for each toe to get its *insertion* over the dorsal surface of the middle and distal phalanges. *Action:* The muscle is a prime mover for dorsiflexion at the ankle, and for extension of all four toes, and also assists the eversion of the foot.

Peroneus Tertius

This little muscle **(Fig. 11.16)** lies lateral to the extensor digitorum longus and is sometimes described as its fifth tendon. The muscle takes *origin* from the anterior surface of lower third of fibula, and *inserts* over the dorsal surface of the base of fifth metatarsal. It is a prime mover for dorsiflexion at the ankle, and also acts as an evertor at the foot.

Extensor Hallucis Longus

This smaller muscle takes *origin* from the anterior surface of the middle half of the fibula and *inserts* over the dorsal surface of the base of the distal phalanx of the great toe **(Fig. 11.17)**. It is primarily an extensor of the great toe, but it does assist in dorsiflexion at the ankle.

Plantarflexors: Prime movers for the plantarflexion are the gastrocnemius and soleus muscles, further assisted by tibialis posterior, peroneus longus, peroneus brevis, plantaris, and long flexors of the toes and hallux.

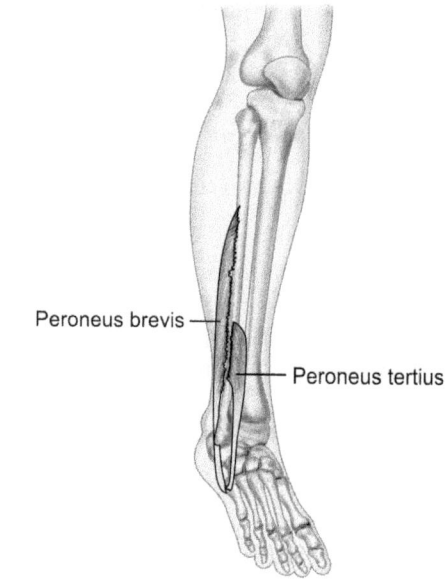

Fig. 11.16: Peroneus brevis and tertius muscles

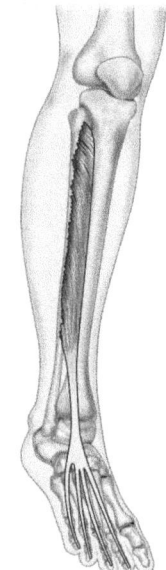

Fig. 11.15: Extensor digitorum longus

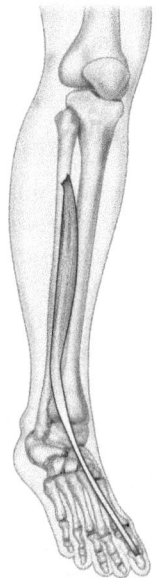

Fig. 11.17: Extensor hallucis longus

Gastrocnemius

This large muscle situated over the posterior aspect of the leg gives it a rounded contour and is also known as the *calf muscle* (**Fig. 11.18**). This muscle takes *origin* by two tendons from the posterior aspect of each femoral condyle, and *inserts* over the posterior surface of the calcaneous by means of tendon of Achilles.

This is a powerful plantarflexor of the ankle and also assists in knee flexion. When the knee is flexed beyond 90°, this muscle is ineffective to produce plantarflexion at the ankle. (Two heads of this muscle along with soleus is also sometimes called as *triceps surae*). The muscle is used in running, jumping, hopping and skipping activities.

Soleus

This muscle lying beneath gastrocnemius takes *origin* from the posterior surface of upper part of tibia and fibula, and *inserts* through the tendon of Achilles into the calcaneous. It is a prime mover in plantarflexion at the ankle. When the knee is straight, plantarflexion is produced by both gastrocnemius and soleus muscles, but when the knee is flexed, the plantarflexion is produced by soleus alone.

Plantaris

The muscle takes *origin* from the linea aspera and oblique popliteal ligament of the knee joint. The muscle gets *inserted* over the posterior part of calcaneous. The muscle is a very weak assistant for the knee flexion, and plantarflexion at ankle.

Both soleus and plantaris muscles are shown in **Figure 11.19**.

Invertors: *Tibialis anterior* and *tibialis posterior* muscles are the prime movers for inversion, assisted by flexor hallucis longus, flexor digitorum longus, and extensor hallucis longus muscles. Tibialis anterior is already discussed as a dorsiflexor, however it may be recalled that it is an invertor of the foot in dorsiflexion position, whereas the tibialis posterior works better as an invertor in the plantarflexion.

Tibialis Posterior

This is the deepest muscle on the posterior aspect of leg (**Fig. 11.20**) which takes *origin* from the posterior surface of upper half of interosseous membrane and adjacent surfaces of tibia and fibula. The muscle gets *insertion* over the inner surface of the navicular, three cuneiforms, and the bases of 2nd, 3rd and 4th metatarsal bones.

This muscle is a prime mover for inversion at the foot, and assists in plantarflexion at the ankle. This is a very strong invertor in the plantarflexed position. One of its important function is to maintain the longitudinal arch of the foot. The muscle must be strengthened in sportsmen by the resistance exercises, and is the most useful in hockey, ice-skating, skiing, etc. The weakness of this muscle may indirectly affect the knee and back causing pain there.

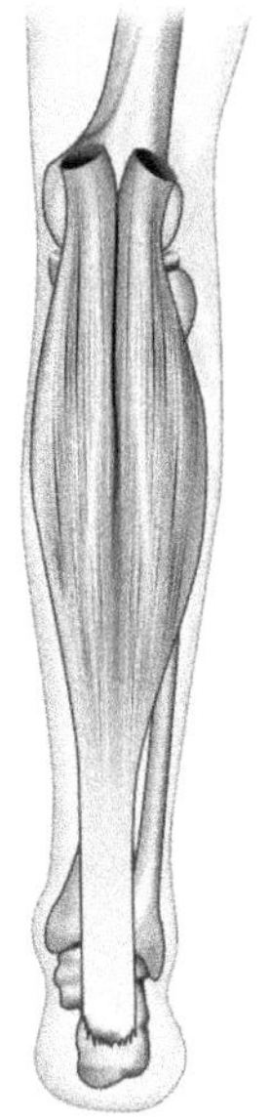

Fig. 11.18: Gastrocnemius muscle (both heads)

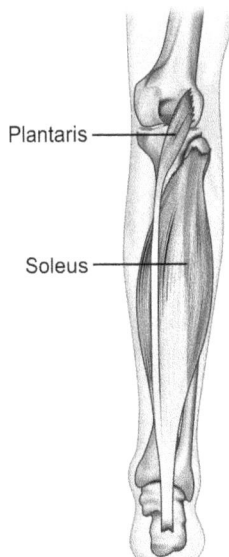

Fig. 11.19: Soleus and plantaris muscles

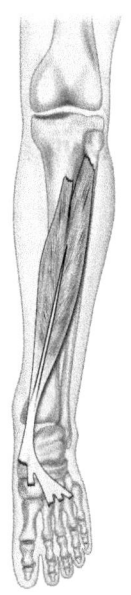

Fig. 11.20: Tibialis posterior muscle

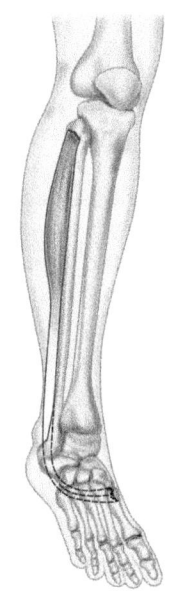

Fig. 11.21: Peroneus longus

Evertors: *Peroneus longus, peroneus brevis* and *peroneus tertius* are the prime movers for the eversion, and the extensor digitorum longus is an assistant mover. Peroneus tertius and extensor digitorum longus are the evertors in the dorsiflexed position, whereas the peroneus longus and peroneus brevis are the evertors better in the plantarflexed position. Peroneus tertius and extensor digitorum longus muscles have already been described above as dorsiflexors.

Peroneus Longus

This is a superficial muscle situated along the fibula on the lateral side of the leg **(Fig. 11.21)**. This muscle has got exceptionally greater power than to its size. The muscle takes *origin* from the upper two-third of the lateral aspect of the fibula. Its tendon passes behind the lateral malleolus and proceeds forwards and downwards. It passes behind the tuberosity of the fifth metatarsal, turns under the foot, then through the peroneal groove of cuboid slants forward on the plantar aspect of the foot to take its *insertion* over the base of 1st metatarsal and 1st cuneiform. *Action:* The muscle is a prime mover is eversion, and an assistant in the plantar flexion. This muscle also helps in the preservation of the transverse or metatarsal arch of the foot, and often used in running, jumping, hopping and skipping activities.

Peroneus Brevis

This small associate muscle of peroneus longus lies under the peroneus longus muscle on the lateral aspect of the lower half of the leg. It takes *origin* from the lateral surface of lower two-third of fibula, and gets *inserted* over the tuberosity at the proximal end of the V metatarsal bone. *Action:* The muscle is a prime mover for eversion movement and an assistant in plantarflexion. This muscle helps to maintain the outer longitudinal arch of the foot. Both peroneus brevis and peroneus tertius muscles are shown in **Figure 11.16**.

Flexors and extensors of the toes mainly act at the metatarsophalangeal (MP) and interphalangeal (IP) joints. The main muscles producing these movements are briefly described hereunder.

Flexor Digitorum Longus

This muscle is situated on the medial aspect of tibia **(Fig. 11.22)** and takes its *origin* from the posterior surface of lower two-third of the tibia, and gets *inserted* over the plantar surface of base of distal phalanx of each of 4 lateral toes. *Action:* This muscle is a prime mover for flexion of the toes, which also assists the plantarflexion and inversion movements.

Inversion movement can be assisted by this muscle only in the non-weight bearing position. This is also

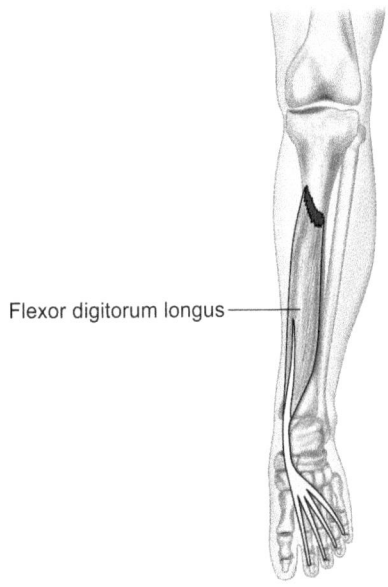

Fig. 11.22: Flexor digitorum longus

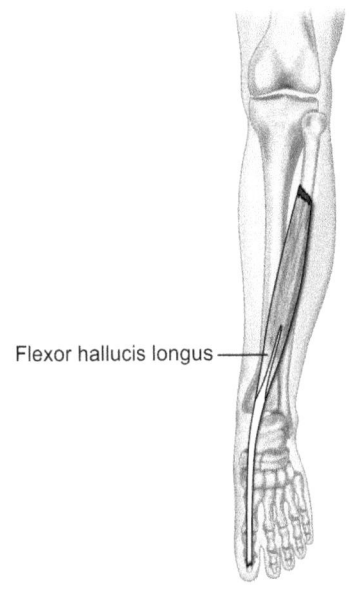

Fig. 11.23: Flexor hallucis longus

an important muscle in the maintenance of the medial longitudinal arch of the foot. When this muscle contracts it helps to lift the arch. Strengthening of this muscle is also very important in sports. Running in sand is a good exercise for developing this muscle.

Flexor Hallucis Longus

This muscle is situated on the lateral aspect of the fibula **(Fig. 11.23)**. The muscle takes *origin* from the lower two-third of posterior surface of the fibula, passes around the medial malleolus, runs forward along the medial side of the plantar aspect of the foot to *insert* over the under surface of base of the distal phalanx of the great toe. *Action:* This muscle is a prime mover for the flexion of great toe, and also assists the plantarflexion and inversion of the foot. This is a very essential muscle to maintain the equilibrium while standing and walking on the toes.

Extensor digitorum longus (EDL) and *extensor hallucis longus* (EHL) muscles have already been described above as the dorsiflexors of the ankle. The extensor digitorum longus is primarily an extensor muscle of all 4 lateral toes, but it does assist the dorsiflexion of the ankle, and eversion at the foot. Similarly, the extensor hallucis longus is a prime mover for the extension of the hallux or big toe, but does assist the dorsiflexion, and inversion at the foot.

Intrinsic Muscles of the Foot

There are about 11 intrinsic muscles in the foot. These small muscles originate on the tarsal bones and are located entirely within the foot. These muscles act as a functional unit and probably have a stabilizing effect on the foot during propulsion. The details of the following main intrinsic muscles may be found in a standard text book on kinesiology:
- Extensor digitorum brevis
- Flexor digitorum brevis
- Quadratus plantae
- Lumbricales
- Abductor hallucis
- Flexor hallucis brevis
- Adductor hallucis
- Abductor digiti minimi
- Dorsal interossei
- Plantar interossei.

Table 11.3: Summary of major muscles acting on ankle and foot

Muscle	Location	Origin	Insertion	Main action	Supporting action
Tibialis anterior	Anterior fibula	Upper lateral tibia	Medial cuneiform and 1st metatarsal	• Ankle dorsiflexion • Foot inversion	
Tibialis posterior	Posterior leg (deep to soleus)	Posterior surface of upper half of interosseous membrane, adjacent tibia and fibula	Navicular, and all three cuneiforms, and 2nd to 5th metatarsals (bases)	Foot inversion	Ankle plantarflexion
Gastrocnemius	Calf of leg (superficial)	*Medial head*: medial condyle of femur (posteriorly) *Lateral head*: lateral condyle of femur (posteriorly)	Posterior calcaneous	Ankle plantarflexion	Knee flexion
Soleus	Calf of leg (deep to gastrocnemius)	Posterior tibia and fibula	Posterior calcaneous	Ankle plantarflexion	
Peroneus longus	Lateral leg	Upper two-third of lateral aspect of fibula	Plantar surface of ist cuneiform and 1st metatarsal	Foot eversion	Ankle plantarflexion
Peroneus brevis	Lateral leg	Lateral distal aspect of fibula	Base of fifth metatarsal	Foot eversion	Ankle plantarflexion
Peroneus tertius	Anteior fibula (lower end)	Distal medial fibula	Base of fifth metatarsal	• Ankle dorsiflexion • Foot eversion	
Extensor hallucis longus	Anterior fibula	Anterior surface of fibula (middle)	Distal phalanx of big toe	Extension of big toe	Ankle dorsiflexion
Extensor digitorum longus	Anterolateral leg	Lateral condyle of tibia, head of fibula	Distal phalanx of four lesser toes (on dorsal surface)	• Extension of all 4 lesser toes • Dorsiflexion of ankle	Foot eversion
Plantaris	Posterior leg (medial)	Posterior lateral epicondyle of femur	Posterior calcaneous	—	• Weak assistant in knee flexion • Ankle plantarflexion
Flexor hallucis longus	Posterior fibula	Posterior surface of fibula (middle two-third)	Distal phalanx of great toe (plantar surface)	Flexion of great toe	• Ankle plantarflexion • Foot inversion
Flexor digitorum longus	Posterior tibia	Posterior surface of tibia (lower two-third)	Distal phalanx of each of four lesser toes (plantar surface)	Flexion of 4 lesser toes	• Ankle plantarflexion • Foot inversion

12 CHAPTER

Spine

The vertebral or spinal column is the main support of the body. It is the spine which enables us stand erect and frees our both the hands for more efficient working. **The spinal column is an organ of stability, mobility, support and protection.** This extends from the lower part of the head to the tail of the back just above the anal opening. The spinal column is also called *the longitudinal bony axis of the trunk.*

The spinal column provides a pivot point for motion and support of the head at the cervical region. The weight of the head, shoulder girdle, upper extremities and trunk is transmitted through the spinal column. Because it encases the spinal cord, the spinal column is able to provide protection to the cord. This multi-jointed structure not only provides movement, but the arrangement of these segments also provides for effective shock absorption and transmission.

The skull which is a bony structure of the head is seated on the top of this long multi-jointed spinal column. The skull contains and protects the brain and the facial bones, and since the sensory organs for sight, hearing, taste and hearing are located within the cranium, it is important that the head be able to move freely. This occurs through movements at various levels of the cervical spine.

The spinal column consists of 33 vertebrae. There are 7 *cervical* (neck region), *12 thoracic* (chest region), *5 lumbar, 5 sacral,* and the *4 lower most vertebrae* (developed partially to form the coccyx). 5 sacral vertebrae are fused to form one sacrum, and similarly 4 lower ones are also fused to form one coccyx. The lateral view of the spine is shown in **Figure 12.1**.

The whole spinal column presents four anteroposterior curves, i.e. cervical, thoracic (dorsal), lumbar and sacrococcygeal while viewed from the side. The cervical and lumbar curves are convex anteriorly, whereas the thoracic

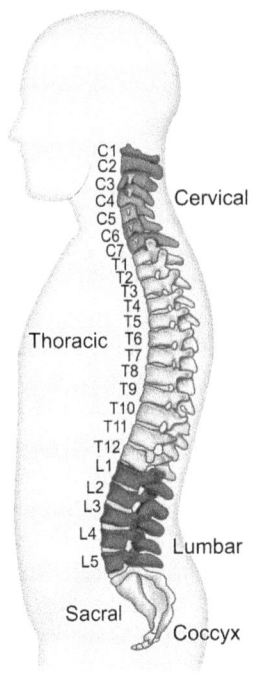

Fig. 12.1: Lateral view of the spine

and sacrococcygeal curves are convex posteriorly when viewed from the side. Out of these four curves, the thoracic and sacrococcygeal curves are called **primary curves** as they exist since birth. The cervical and lumbar curves are called **secondary curves** as they develop after the birth in the early childhood. The spinal column is flexible from cervical to lumbar region, and because the sacrum and coccyx are the rigid units, the flexible part of the spine rests over the rigid one. These curves provide the spinal column with much more strength and resilience, approximately 10 times more than if it were a straight structure.

The spinal column is about 71 cm. long in a man of average height. While descending from the cervical spine down, we see that size of the vertebral bodies become progressively more larger. This very well explains the weight bearing function of the spine. Each vertebra has to bear the whole weight of the body parts above it, and since the lower vertebrae have to bear comparatively much more weight than the upper ones, they are much larger in size to bear more weight. The vertebrae of the upper three regions, i.e. cervical, thoracic and lumbar are also called *movable* or *true vertebrae*, whereas those of sacral and coccygeal regions as false or fixed vertebrae.

Apart from 33 bones of spinal column, the *sternum* that is developed of 3 parts (or bones) and *12 pairs of ribs*, altogether form the bony structure and shape of the trunk.

Though almost all the vertebrae differ in size and shape, however their general structure is more or less the same and is briefly described below.

STRUCTURE

A verterbra consists of its *body* which is the largest and the most important portion because of its weight transmitting function. From the body, two short thick *pedicles* project backward, one on each side to join the *laminae* which unite posteriorly. These five parts, e.g. the body, two pedicles and two laminae enclose the vertebral foramen. A *spinous process* extends posteriorly from the junction of both the laminae. Two *transverse processes*, one on each side also extends laterally from the place of junction of laminae and pedicles. Each vertebra has 4 more articular processes, two to connect with the upper vertebra (i.e. *superior articular processes*), and two for connection with lower vertebra (i.e. *inferior articular processes*). The vertebral or spinal foramen, along with other vertebrae forms a *vertebral canal* for the lodgement of the spinal cord. Beneath each pedicle is an intervertebral notch which provides place for the nerves to come out of the spinal cord.

There are two types of articulations between the vertebrae:
1. Between the vertebral bodies, and
2. Between the articular processes of the adjacent vertebrae (or vertebral arches).

The joints between the vertebral bodies are slightly movable and are non-synovial, cartilaginous type. The vertebral bodies are united to the adjacent vertebrae by means of fibrocartilaginous intervertebral discs. Each disc consists of two parts, a centrally located gel like mass, *nucleus pulposus* surrounded by a strong fibrous layer, *annulus fibrosus*. There are *23 intervertebral discs* (IVD) that permit movement between the vertebral bodies due to the compression, and also act as shock absorber and a cushion between the vertebral bodies. The intervertebral discs make up approximately 25% of the total length of the spinal column.

The joints between the articular processes of the adjacent vertebrae are called *facet joints* which are freely movable synovial joints permitting gliding movements. On the posterior side of the vertebrae, facet joints are formed by the articulation between the superior articular processes of the vertebra below and with the inferior articular processes of the vertebra above, on each side. Since each vertebra has two superior, and two inferior articular processes, each vertebra is involved with formation of two facet joints. The lateral view of two adjacent vertebrae showing the intervertebral foramen and the facet joint is shown in **Figure 12.2.**

Ligaments

There are a number of ligaments that hold these vertebrae together. The *anterior longitudinal ligament* extends from skull to the sacrum on the anterior surface of vertebral bodies, and tends to prevent excessive hyperextension.

- *The posterior longitudinal ligament* runs along the vertebral bodies posteriorly inside the vertebral foramen, and prevents excessive flexion of the spine.
- The laminae of adjacent vertebrae are joined by the *ligamentum flava* which encloses the spinal canal.
- Adjacent spinous processes are connected by *interspinous ligament* which extends from the root of the process to its apex and meets the ligamentum flava in front, and supraspinal ligament behind.
- The spinous processes at the tips or apex are connected by the *supraspinal ligament* which extends from the 7th cervical vertebra to the sacrum. In the neck, this ligament continues upward up to the occiput and is known as *nuchal ligament* (or the ligament of neck) **(Fig. 12.3)**.

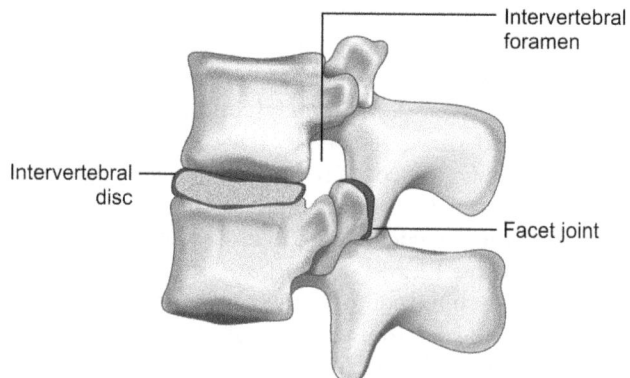

Fig. 12.2: Lateral view of two adjacent vertebrae showing the intervertebral foramen and the facet joint

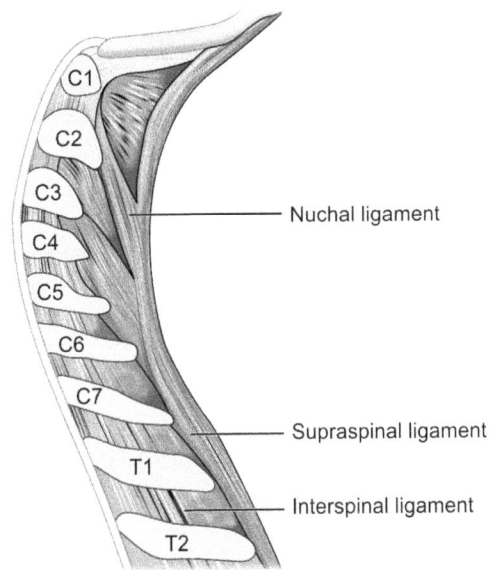

Fig. 12.3: Nuchal ligament, becomes supraspinal ligament in the cervical region

- The transverse processes are connected by the *intertransverse ligaments* (on each side) to the process of the adjacent vertebrae, in the thoracic and lumbar regions.

MOVEMENTS

Movements between any two adjacent vertebrae is very limited but the sum of these small movements results in the production of a considerable range of motion in the spine as a whole. The movements possible in a particular region of the spine depends upon, and varies according to the shape and thickness of intervertebral discs, direction of the articular surfaces, and shape of the spinous processes.

The summary of freedom of various spinal movements according to the region is given in **Table 12.1**.

The cervical spine allows movement and positioning of the head. The first two cervical vertebrae (i.e. the *atlas* and *axis*) are unique in their shape and allow extensive movements of the head on the neck. Movement of the head i.e. at the *atlanto-occipital joint*, and that of neck at the *atlanto-axial joint* are virtually inseparable. The atlanto-occipital joint is formed between the condyles of the occiput (head) and the first cervical vertebra (i.e. atlas) which lies between the head and neck, and permits the flexion-extension movements of the head as in nodding. This nodding movement is in sagittal plane around a frontal axis. The joint formed between the 1st and 2nd cervical vertebrae (the atlas and axis), is *atlanto-axial joint* which is peculiar as it permits the free rotation movement. Therefore, the rotation movement of the atlas on the axis largely attributes to the rotation movement of the head on the trunk.

In the cervical spine, in general, the cumulative movements of each of flexion, extension, lateral flexion and rotation are possible in free range, permitting range of motion of these movements, i.e. 45°, 45°, 45°, and 60° respectively. Various movements that occur at neck or cervical spine are shown in **Figures 12.4A to E.**

Generally speaking, the spine as a whole permits the following movements. The movements that occur at the thoracolumbar spine or trunk are shown in **Figures 12.5A to E.**

Flexion: This is forward bending of the trunk, in which the anterior surfaces tend to come together. This movement is free in cervical and lumbar regions, but greatly limited in the thoracic region due to the presence of rib cage. This movement involves a compression of anterior parts of intervertebral discs and gliding motion of the articular processes.

Extension: It is the return movement from the flexion. The backward movement of the spine farther from the anatomical position is referred to as hyperextension. It is also free in the cervical and lumbar regions (particularly lumbosacral junction). In the thoracic region, the hyperextension is limited due to the overlapping of the spinous processes.

Lateral Flexion: It is the sideward bending of the spine in the lateral plane on either side. In this movement the shoulder tries to approach the hip joint laterally. This is also free in the cervical and lumbar regions, and limited in thoracic region due to the ribs.

Rotation: This is a twisting movement of the spine around the long axis of the spine. In standing, it takes place in the horizontal plane about a vertical axis. Spinal rotation is named according to the direction the upper spine moves in relation to the lower part. For example, when the head is turned to right in relation to the fixed pelvis – it is the rotation movement to the right, or when the lower part,

Table 12.1: Summary of freedom of spinal movements according to the region

Region	Flexion	Extension	Lateral Flexion	Rotation
Cervical	Free	Free	Free	Free
Thoracic	Free	Free	Limited (due to rib attachments)	Free
Lumbar	Free	Free	Free	Limited (due to positioning of articular processes)

146 | SECTION II: Kinesiology of Body Regions

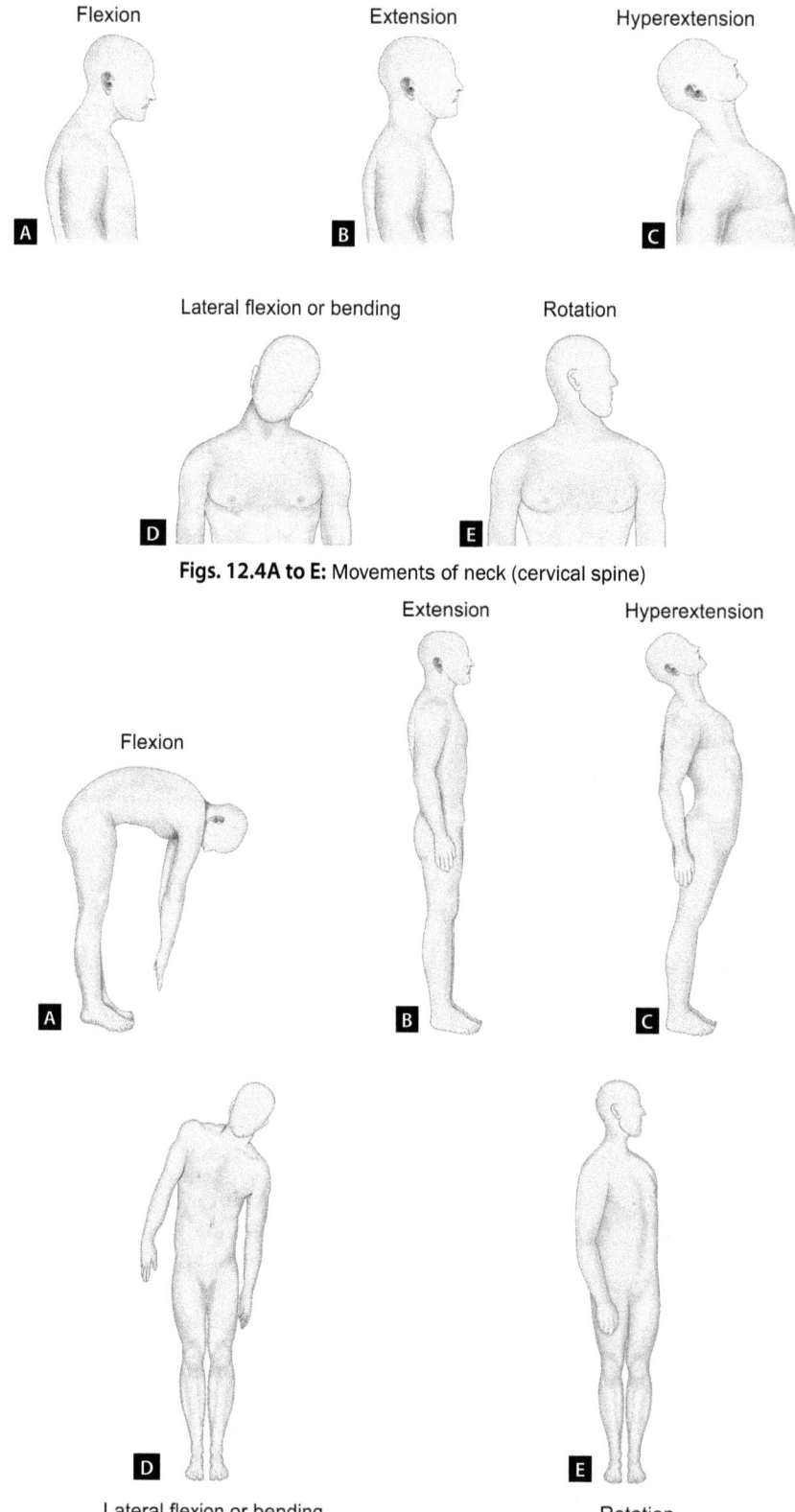

Figs. 12.4A to E: Movements of neck (cervical spine)

Figs. 12.5A to E: Movements of thoracolumbar spine (trunk)

i.e. the pelvis and legs move to the left in relation to the fixed head, then also it is termed as the rotation to the right. Because in this case also, the anatomic relationships are the same as in the former case. Rotation is also accompanied by a slight lateral flexion. This movement is most free in cervical and thoracic regions, and restricted in lumbar region due to the articular processes.

- Sometimes a movement of **circumduction** is also described in which the upper part of the trunk makes a circular movement over the lower one. The movement consists of flexion, lateral flexion and hyperextension. Rotation does not combine the circumduction.
- Generally speaking, most of the spinal movements occur freely in the cervical and lumbar regions, however in thoracic region, these movements are slight as compared to the cervical and lumbar regions.

The summary of muscular analysis of various spine movements is given in **Table 12.2**.

MUSCLES

The muscles producing movements in the head and trunk are many and are bilateral. They, at times can also contract independently. There are many small muscles also found in the spinal column region. Most of them start from one vertebra to the next one, and sometimes from the spinous process of one vertebra to the transverse process of another. Though these muscles are important in spinal functions, however for the knowledge of these about 140 small individual muscles, a physical education student or any other interested reader may refer to a standard textbook on anatomy or kinesiology. Therefore, here only a review of the major (larger) muscle groups is described for the general analysis of the movements. The summary of muscles of spinal column according to various aspects and regions is given in **Table 12.3** and the summary of major muscles acting on spine is given in **Table 12.4**.

Table 12.2: Summary of muscular analysis of spine movements

Movement	Prime movers	Assistant movers
Cervical spine (neck)		
Flexion	• Sternocleidomastoid • Prevertebral group of muscles (4)	Scaleni (three)
Extension	• Splenius capitis and cervicis • Erector spinae • Deep posterior spinal muscles	
Lateral flexion	• Sternocleidomastoid • Scaleni (3) • Splenius capitis, and cervicis	• Prevertebral group of muscles • Levator scapulae
Rotation to right	• Right sided splenius capitis and cervicis • Left sided sternocleidomastoid • Left sided semispinalis cervicis	• Prevertebral group of muscles • Erector spinae
Thoraco lumbar spine (trunk)		
Flexion	*Abdominal muscles* • Rectus abdominis • Internal oblique (both sided) • External oblique (both sided)	Psoas
Extension	• Erector spinae (sacrospinalis) • Semispinalis thoracic (transversospinalis)	Deep posterior spinal muscles
Lateral flexion	• Quadratus lumborum • Internal oblique (same sided) • External oblique (opposite sided)	
Rotation to right	• Right sided internal oblique • Right erector spinae • Left sided external oblique • Left sided semispinalis thoracic (transversospinalis)	Left sided deep posterior spinal muscles

Table 12.3: Summary of muscles of spinal column according to the aspect and region

Anterior aspect	Posterior aspect	Lateral aspect
Cervical region only • Prevertebral muscles (longus capitis and colli, rectus capitis anterior, and lateralis)	*Cervical region only* • Splenius capitis, and cervicis • Suboccipitals (rectus capitis posterior major and minor, obliques capitis superior and inferior)	*Cervical region* • 3 scalene muscles (scalenus anterior, posterior, and medius) (Fig. 12.8) • Sternocleidomastoid • Levator scapulae
Thoracic and lumbar regions • Abdominal muscles (rectus abdominis, internal oblique, external oblique, transversus abdominis)	*Cervical, thoracic and lumbar regions* • Erector spinae (iliocostalis, longissimus, and spinalis) • Deep posterior spinal muscles (multifidi, rotatores, interspinales, intertransversarii and levatores costarum) • Semispinalis thoracis, cervicis, and capitis	*Lumbar region* • Psoas major • Quadratus lumborum

We can classify these muscles into two headings:
1. Those acting to produce movements in the cervical spine (neck), and
2. Those acting and producing movements in the trunk or thoracic and lumbar spines together.

Cervical Spine (Neck)

Flexors: Flexion at the neck is produced by the *sternocleidomastoid* and the *prevertebral* group of muscles, assisted by 3 scaleni muscles.

Sternocleidomastoid

A prominent pair of these muscles is located at the front and sides of the neck (**Fig. 12.6**). This muscle is named from its attachment over the bones (*cleido refers to clavicle*). This muscle takes *origin* by two heads, one arises from anterior aspect of the sternum, and the other one from the medial third of the clavicle (anterior and superior surfaces). The muscle gets *insertion* by a strong tendon to the mastoid process of the temporal bone. When both sided muscles contract, it causes flexion at the head and neck. When the sternocleidomastoid of one side contracts, it causes lateral flexion of the same side with rotation to the opposite side. As such, *this muscle is a prime mover for flexion, lateral flexion, and rotation to opposite side* and (on the head & cervical spine). When this muscle is paralyzed or abnormally contracted, *wry neck* or *torticollis* results.

Prevertebral Muscles

The prevertebral group of muscles (**Fig. 12.7**) are the deep anterior muscles of the head and neck. The prevertebral group consists of *longus capitis, longus colli, rectus capitis anterior* and *lateralis muscles*. They produce flexion of head and neck when the muscles of both sides contract. One sided muscles give assistance to lateral flexion. They are also prime movers for rotation to the same side.

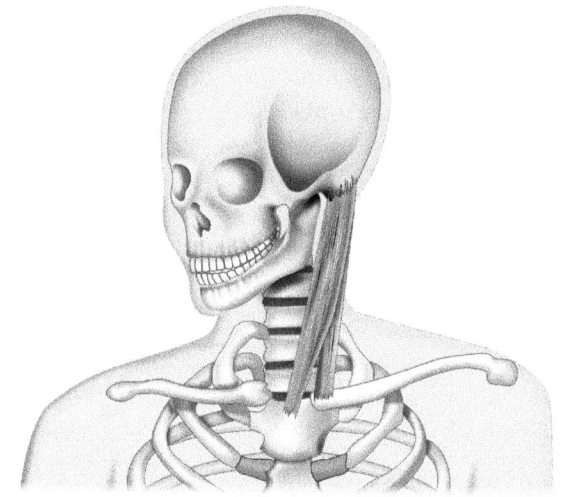

Fig. 12.6: Sternocleidomastoid muscle

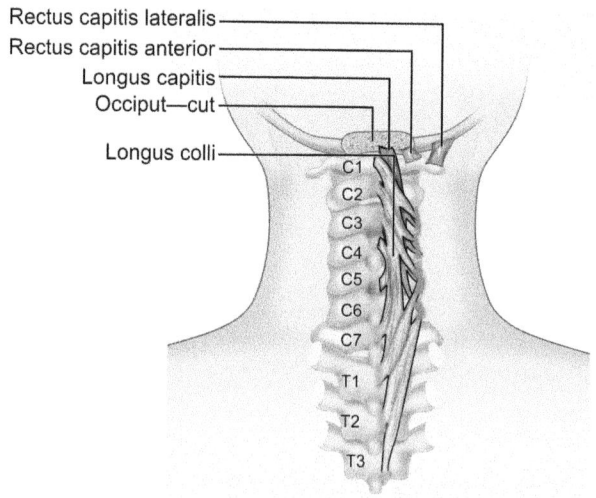

Fig. 12.7: Prevertebral muscles (anterior view)

Lateral Flexors: Lateral flexion of the cervical spine is produced primarily by the *sternocleidomastoid,3 scaleni, splenius capitis,* and *cervicis muscles.* The movement is assisted by the muscles of prevertebral group which are true flexors and extensors of the neck, and also by levator scapulae muscles.

Scaleni

These three muscles (or the parts of scalene muscle shown in **Figure 12.8**) scalenus anterior, medius and posterior are named from their relative positions. They take *origin* from the transverse processes of cervical spine and *insert* over the first 2 ribs. *These muscles are the prime movers for lateral flexion*, and when muscles of both sides contract they assist the flexion at the cervical spine.

The *splenius capitis* and *cervicis* muscles are described in the extensor group of muscles.

Extensors: Extension of the neck is produced by the *splenius capitis* and *splenius cervicis, erector spinae* group of muscles, and *deep posterior spinal muscles.*

Splenius Capitis and Cervicis

The muscle *splenius capitis takes origin* from the spinous processes of 6th cervical and upper 4 thoracic vertebrae and *inserts* at the occipital and temporal bones. *The splenius cervicis arises* from the spinous processes of 3rd to 6th thoracic vertebrae, and *inserts* at the transverse processes of upper 2-3 cervical vertebrae. When these two muscles of both sides work together, they are the prime movers for the extension and hyperextension movements. And when both of these muscles work on one side only, they are the prime movers for the lateral flexion and rotation of the head and neck to the same side. These are the important muscles which hold the head erect in standing and sitting postures. These muscles are shown in **Figure 12.9**.

Suboccipitals

This is a group of four short muscles, i.e. *obliquus capitis superior,* and *inferior*; and the *rectus capitis posterior major* and *minor*. These muscles (**Fig. 12.10**) are situated at the back of the lower skull (occipital bone), and upper two cervical vertebrae (atlas and axis). Acting together on both sides, this group of muscles extends and hyperextends the head. When one sided group of muscles acts alone, it flexes the head laterally or rotates it to the same side.

Rotators: Rotation to the right side is produced by right sided *splenius* muscles assisted by the prevertebral group and erector spinae group of muscles, whereas the sternocleidomastoid and semispinalis cervicis of the left side also produce the rotation to the right side.

Trunk

Thoracic and lumbar spines together comprise the trunk.

Flexors: Flexion is mainly produced by *abdominal muscles i.e. rectus abdominis, external oblique,* and *internal oblique,* assisted by psoas muscle.

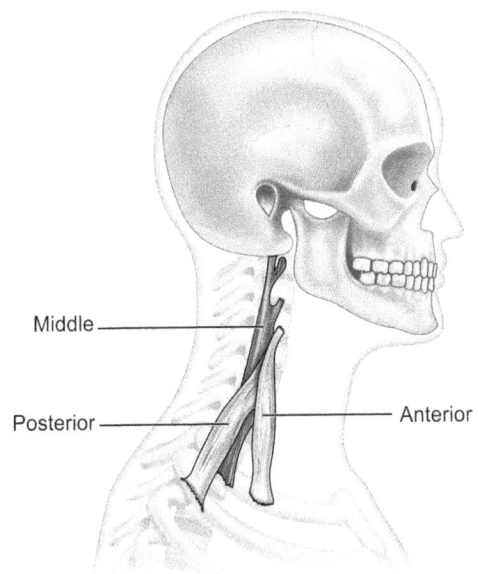

Fig. 12.8: Scalene muscles—three parts

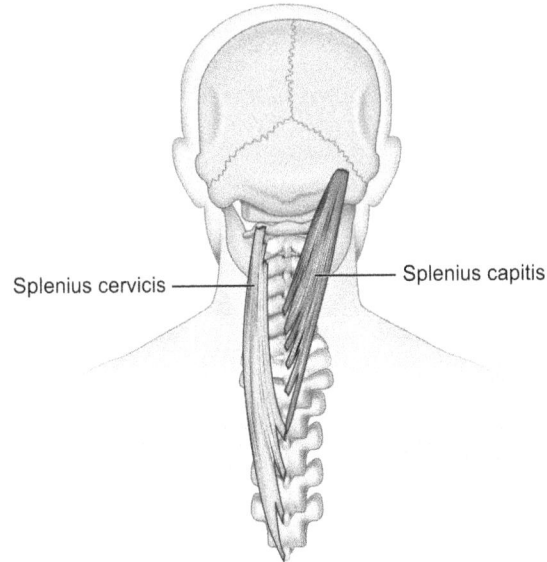

Fig. 12.9: Splenius capitis and cervicis muscles

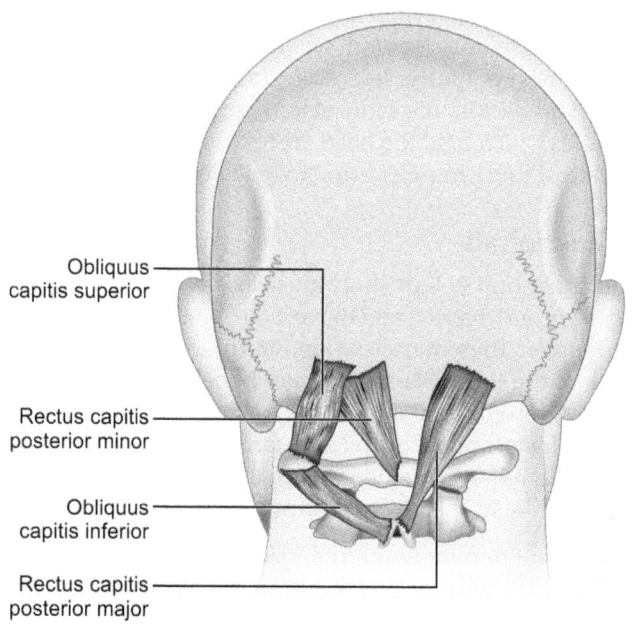

Fig. 12.10: Suboccipital muscles (posterior view)

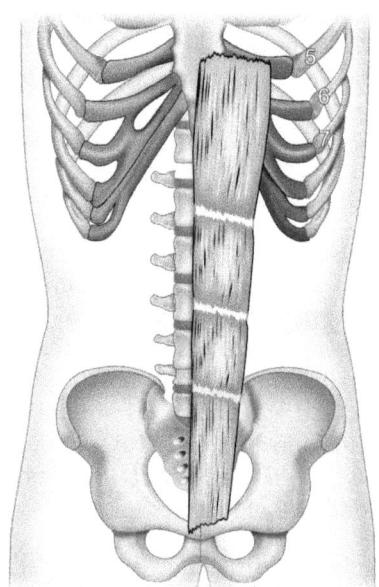

Fig. 12.11: Rectus abdominis muscle

Rectus Abdominis

It is a slender muscle which extends vertically in front of the abdominal wall (**Fig. 12.11**). This takes *origin* from the crest of the pubis and *inserts* over the cartilages of 5th, 6th and 7th ribs. Action: This muscle is a prime mover for the flexion of thoracic and lumbar spines when both sided muscles work together. Contraction of one sided muscle only assists lateral flexion to that side. The recent studies have shown that the upper rectus is more active in the flexion of the spine from the supine lying position, whereas the lower rectus is found to be more active in the bending movements of the hip and knee towards the face when the pelvic tilt is also decreased.

External Oblique

External oblique muscle covers the front and side of the abdomen and extends from the lower border of the lower eight ribs to the distal attachment at the anterior half of iliac crest and crest of pubis (**Fig. 12.12A**). Because the fibers run diagonally upward and outward from the lower part of the abdomen, *it is a prime mover for the flexion of the trunk when both the muscles contract*. One sided external oblique is a prime mover for the lateral flexion and rotation of the trunk to the opposite side.

Internal Oblique

This muscle lies beneath the external oblique and extends from the crest of ilium and inguinal ligament to the cartilages of 8th, 9th, and 10th ribs (**Fig. 12.12B**). *This muscle is a prime mover for flexion, lateral flexion and rotation to the same side.* When muscles of both sides work together it causes flexion at thoracolumbar spine.

Extensors: Extension of the trunk is mainly produced by *erector spinae group of muscles* and transversospinalis, assisted by deep posterior spinal muscles.

Erector Spinae

This extensive muscle of the back consists of three branches; *lateral*-the iliocostalis, *middle*-longissimus, and *medial*-spinalis (**Fig. 12.13**). The muscle takes *origin* from the thoracolumbar fascia, posterior portion of lumbar, thoracic and lower cervical vertebrae, and angles of ribs, and *inserts* at the angles of ribs, posterior portions of cervical and thoracic vertebrae, and to the mastoid process of temporal bone. *When these muscles of both sides contract, they cause extension and hyperextension of the head and entire spine.* However, one sided muscle produces lateral flexion and rotation of head and spine to the same side. Recent studies have shown that this muscle contributes very little in the maintenance of erect posture in standing unless one makes efforts to extend the thoracic spine more, or when some weight is carried in front, and in that case the muscle contracts statically. In forward flexion from standing position the muscle works eccentrically.

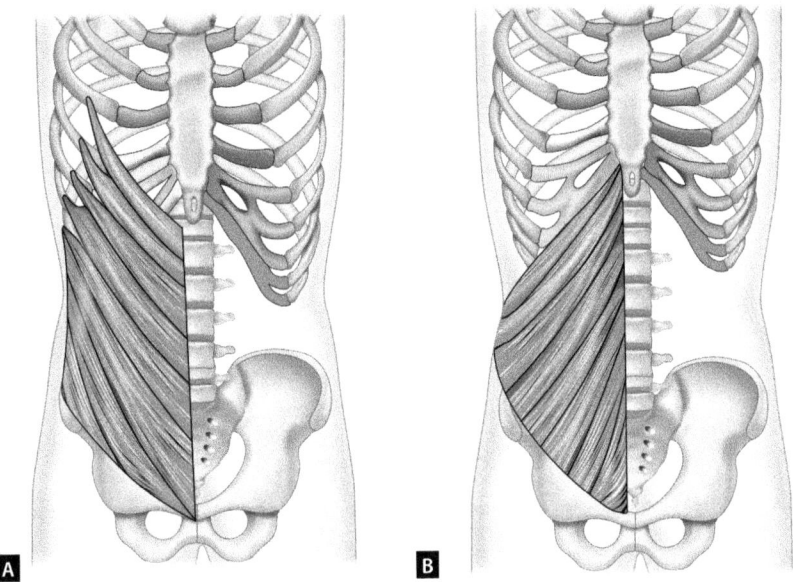

Figs. 12.12A and B: (A) External oblique; (B) Internal oblique

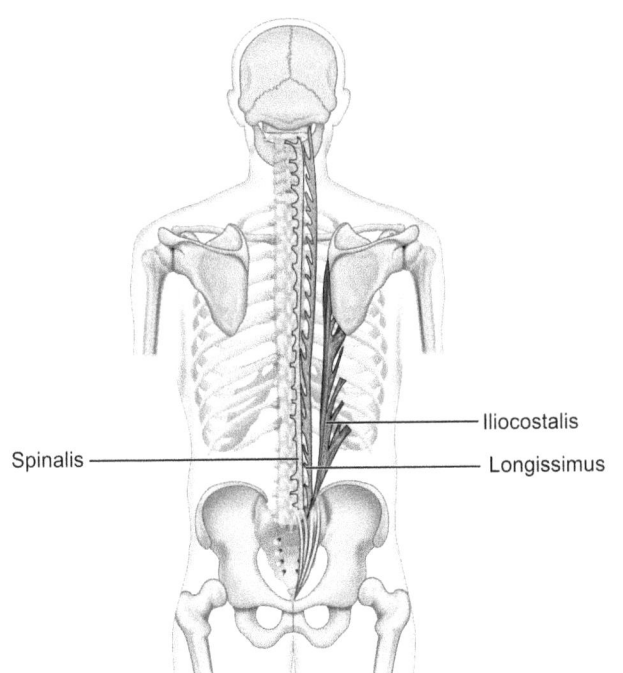

Fig. 12.13: Erector spinae muscles

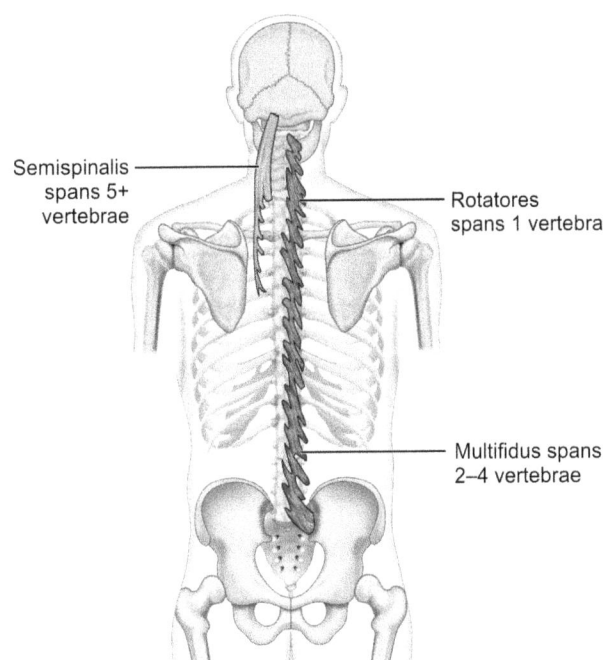

Fig. 12.14: Transversospinalis muscle group

Transversospinalis

This is actually a group of three muscles located to the deepest of the spine **(Fig. 12.14)**. These muscles have an oblique line of pull, essentially attaching from a transverse process to the spinous process of a vertebra above and therefore, are very effective in rotation of trunk (spine). One of the three muscles is *semispinalis* which extends from 6 to 10th thoracic vertebrae to two lower cervical and four upper thoracic vertebrae. The other muscle *multifidus* extends to two to four vertebrae, and the remainder muscle *rotatores*, is the shortest and deepest muscle of the group which extends only to one vertebra. *These muscles extend, and rotate the trunk (spine) to the opposite side.* The semispinalis is the most superficial muscle of this group,

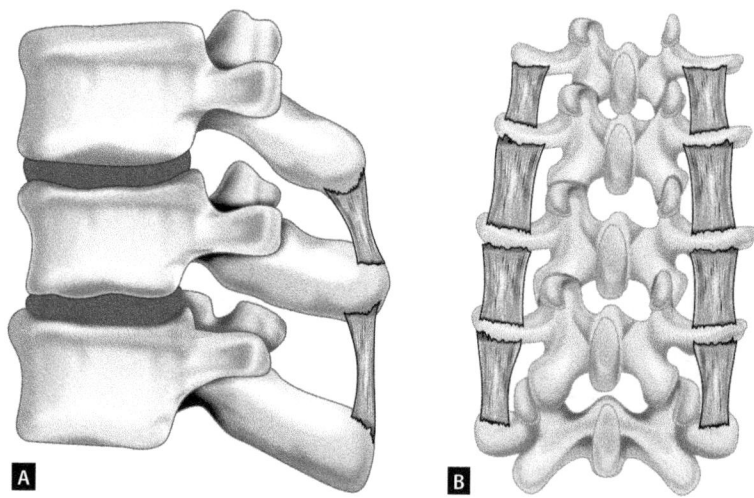

Figs. 12.15A and B: (A) Interspinales muscles (lateral view); (B) Intertransversarii muscles (posterior view)

the multifidus lying underneath it, and the rotatores being the deepest of these muscles.

Deep Posterior Muscles of the Spine

Other than transverso-spinalis group of deep muscles described above, there are many other deep muscles (i.e. *interspinales, intertransversarii* etc.) **(Figs. 12.15A and B)** over the back which extend from the posterior surface of sacrum and spinous processes of all the vertebrae to the spinous and transverse processes of slightly higher vertebrae. *Muscles of both sides cause extension and hyperextension of spine and, one sided muscles mainly assist rotation and lateral flexion to the opposite side.*

Side Flexors or Lateral Flexors: Side flexion or lateral flexion is produced mainly by the *quadratus lumborum*, and the *extensors* and *flexors* of the spine particularly, the external and internal obliques.

Quadratus Lumborum

It is a pure lateral flexor. It is a flat sheet of fibers lying on each side of the spinal column **(Fig. 12.16)**. This muscle takes *origin* from the posterior surface of iliac crest and transverse processes of lower 4 lumbar vertebrae, and *inserts* over the transverse processes of the upper 2 lumbar vertebrae and lower border of the last rib. *This muscle is a prime mover for lateral flexion to the same side,* and both sided muscles may help in extension of the lower back, and assist the stability of the lower spine and pelvis. The flexors

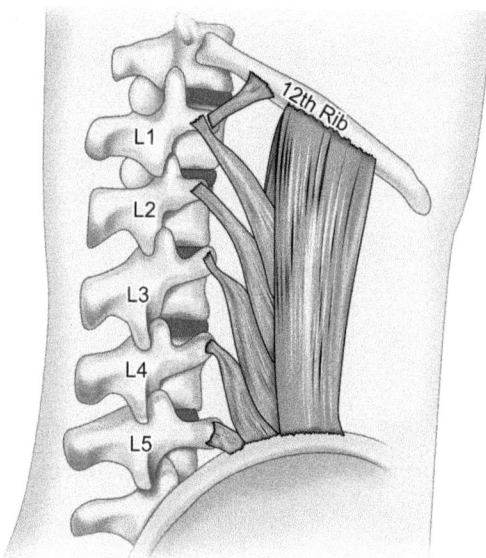

Fig. 12.16: Quadratus lumborum muscle (lateral view)

(external and internal obliques) and extensors of one side, while working together may produce lateral flexion to that side, as described earlier.

Rotators of the Spine: Prime movers for the rotation of the spine to the same side are the *internal oblique* and the *erector spinae* group of muscles, and the *external oblique* and *semispinalis thoracic* muscles are the prime movers for the rotation to the opposite side. Some of the deep posterior spinal muscles of the opposite side also help in the rotation movement.

Table 12.4: Summary of major muscles acting on spine

Muscle	Location	Origin	Insertion	Main action	Supporting action
Sternocleido-mastoid	Lateral neck	Anterior surface of sternum and clavicle	Mastoid process of temporal bone	• Both sided muscles: Flexion of head and neck • Lateral flexion of neck to the same side • Rotation of neck to the opposite side	
Prevertebrals	Anterior neck (deep)	*Longus capitis:* Transverse processes of C_3–C_6	Occipital bone	Both sided muscles: Flexion of head	
		Longus colli: Bodies and transverse processes of C_3–T_2	Transverse processes and bodies of C_1–C_6	Both sided muscles: Flexion of neck	
		Rectus capitis anterior: atlas	Occipital bone	Both sided muscles: Flexion of head	
		Rectus capitis lateralis: Transverse process of Atlas	Occipital bone	Lateral flexion of neck to the same side	
Scaleni (anterior, medius and posterior)	Lateral neck (deep to sternocleido-mastoid)	Transverse processes of the cervical vertebrae	First and second ribs	• Neck flexion • Lateral flexion of the neck to the same side	
Splenius capitis	Posterior neck	Spinous processes of 6th cervical and upper 4 thoracic vertebrae	Lateral occipital and temporal bones	• Both sided muscles: Extension of head and neck • One sided muscle: Lateral flexion and rotation of head to the same side	
Splenius cervicis	Posterior neck	Spinous processes of 3rd to 6th thoracic vertebrae	Transverse processes of upper three cervical vertebrae	• Both sided muscles: Extension of neck • One sided muscle: lateral flexion and rotation of neck to the same side	
Rectus Abdominis	Anterior mid line of abdomen (superficial)	Front of the pubis	Cartilages of 5th, 6th and 7th ribs	Both sided muscles: Flexion of trunk (thoracic and lumbar spine)	One sided muscle: Lateral flexion of trunk to the same side
Internal oblique	Anterior and lateral abdominal (deep to external oblique)	Iliac crest, Inguinal ligament	Cartilages of 8th, 9th and 10th ribs	• Both sided muscles: Flexion of trunk • One sided muscle: Lateral flexion and rotation of the trunk to the same side	
External oblique	Anterior abdominal (superficial)	Lower eight ribs laterally	Iliac crest and pubis	• Both sided muscles: Flexion of trunk • One sided muscle: Lateral flexion and rotation of the trunk to the opposite side	
Erector spinae Three parts: 1. *Iliocostalis* 2. *Longissimus* 3. *Spinalis*	Length of back (sacrum to skull)	Posterior portion of lumbar, thoracic, and lower cervical vertebrae, angles of ribs	Posterior portion of cervical and thoracic vertebrae, mastoid process of temporal bone	• Both sided muscles: Extension of trunk • One sided muscle: Lateral flexion of trunk to the same side	
Transversospinalis (Group of three muscles) 1. *Semispinalis* 2. *Multifidus* 3. *Rotatores*	Posterior spine	Transverse processes	Spinous processes of vertebra above	• Both sided muscles: Extension of trunk • One sided muscle: Rotation of trunk to opposite side	
Quadratus lumborum	Between iliac crest and lower rib (laterally)	Iliac crest	12th rib, transverse processes of all five lumbar vertebrae	Lateral bending of trunk	

SECTION III
Motor Skills: Principles and Applications

- Posture
- Locomotion
- Application of Kinesiology to Selected Daily Life Activities and Sports Skills
- Prevention of Sports Injuries: The Mechanical and Kinesiological Viewpoint

13
CHAPTER

Posture

INTRODUCTION

Posture means body position. By posture we mean the relative alignment of various body parts such as head, neck, shoulders, trunk, pelvis and limbs, with one another and also with respect to the surroundings. In general, posture is the position of body parts in relation to each other at any given time. The term *posture* is probably derived from the Latin word *ponere* which means 'to put or place'. The way our body is held when we sit, walk, lie, play or the way we carry things on arms, on back, hang on the shoulders, or on head or perform an activity constitutes our posture.

Posture and movements are intimately associated. Every movement of the human body begins in a posture and ends in a posture. Posture expresses physical as well as mental status of an individual. Posture is also an index of one's personality. It is very rightly said that *'who stands erect with a well poised, controlled and therefore a graceful body will feel that he is a master of himself and leader of men.'*

The development of good posture in children and adolescents is one of the aims of physical education teachers. Good posture means more than merely the ability to assume an erect standing position when it is desired. It is the ability to handle the body easily, gracefully, and efficiently under all circumstances.

Definitions

- **Human posture refers to the carriage of the body as a whole, the attitude of the body or the position of the limbs (the arms and legs).**
- **Posture is a general term that is defined as a position or attitude of the body, the relative arrangement of body parts for a specific activity or a characteristic manner of bearing one's body.**

From the viewpoint of kinesiologically oriented physical educator,

Posture can be defined as a gauge or an instrument of mechanical efficiency, kinesthetic sense, muscle-balance, and neuromuscular coordination.

Simply speaking, posture is a position or attitude, assumed by the body by means of co-ordinated action of many muscles which are working to maintain stability.

Evolution of Erect Posture

When we talk of posture, it generally refers to the erect posture of man which is a unique development of evolution from ape to man. This erect or upright standing posture represents an interesting stage in the history of human structure.

The erect posture of a human-being distinguishes him from all other animals. It is postulated by the anthropologists that erect or upright posture is the product of perhaps more than 35 crore years of evolution. In the evolutionary process, the paired fins of certain tetrapods developed into limbs and provided for locomotion. In the course of time, possibly some 15 crore years ago, the first mammals came into existence. By 7 crore years ago, quadruped primates about the size of rats were in existence. Over the millennia, certain changes in body form adapted them to assume a vertical position. Gradually some 3 crore years ago, the forests began to recede forcing man's predecessors to become ground dwellers. Among the higher primates, bipedalism appears to a rise whenever it improves the organism's chances of survival. Over the process legs lengthened and straightened. The foot lost most of its grasping abilities and became specialized for bipedalism. As such, the erect posture of man was perhaps evolved to suit the way of life and the survival of man in this universe.

The evolution of erect posture has given man a tremendous advantage since his upper extremities are freed from the burden of supporting the body, which later evolved into the instruments of very fine and delicate movements and work. Manipulative hand is a very important and distinct feature of the human beings, and most of the present day progress and development made in the areas of science, technology, medicine, engineering, art etc, can be attributed to the manipulative hand, apart from a large brain possessed by the man. The erect posture has also resulted into the raising of the head, and the line of vision higher which may be used for better advantage. A graceful appearance and different type of gait with greater speed are other advantages of the erect posture.

However, the evolution of the erect posture has resulted into certain disadvantages also because the upper limbs, instead of supporting the upper part of body have now to be supported by the upper part of the trunk. The raising of body's center of gravity over a decreased base of support, change in the alignment of the spine, and its relationship to pelvis are some of the issues that present a continuous threat to the maintenance of balance and stability in the humans.

During the process of evolution, certain structural changes and modifications were necessary for effectively bearing the stresses of body weight. The lower extremities have greatly been modified, as earlier all the body weight which was supported by the four limbs is now to be borne by the two hind limbs alone. However, the pelvis by which the lower extremities are attached to the spinal column has remained essentially that of a quadruped. Likewise, the spine has also not adapted fully to the demands of the upright posture. Though the development of various curves in the spine helps in balancing and maintenance of an upright posture, however the human spine remains mechanically weak and can become deformed under pressure.

Effects of Erect Posture

In the evolutionary process, from a quadruped to an erect bipedal posture, certain changes have resulted:

- *Muscular Development:* Because of greater body weight being borne by the lower limbs, they have now become increased in size and strength of extensor muscles. The trunk flexors or the abdominal muscles relieved of much of the strain that they were subjected to in quadruped position, have now a tendency for the weakness, and sometimes even allows the sagging of the abdominal organs.
- *Co-ordination:* Due to the increased difficulty in balance and poise in the erect posture, a corresponding development of certain nervous reflexes occurs to maintain exact balance under all conditions. The upper limbs, relieved from the burden of support and locomotion, are now employed in skilled occupations under the guidance of the eyes.
- *Breathing:* In quadrupedal position, the ribs hang down below the spinal column and swing back and forth like a pendulum in breathing, requiring very little muscle work. However, in the erect position, the entire weight of the chest wall must be lifted with each inspiration and held up to proper level continuously. The pull of gravity is also great enough on the chest, neck and spine that the ribs usually sink gradually as age increases, along with the sinking of the internal organs as well.
- *Circulation:* In the horizontal position, the blood returns easily and evenly from the anterior and posterior portions of the body. However, the erect posture has created a column of blood that greatly increases the problem of blood return to the heart. Therefore, the circulatory mechanisms are subjected to postural stress.
- *Position of Internal Organs:* In the horizontal position, the abdominal organs are suspended from the spine and are supported from below by the contracting abdominal muscles. However, with the erect posture these organs sag downward into the pelvic region creating pressure and congestion. This has therefore altered the weight centers of various segments of the trunk, and poses a continuous threat to maintenance of balance.

TYPES OF POSTURE

Postures can be *inactive* or *active*. Inactive postures are adopted for resting, sleeping or training general relaxation. In inactive postures, the essential muscular activity required to maintain life is reduced to a minimum, whereas active postures require an integrated action of many muscles to maintain these.

Active postures may be either *static* or *dynamic*. In a static posture the whole body is held practically in a static or stationary position such as standing, sitting or lying, without any appreciable motion of any body part, whereas in a dynamic posture the whole body or at least some body parts are moving from one position to another, as in walking the arms and legs are moving as against the head and back which are relatively in static position. Likewise, when a person writes, the hands and eyes are moving but the back and legs are almost held stationary. As such, in these mobile or dynamic states, the posture of the static body parts is a comparative state over the moving body parts.

A static posture is maintained by the interaction of group of muscles which work more or less statically to stabilize the joints, and in opposition to gravity or other forces, whereas a dynamic posture is required to form an efficient basis for movement, and the pattern of posture is constantly modified and adjusted to meet the changing circumstances which arise as a result of movement.

STATIC VERSUS DYNAMIC POSTURE

Posture is a dynamic, not a static concept. The body is seldom stationary for more than few moments, even in its apparently still positions. It is found out that even the static upright posture also generally involves of about 4 centimetres of swaying of the body in the anterior posterior direction. The body is often engaged in the movements which are greatly varying in the extent and direction.

The human body assumes many postures and hardly holds any of them for an appreciable length of time. Although characteristic postural patterns become apparent as we observe an individual over an extended period. However, it is very difficult to accurately measure or record such patterns in the absence of comprehensive series of motion pictures of an individual's varied stance and movement patterns. This is one reason because of which most posture research perhaps has been related to the active static standing position. Although people seldom standstill, however, *static standing posture is used as the reference for posture evaluation*. It is the static posture that can reveal abnormalities in relative balance and alignment of body parts that affect structure and function.

From the point of view of physical education, dynamic postures should be of greater concern than the static postures of the individuals. However, it is a convenient custom to accept the standing posture as the individual's basic posture from which all other postures are derived. The static standing posture serves as a reflection of individual's characteristic postural patterns. However, it should be kept in the mind that the importance of static standing posture lies only when it represents an individual's habitual postural pattern held for long periods of time and the understandings derived from the knowledge of the static postures should not be always applied to all the dynamic situations of the body.

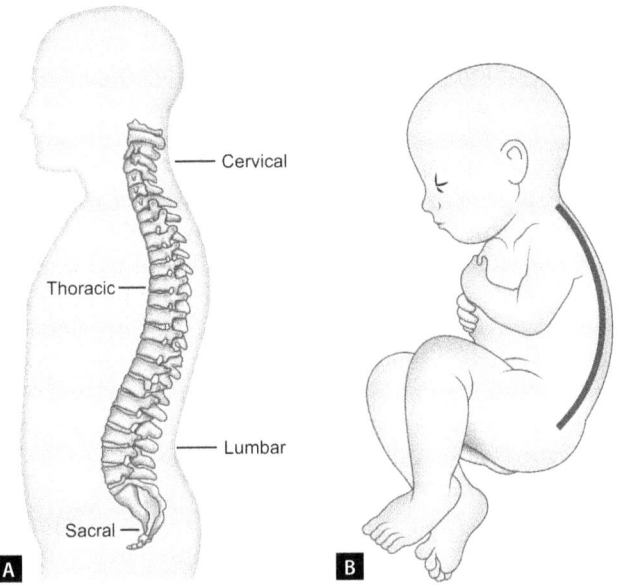

Figs. 13.1A and B: (A) Four major curves of vertebral column; (B) A single primary curve (flexion) of a new born

Vertebral Alignment/Postural Curves

The vertebral column can be compared to the column of blocks. The vertebral column is not completely straight (as generally appears when viewed from the front or back of the body), rather it has a series of counterbalancing curves in anterior-posterior direction. In a normal human adult, the spine adopts four curves (**Fig. 13.1A**). The flexion curves in the thoracic and sacral regions are convex posteriorly, and the hyperextension curves present in the cervical and lumbar regions are convex anteriorly.

As shown in **Figure 13.1B**, at birth the entire vertebral column presents a convex curve posteriorly (flexion). Since the thoracic and sacral flexion curves are present at the time of birth, they are considered as **primary curves**. The cervical and lumbar curves are developed after birth as the child grows, and hence these are referred to as the **secondary curves**. These secondary curves in cervical and lumbar regions counterbalance the already existing primary curves in thoracic and sacral regions. These curves, which must be maintained during rest and activity, act as shock absorbers and reduce the amount of injury (than it the spine were straight). Mechanically, these curves enable the three masses of weight in the body (i.e. the head, thorax, and the pelvis) kept balanced one above the other. When one or more of these vertebral curves either increases or decreases significantly from what is considered good posture, poor posture results. For example, a kyphosis or lordosis represents an increased thoracic or lumbar curve respectively, whereas a flatback shows a decreased thoracic curve. No lateral curves should exist in a normal spine. Any lateral curvature of the spine is a pathological condition called *scoliosis*.

The position of pelvis is very important in maintaining the curves of the spinal column, especially the curve of the lumbar region. Since the pelvis functions as a base for the entire trunk/spine, it should be held in a neutral position. The neutral position of the pelvis is defined as when the anterior superior iliac spine (ASIS) and posterior superior iliac spine (PSIS) are level with each other in a transverse plane; and when the anterior superior iliac spine is in the same vertical plane as the pubic symphysis. When the pelvis is in a neutral position, the lumbar curve has the desired amount of curvature. Any tilt or inclination of the pelvis beyond the normal will cause imbalance of the whole structure above it. When the pelvis is tilted anteriorly, there is an increased amount of lumbar curvature, or lordosis; and when the pelvis is tilted posteriorly, there is a decreased amount of lumbar curve, or flatback.

Importance of a Strong Abdominal Wall

The importance of a strong abdominal wall is often ignored. Thought the main functions of the abdominal muscles are to hold the contents of the abdominal region (viscera) in place and assist in reflex actions like coughing, sneezing and defecating, however the lower abdominal muscles do have some role in reducing the tilt of the pelvis, which is primarily the action of gluteal muscles. The abdominal muscles, though reflexly take part in trunk movements, however, they are not the main muscles to perform the bendings and twistings of the trunk.

There may be sagging of the abdominal wall when the normal thoracic curve is increased or when the lumbar curve is flattened markedly. However, when both of these curves are increased moderately, then the abdominal wall does not sag much, so long as the body parts are in balance, as the abdominal wall is fairly retracted in this position.

Although the health of many people with sagging abdominal wall and dropping of its internal organs may not be affected, there is no advantage in the low position of abdominal organs. However, when the external skeletal muscles lose their tonus, their associated smooth muscles are also affected and constipation or other such conditions may develop due to it. In a condition when the lower abdominal wall is weak, the organs resting against it for support are permitted to sag, and sometimes, the discomfort results along with the improper functioning of these organs. Anyhow, a person feels better and functions better when his internal organs are properly supported by a strong abdominal wall. And this is what all the physical educators must bear in the mind while educating and giving conditioning exercises to their students.

Maintenance of Posture

Posture is maintained by integrated action of effector organs, i.e. muscles with required precision, and it is obtained as a result of neuromuscular co-ordination. It means, it requires co-ordination between muscular system and central nervous system. Maintenance or regulation of posture (which can be either static or dynamic) refers to a person's ability to maintain stability of the body and body segments in response to forces that threaten or disturb the body's equilibrium.

Maintenance of posture mainly depends on:
- Information reaching the spinal cord and higher centers of the brain from various receptor organs (afferent stimuli).
- Efferent impulses reaching the muscles or the effector organs from higher centers and spinal cord.
- Various postural reflexes also play an important role in regulating the posture. Reflex is an efferent response to an afferent stimulation. Here the efferent response to antigravity muscles is elicited via afferent stimulation that arises from variety of sources all over the body like muscles, ears, proprioceptors and exteroceptors (eyes, skin, etc). These are all known as receptor organs.

Mechanism played by these receptor organs is as followed:
- *Muscles:* Muscle spindles present within the muscles record change in tension and increased tension causes stimulation and results in reflex contraction of the said muscles.
- *Eyes:* Visual sensation records any change in the position of the body with regard to its surroundings. Eyes are one of the receptors for righting reflexes and enable the head and body to restore to the erect position.
- *Ears*: Stimulation of receptors of the vestibular nerve results from the movement of fluid in the semicircular canals of the internal ear. Any movement of the head disturbs this fluid, and thus the knowledge of the movement and the direction in which it takes place are recorded for further regulation of the posture.
- *Joint Structures:* In weight bearing position, approximation of bones, stretching or traction of internal structures stimulates proprioceptors in the

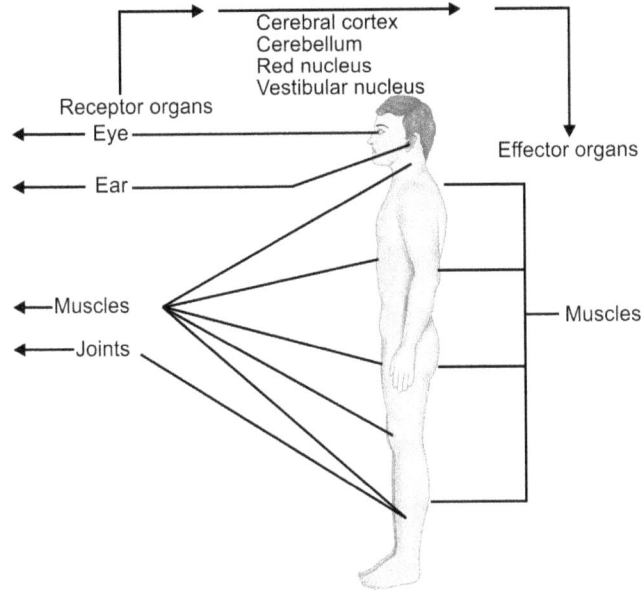

Fig. 13.2: Nervous control of posture

joints, and the joints reflexly presume position of stability.
- *Skin:* Skin of sole of the foot in particular is very sensitive, and its stimulation helps in standing position.

Impulses from all these receptors are conveyed and coordinated in the higher centers of brain, and through the spinal cord, the effector organs (the muscles) maintain or develop a new posture. The nervous control of posture is shown in **Figure 13.2**. Thus, posture is regulated in part by reflex action, and in part by volitional effort. Regulation of posture is briefly summarized here:
- Stimuli arising from eyes, ears, muscles and proprioceptors, and from pressure on the soles of the feet elicit an increase in the tonus of the antigravity muscles, and thus provide the body with a means of remaining upright without the necessity for conscious control.
- Temporary postural adjustments can be made volitionally in the same way as all voluntary movements are performed.
- The gradual changing of habitual postural patterns can be brought about by reconditioning the neuromuscular pathways.
- Ligaments, fascia, bones and joints are inert structures that support the body, whereas the muscles and their tendinous attachments are the dynamic structures which maintain the body in a posture or move it from one posture to another.

GOOD POSTURE

A posture is said to be good when it fulfils the purpose for which it is used with maximum efficiency and minimum effort. In a good posture, the body segments are properly aligned in perfect balance with minimum possible stress and strain on the body parts.

It is only in good posture that the body functions best. Ease and grace of body movements are promoted by correct posture which in turn contributes to health by promoting good body-mechanics. An efficient posture in standing and sitting is to be esteemed as much for its social and psychological values as for its direct hygienic value.

Good posture is characterized by best mechanical efficiency, the least interference with organic function, and the greatest freedom from strain. Since good posture has aesthetic appeal, it is considered as a desirable social asset and one makes better impression, and can thereby impress more people in business and professional life. In the physical education and sports, good posture is even associated with better execution of their sport activity, with improved efficiency and reduced risk of injury.

Factors Essential for Good Posture

- Knowing what good posture is,
- Being able to assume good posture (good health and hygienic conditions, i.e. nutrition, sleep),
- Possessing strong muscles and fitness to maintain good posture for reasonable lengths of time without fatigue,
- Having the desire to practice good posture consistently (a stable psychological background), and
- Plenty of opportunity for free, natural movements (as lack of ventilated and spacious place for study or sleeping, play facilities or equipments at home or school, attitude of the parents toward games and sports can all not let the children engage in their overall body building, and thus the poor posture or postural defects may develop in them).

Values and Advantages Associated with Good Posture

- **Movement Efficiency**: In a good posture all the muscles and bones are properly adjusted and function at their best. An individual is in a better position of readiness for movement when his posture is good. Spinal column has the normal curves permitting spinal cord and nervous system control body movements to function freely. It is logical to expect that a person will be more effective, graceful and co-ordinated in his movements and in all the activities of standing, walking, running, climbing, etc. The athletes are very well familiar with the importance of proper stance and posture for performing the skills necessary for the respective event. A good posture always requires an optimal energy expenditure to maintain it.

- **Improved Health and Fitness**: Good posture denotes general health, fitness and strength status of an individual. Various experts have explained the health values of good posture training; and it is well proved that good posture is associated with the better functioning of all the internal organs, i.e. better expansion of the lungs and normal functioning of the heart. When the trunk alignment and the abdominal muscles are good, the abdominal organs, kidneys, spleen, etc. are all given proper support and therefore, function better.
- **Aesthetic (Cosmetic) Values**: Good general appearance is an asset physically, commercially and socially. Good posture has an advantage of a pleasing appearance. A well co-ordinated human body with grace and poise is definitely more attractive physically, rather than when a person is having sagging abdomen, rounded shoulders or protruding head.

POOR OR BAD POSTURE

Posture is poor or bad when it is inefficient and fails to serve the purpose for which it is designed, and an unnecessary amount of muscular effort is used to maintain it. A bad or poor posture is a faulty relationship of various parts of the body which produces increased strain on supporting structures, and in which there is less efficient balance of body over its base of support. The human body is just like a machine. Like any machine human machine also functions efficiently when all of its parts are maintained in proper alignment. Many people become biologically older than their chronological age just because of neglecting some simple rules of living and of body mechanics.

Poor or bad posture has got its physical, physiological, social and psychological ill effects. Bad posture with poor body mechanics is accompanied by lack of muscle-tone, lowered threshold to fatigue, and lessened available mechanical energy. Especially, in older people exaggeration of normal curves tends to become set-in-rigid patterns and interfere with normal physiology. If a child or youth has had the proper developmental experiences in his previous years, it will be possible for all of his joints to function correctly and his movements will be smooth and efficient. Lack of development of erect carriage predisposes toward inefficient movements, lack of success in competition with one's peers, loss of interest in the physical activities characteristic of his age and a lessening of vitality as well as joy of living. Thus, the results are psychophysical.

A poor posture also shows a displeasing appearance and a state of unwelcome psychological status. The clothes do not fit well these subjects, and the person lacks in vigor and poise.

Disadvantages/Drawbacks of poor or bad posture can be summarized as following:
- Muscles, bones and ligaments are poorly adjusted tending to cause lack of strength and tone producing strain on certain body parts.
- Flat chest impairs full use of lungs, relaxed and protruding abdomen tends to cause sagging of the viscera, constipation or auto-intoxication. In severe cases the heart may be cramped.
- Exaggerated spinal curves tend to cause weakness of the supporting ligaments and muscles.
- Poor general appearance shows lacks of vigor, tone, poise and aesthetic quality.

STATIC POSITIONS

Thought of the term *good posture* usually draws us to think about static standing position that fulfills certain mechanical and aesthetic characteristics. However, today posture is not just related to the standing position alone, as was generally the practice in the past. Rather, it involves many aspects of a person's stances like sitting, lying, etc. and gives a greater scope to posture considerations. Though there is a wide variation in the human physique, and it is almost impossible to define a uniformly standard posture for all body types, however, it is desirable to have certain criteria for guidance. Out of all static body positions, standing, sitting and lying postures are important ones having more practical relevance, hence they are described here.

Standing

Posture is easier to describe in a static standing position because, except for a slight amount of sway when standing, the body is not moving. Standing position is a unique extended human position and this is the only position from which vigorous body movements can be performed freely, and without the danger of strain. In this position poise of the body should be such that provides conditions of efficient balance while permitting favorable functioning of the internal organs the same time. This should be an easy, balanced extension and should not be accompanied by strain or tension. However, this is the most difficult position to maintain as the whole body must be balanced and stabilized in correct alignment on a small base by the co-ordinated work of many muscle-groups.

Characteristics and Standards of Static Standing Posture

In this posture, the body is held as tall as possible; the head is erect, the chin is slightly drawn in; the shoulders are slightly posterior to the center of gravity, chest is high; abdomen is in, spinal curves are not exaggerated; pelvis is tilted slightly backwards; knees are straight but not stiff; the feet are so placed that their inner borders are almost parallel and 2″-3″ apart. The arms are loosely hanging to the sides of the body, and the elbows and fingers are only very slightly flexed **(Figs. 13.3A and B)**.

When this position is maintained, the line of gravity which is drawn from head to feet, divides the body into two equal halves, and distributes the body weight equally on either side, and also in front and back of the body.

- For proper and perfect balance of the body in erect standing position, the line of gravity must fall well within the supporting base, preferably nearer to the center of the base.
- The entire body is held in erect position by the ligaments which bind the skeleton together, and by the muscles which maintain the balance throughout. The muscles of the front of the body are opposed in their action by the muscles of the back of the body. Similarly, the muscles on the right side of the body are balanced by this on the left side of the body. The antigravity muscles should have adequate strength and muscle-tone to resist the pull of gravity successfully.
- The erect standing posture is maintained by the co-ordinated work of many antigravity muscles, many postural reflexes, eyes, ears, muscles, joints and skin that are all under the control of the nervous system.
- The center of gravity of each body segment (i.e. head, thorax and pelvis) should be vertically above the center of the supporting base. In **Figures 13.4A and B**, the human body segments are shown respectively when they are properly balanced, and imbalanced.
- In a good standing posture, there should be minimum friction in the joints for their efficient functioning. This will also prevent unnecessary wear and tear of the joints.
- A good standing posture as also the other postures should favor the organic functions, and in no way their activities should be restricted.
- A good posture also requires a sufficient flexibility in the structures of the weight bearing joints to permit

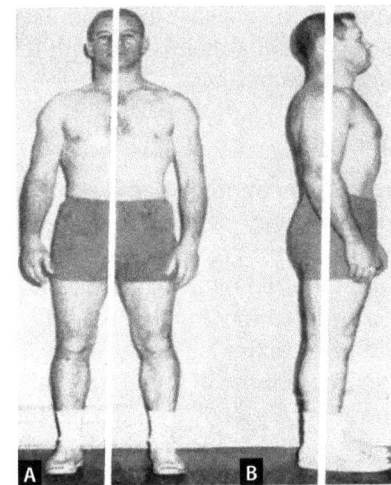

Figs. 13.3A and B: Good static standing posture: (A) Front view; (B) Side view

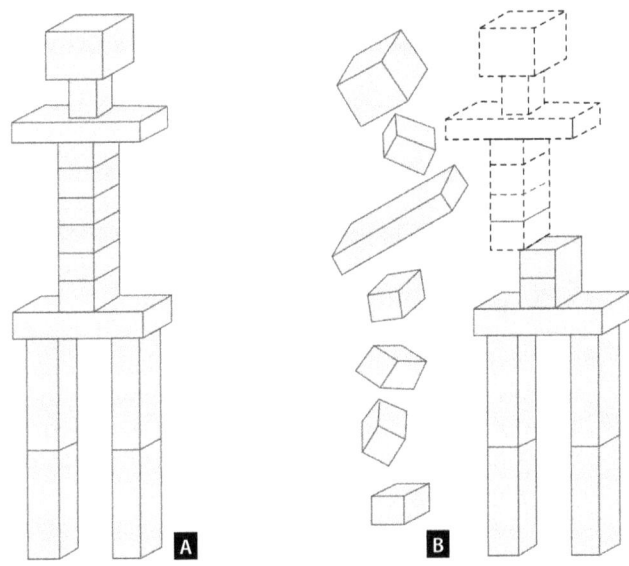

Figs. 13.4A and B: Human body segments: (A) Properly balanced; (B) Imbalanced

alignment without much strain. However, the flexibility should not be too great nor too less.

Energy Cost

A good posture requires an optimal energy expenditure for the maintenance of proper alignment. Rigid military or NCC type of standing postures in which the muscles are hypertensed, require more energy expenditure are usually fatiguing. Similarly, the fatigue slump posture requires only some less energy, but it places too much strain on the ligaments.

When standing, the increase in metabolic rate over the basal rate is very small as compared with the metabolic cost of moving and exercising.

Muscular Activity

Humans have very economical antigravity mechanisms and the muscle energy needed to maintain the erect standing position is not great. A reason for the economy of muscle effort required is the major role of the ligaments in supporting and maintaining the integrity of the joints. The muscles that are active are those which aid in keeping the weight-bearing column of bones in relative alignment and that oppose gravity's downward pull. The significant muscular activity recorded during the erect standing position on the electromyography (EMG) by some scholars is briefly presented here.

- *Foot:* The intrinsic muscles are not active during normal standing, but become active in the push off for walking or rising on toes.
- *Leg:* The posterior calf muscles are more active than the anterior muscles. Rising on the toes or wearing high heels increases the activity of both anterior and posterior muscles.
- *Thigh and hip:* There is very little muscle activity in thigh muscles during relaxed standing. The iliopsoas is constantly active to prevent hyperextension at the hip joint. In swaying, there is some alternating activity of gluteus medius and tensor fascia lata muscles.
- *Spine:* There is very slight activity either in the lower erector spinae or abdominal muscles, depending on the relation of the line of gravity to the spinal column. However, it is reported that slight to moderate activity in back muscles is at least three times as likely as the activity in the abdominal muscles.
- *Upper extremity:* Since the arms are hanging passively, the integrity of the joints of upper extremity is assisted by low grade activity of some muscles. The serratus anterior and fibers of trapezius support the shoulder girdle. The supraspinatus muscle resists downward dislocation of the elbow or wrist joints due to the arms hanging passively.

Neuromuscular Mechanism for Maintaining Erect Posture

The proprioceptors are mainly responsible for most of the reflex movements necessary for the maintenance of the erect standing position, and for the adjustments that must be made to meet changing conditions. They include the receptors in the muscles, joints and the ears, accompanied by two exteroceptors, i.e. visual and cutaneous. Volitional postural adjustments are made the same way as all the voluntary movements take place.

Static standing position is considered as an individual's basic posture from which most of all other body postures are derived. It is the standing position that allows a large number of varied activities and exercises to be performed from it. Therefore, *Metheny's* following quote very correctly lays the practical significance on accepting and analyzing an individual's static standing posture.

Sitting

In our sedentary life most of us actually spend more time sitting than standing. Good postural alignment while sitting is important because sitting is an activity that can place a great deal of pressure on the intervertebral discs. Studies have shown that intervertebral disc pressure is least when lying supine and it increases as one stands, and decreases still more as one sits. Long continuous sitting habits combined with poor furniture may force the body to progressive degenerative tissue changes, and pain. However, it is only recently that the furniture is being designed scientifically along the knowledge of anatomical and physiological principles.

From the viewpoint of kinesiology, it may be more important to pay greater attention to one's habits of sitting than of standing. A person rarely stands in one position for more than a few moments at a time, but most of us spend many minutes at a stretch in one sitting attitude. Plenty of time is spent sitting to establish unfortunate postural habits.

With sitting postures, a chair that has a lumbar support that maintains the lumbar lordosis will place the least amount of pressure on the intervertebral discs. Maintaining the vertebral curves, keeping the feet flat on the floor, having the low back supported, and keeping the upper body in good alignment are key elements of good sitting posture.

Lying

Lying down is considered a resting position. Generally speaking, lying supine position should only be taken when prescribed, as otherwise, the individual while sleeping constantly changes position during the night. This is the easiest position as the body can be completely supported in the supine position and is as stable as possible. A good resting surface should be firm enough to avoid loss of the lumbar curve, yet soft enough to conform and give support to the normal curves of the body. In this position, the spine is relieved of the burden of transmitting the weight of the head and shoulders when it is in the horizontal position. Therefore, it tends to elongate and straighten out.

The muscle work required to maintain this position is minimal. The supine position on a completely supporting surface permits maximum relaxation. However, lying prone is generally not recommended because of the increased pressure placed on the neck. It should also be remembered that many positions which are commonly considered to be resting positions do not completely eliminate the need for muscular control.

POSTURAL DEFECTS

Faulty posture is a universal finding among children and untrained individuals, which if neglected may later turn into postural defects. Various research studies have indicated that the incidence of postural defects among children have rapidly increased over the years. *When a derivation from the normal posture becomes habitual and produces ill effects, it becomes a postural fault.* There are some postural deviations due to individual variations in body composition and physique type, and therefore, should not be considered postural faults. While making postural examination, it is best to note all the postural deviations and faults regardless of their degree and cause. Each type of postural deviation or defect has its particular cause and effects, consideration of which simplifies the process of selecting proper activities for correcting these conditions.

From the point of view of postural defects, childhood is considered to be the most crucial and important stage of life when the defects can be easily prevented and corrected. The maintenance of an erect posture is a distinct problem to the humans since the skeleton is fundamentally unstable in this position. The two legged human body presents a continuous problem in maintaining balance because the feet are very small base of support for a multisegmented towering superstructure.

There are various types of postural defects ranging from mild to severe in degree. Some of the defects affect the spine and the others the lower extremities and shoulder girdle. Quite often, combinations of two or more defects are found together among the children and youngsters. Sometimes, the presence of one postural defect can lead to the development of another defect at a higher or lower level to counterbalance the already existing one.

Causes of Poor Posture and Postural Defects

There are many and diverse causes of poor posture and postural faults. Quite often, more than one cause is present in the same situation. Sometimes, the general factors such as poor hygienic conditions, general debility or the mental attitude are responsible for the poor posture and specific postural defects, whereas the local factors such as muscular weakness or occupational stresses can also alter the postural patterns and lead to the poor posture or postural faults. Some of the important and significant causes of poor posture and postural defects are described here.

Lack of Posture Knowledge

Generally, boys and girls do not know what good posture is unless they have been given instruction in proper posture and body-mechanics. They are not aware of proper standing, sitting, walking or lying postures, and as such develop poor posture. Poor posture is usually acquired in the childhood and is continued unconsciously in later life also. These persons are not aware of their poor posture and feel the wrong posture as most natural. The correct position seems strange to them. In a large percentage of the cases of faulty posture found among school children and college students, the bones, joints, muscles and ligaments are in normal condition, the fault is mainly a wrong habit of co-ordination.

Poor Muscular Development

Poor muscular development is often a cause of poor posture. Unfit boys and girls are prone to poor posture, as they do not have the muscular strength to maintain good posture for any length of time. Their antigravity muscles lack muscular endurance. The usual 'fatigue slump' reveals a relaxation of the whole upright stance, resulting in pelvic tilt, protruding abdomen, lumbar lordosis, thoracic kyphosis, round shoulders and forward head. A major effort is needed with these individuals to improve their general strength and endurance before any posture improvement program can be effective.

Poor Postural Habits

Poor postural habits are closely associated with the presence of bad posture and postural defects. Such habits may involve faulty standing or sitting positions, in which the boy or the girl permits the spine to sag, the abdomen to protrude, and the shoulders to round. These defects most likely appear when the child assumes a particular posture for a greater length of time. Poor muscular strength and endurance often aggravate this situation.

Poor postural habits are generally caused by an injury, disease and environment. Improper shoes and tight clothing hold one in a faulty position. Defects of vision and hearing, improper chairs or bed, excessive height, overdeveloped breasts in girls can all lead to developing poor posture. A boy carrying a heavy school bag consistently over the same shoulder promotes lateral deviation of spine called *scoliosis*. Even overspecialization in some sports, despite an excellent body development can lead to postural deviations. Likewise, the practices of some occupations are conducive to poor posture or postural faults. For example, an accountant who spends many hours at a desk is prone to develop round upper back.

Injury

When a bone, ligament or muscle is injured, it weakens the support at that point and throws the body frame workout of balance. As long as this condition is present, perfect posture is impossible. Sometimes, even after the injury has been fully repaired, a wrong habit set up may persist and faulty posture continues to exist for a long time. For example, even a minor injury like right ankle sprain causes most of weight bearing on his left foot. This may develop into a habit which remains with him for years. A malunited fracture of femur producing shortening can lead to causing poor posture for the whole life unless proper shoe-modification is provided.

Diseases

Diseases that weaken muscles, bones or cause joints to lose their strength or freedom of action often upset the control of posture. Rickets, tubercular disease of joints or vertebrae, poliomyelitis, cerebral palsy etc. are the examples of the diseases that produce poor posture. Sometimes, untreated chronic diseases of the lungs and other internal organs may also not allow an individual to hold his body in normal posture and thus, poor posture is the result.

General Weakness

Erect posture can not be maintained without energy expenditure and therefore, requires some strength and endurance. It has been shown that a poor and slouched posture can be maintained at less metabolic rate than the erect alert position, largely because the individual is mainly supported by the ligaments of the hyperextended joints, rather than the action of muscles. Any factor that affects the general health and vitality of an individual, i.e. a chronic debilitating disease, general muscular or nervous weakness, can cause poor posture to develop as a means of energy conservation. Undernutrition or other dietary deficiencies also result in poor posture as the whole body is weak and it is very difficult to hold the body in correct posture. Overwork or fatigue produced due to any reason can also account for developing a poor posture.

Mental Attitude

Posture frequently reflects the mental attitude of a person. Feelings of confidence, satisfaction and happiness help in the maintenance of erect posture, whereas the depression, humility and inferiority complex all produce poor posture. As such, a good mental attitude is reflected in an erect, alert posture. The famous saying, *'as you feel, so you stand'*, rightly expresses the effect of mental attitude on the posture. Any ill or negative mental status of an individual will lead to developing his poor posture.

Heredity and Others

Sometimes, the poor posture or certain postural defects may be hereditary. It is believed that kyphosis and flat foot are hereditary, and other postural defects may also have genetic basis. Similarly certain anatomical congenital anomalies present since birth like hemivertebrae, cervical rib, asymmetrical sacralization, etc. can also contribute to the development of poor posture. Excessive height of an individual may also lead him to stooping posture.

Improper Clothings, Footwear etc.

Too tight or too loose clothings, improper footwear or shoes, etc. that are often used can also cause poor posture. Such clothing does not allow the person to hold his body in correct position and permit free movements. The high heels often worn shift body's center of gravity forward, and posterior calf and back muscles over-contract to prevent the body from falling, producing lordotic posture. The footwear not having proper arch support also sometimes causes flatfootedness, and so on.

General Prevention and Correction of Poor Posture and Postural Defects

The physical educators must learn to recognize the postures which are in need of improvement, and which can be improved with guidance and corrective exercises, from those children who are merely different because of variations in their individual structures. Fortunately, the human body has a remarkable ability to compensate for deviations from the normal. It is also not necessary that the poor posture always leads to functional impairments. Physiologically speaking, according to some researchers there is no clear cut evidence that correction of common functional postural defects will conserve energy expenditure. However, it is generally established fact that the poor posture and the postural defects hamper the normal growth and development in the children, and these children may suffer from various types of sickness and diseases that ultimately interfere with their overall physical, functional and mental efficiency.

It is the duty of every physical education teacher to pay attention to whether or not every one of his students can carry himself beautifully in erect standing. It is the physical education teachers who are supposed to have received better and enough training in anatomy, physiology, neurology, kinesiology etc. so as to helping normal children or youths attain good postures in standing, sitting and all other sports activities

To detect poor posture or any postural fault in the children, their regular physical and health examination must be conducted in the schools. As soon as any deviation from the normal posture or any such health condition contributing to poor posture is noticed, the parents and the school health personnel must be intimated, and immediate appropriate remedial measures should be taken. The school authorities and the physical educators also have a great responsibility in the correction of the poor posture and postural faults.

The important general corrective measures often employed for most of the postural defects are briefly described here.

Removal of Cause

Removal of underlying factor that produced poor posture or postural defect is very important if permanent correction is to be effective. If faulty posture is due to any injury or disease, it should be properly treated before anything else is attempted. A fault in posture due to improper footwear or high heels cannot be much improved so long as the improper footwear or high heels are used, no matter how much proper and supervised corrective exercises are performed. Likewise, if the poor posture is due to any fatigue, nutritional disorder, mental strain, improper digestion, hearing or visual defect, etc. it should be treated for the said cause by the respective expert doctor. Any environmental factor whether it is the wrong furniture or otherwise at home or school, needs improvement. Rest and proper nutrition are very important in most of the children suffering from poor posture or postural defects.

Postural Training and Re-Education

Habit of good posture patterns are naturally trained in the nervous system of the children at each stage of their development. When a child has not had a training of proper postural habits, there is only one sure way of re-establishing the proper habits, i.e. many repetitions of a specific act until the required pattern becomes conditioned to work without conscious control. It has repeatedly been observed that if a posture is maintained and practiced long enough by conscious effort, then it serves to maintain the same position without the need of further thought or attention. However, for such postural re-training, all the conditions of health, general physical strength, and emotional desire must be optimal. The starting point of any posture improvement program is to teach students proper standing, sitting and walking stances.

Cardinal Principles of Postural Training

There are at least four cardinal principles that are important to follow for a program of postural education or re-education, particularly in younger individuals:

1. *Use the entire body in a great variety of activities in order to maintain the muscles and joints at their highest peak of efficiency*: Use of the muscles is the best stimulation to their growth, repair and regeneration, and the most important factor in normal development of bone. It is only when the body is permitted to move freely and vigorously since birth that it will develop normally.

2. *Choose activities that afford a sufficient amount of extension*: The extension is the main characteristic of the human adult, and flexion, as a maintained attitude, must be discouraged. It is very interesting to note that the emotions of joy, pride, happiness all express themselves in postures of extension. Therefore, it is important for a physical educator to keep this fact in the mind that, if a child's life is filled with happiness, his body will most likely develop into the full stature of a man. The bodies handicapped physically or psychologically need to be corrected, prevented or overcome.

3. *Activities that train desirable reflexes must be considered*: Creeping, hanging and balancing are some of the activities that condition the reflexes which encourage normal bodily development. Children who do not creep miss an important phase of development. Their spinal and abdominal muscles do not strengthen properly. Hanging also makes the abdominal muscles contract reflexly. Balancing is also of great value as it makes the body to adjust its weight statically near center and thereby helps to train the reflexes necessary for standing.

4. *Do not neglect relaxation*: Relaxation is very important means of further continuation of normal action in specific muscles, and also for the general benefits to

the body and mind, particularly when the muscles and the body are exhausted due to the exercises or activities.

Corrective Exercises

Once the cause of the poor posture or postural fault has been determined, corrective exercises are used to reduce or relieve the fault. In many instances corrective exercises which are specially designed to permit the individual to assume proper body positions easily are very essential. For this purpose, two types of exercises are generally needed:
1. Where muscles and ligaments have shortened and restrict joint-movements, stretching exercises will be necessary.
2. Where muscles are weak and lengthened, muscular strength and endurance must be developed.

These exercises are usually complimentary to each other, for example, for giving postural training in lordosis, the lumbar spine must be stretched anteriorly and the abdominal muscles are shortened and strengthened.

Relaxation

To relax an individual is very important, as some degree of useless and unnecessary muscular and nervous tension is almost always associated with poor posture. To begin with, general relaxation with the body in horizontal positions reduces muscular tension and gives a feeling of alignment. Later, voluntary relaxation of specific muscle groups can be taught and practiced so that the individual learns to recognize tension and is able to relax at will, and also if and when the tension develops during the maintenance of static or dynamic postures.

The deep breathing exercises and relaxing music can also be helpful in relaxing a tensed individual and as such, these should be appropriately used in the corrective program.

CLASSIFICATION OF POSTURAL DEFECTS

Most of the postural defects can be classified into *postural (or functional)* and *structural* types. A postural or functional defect is usually reversible and can be corrected with physical measures and corrective exercises, whereas in a structural defect definite changes in the soft tissues and bones have occurred, and such cases usually do not respond to the physical measures or corrective exercises alone. However, according to most of the experts, all the postural defects can also be classified into following three grades on the basis of their severity:

1. *First Degree (Mild):* The postural defect is mild and can be corrected completely by active or passive stretching or by positioning of the body part in the opposite direction.
2. *Second Degree (Moderate):* Shortening in the soft tissues prevents full active or passive correction of the defect. Repeated attempts to passive correction result in pain. Some degree of bony changes may be present in this category of postural defects.
3. *Third Degree (Severe):* The postural defect is rigid or irreversible, and can not be corrected by passive maneuvers. Most often there are significant associated bony changes.

Most of the postural defects either occur in an antero-posterior or lateral direction:
- *Anteroposterior Defects:* The defects of posture occurring in the anteroposterior direction or sagittal plane are called anteroposterior defects. Examples of most common postural defects are kyphosis, lordosis and flatback.
- *Lateral Defects*: These postural defects occur in the lateral direction or frontal plane. Scoliosis is a classical example of a postural defect in lateral direction.

Anteroposterior Defects

Effect of Pelvic Tilt

The position of the pelvis is very important for maintaining the normal anteroposterior curves in the spine as the pelvis functions as a base for the entire trunk. Any tilt or inclination of the pelvis beyond the normal will cause imbalance of the structure above it. Therefore, it is very essential to understand the mechanism of normal pelvic tilt, particularly in the anteroposterior direction.

The anteroposterior tilt of the pelvis is normally controlled by the muscular activity of the hip (glutei) and the trunk (abdominals). The normal inclination of the pelvis is about 30°, and at this angle the anterior superior iliac spines and public symphysis are on the same vertical plane. The contraction of abdominals and/or the glutei (hip extensors) will tilt the pelvis backward, cause pelvic inclination (about 20°), and flatten the lumbar spine, whereas the contraction of the lower back extensors and/or the hip flexors will tilt the pelvis forward, increase the lumbar lordosis and the pelvic inclination is increased about to 40° or more. The spinal deformities in the anteroposterior direction are produced by the individual in an attempt to keep body's centre of gravity above the feet. Thus, any pelvic tilt beyond normal range produces compensatory adjustments in the lumbar, thoracic and

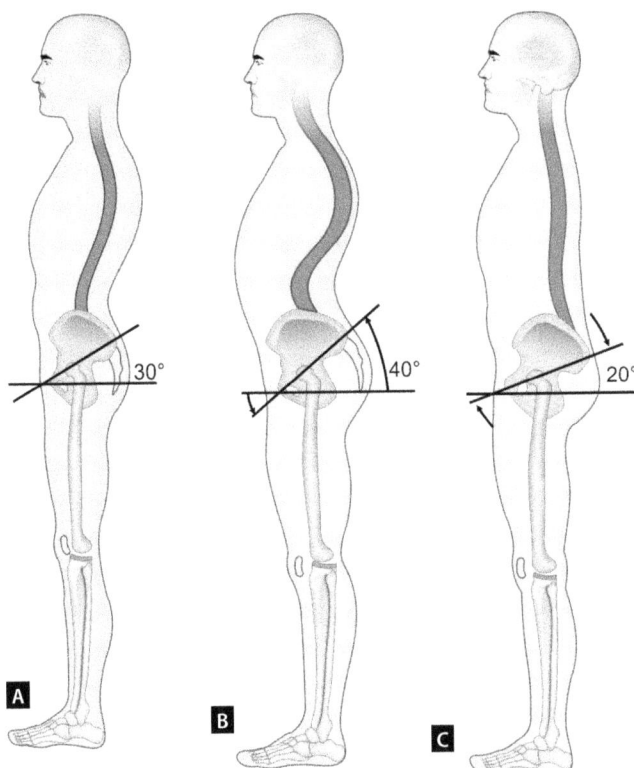

Figs. 13.5A to C: Pelvic inclination:
(A) Normal pelvis; (B) Lordosis; (C) Flatback

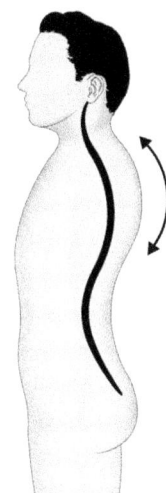

Fig. 13.6: Kyphosis

cervical curves, and possibly downwards to the knees and ankles also. The normal pelvic tilt can also be altered as a result of the deformities of the lower limbs, particularly flexion deformities of the hips or knees. The pelvic inclination in normal pelvis, lordosis, and flatback are shown in **Figures 13.5A to C** respectively.

When there is a forward or anterior tilt of the pelvis beyond normal, by keeping both hips fixed in a slightly flexed position, the body assumes a lumbar curve with its convexity backward. In order to keep his body upright, he compensates by bending his spine backward, thus producing a curve with increased lumbar lordosis. This, in turn is compensated by the increased curve in thoracic region (kyphosis). Sometimes, there may even be a final increased curve in the cervical region (cervical lordosis). Alternatively, anteroposterior curvatures may begin in the upper part of the spine producing compensatory curves below them. In all cases of spinal deformity, the first curve developed initially whether at a higher or lower level is known as *primary curve* and the compensatory curve that develops afterward is called as *secondary curve*.

There are a number of postural defects that occur in antero-posterior direction. The most common and familiar anteroposterior postural defects are kyphosis and lordosis, though flatback, kypholordosis, forward head, swayback, etc. are also the examples of postural defects affecting the spine in this plane.

Kyphosis

Kyphosis is the exaggeration of the normal posterior curve of the spine (**Fig. 13.6**). This deformity generally occurs in the thoracic region where the normal flexion or posterior curve becomes increased, as such it should be called by the term *thoracic kyphosis*. However, in practice thoracic kyphosis and kyphosis have become so synonymous to each other that kyphosis is regarded as a postural deformity of increase in the amount of normal convexity of the thoracic region of the spine itself. In the cervical and lumbar regions of the spine an anterior normal curvature exists, however, if this anterior curve is reversed to posterior convexity, it will result in cervical or lumbar kyphosis respectively.

This deformity is often termed as *'round back'* when it takes the form of a long rounded curve. If there is a localized sharp posterior curvature or angulation, it is called as a *'hump back'*.

Lordosis

Lordosis is the exaggeration of the normal anterior curvature of the spine (**Fig. 13.7**). Theoretically speaking, the cervical and the lumbar regions of the spine have normal anterior convexity, as such these should be called *cervical lordosis* and *lumbar lordosis* respectively. However, lordosis in lumbar region is more common where a slight

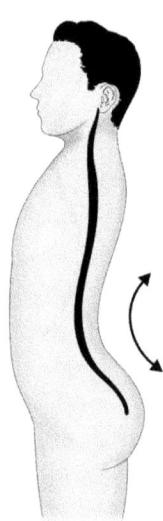

Fig. 13.7: Lordosis

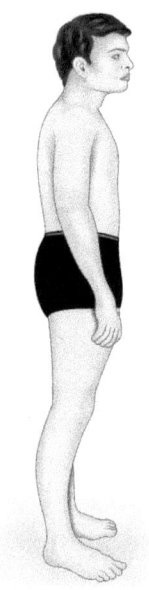

Fig. 13.8: Flat back

anterior curve is normally present. Therefore, the term lordosis should only be used when the curve is increased well beyond the normal. This is more often a postural than a structural deformity.

Kypho-lordosis (Round-Hollow Back)

This postural defect is a combination of two postural defects described above, i.e. kyphosis and lordosis. This is a more common postural defect than any other type of anteroposterior postural defects. It is a definite fatigue slump posture.

In kypholorosis, either kyphosis or lorosis is the primary curve and occurs before the other secondary curve that develops to compensate the existing primary curve. However, according to certain experts, lumber lordosis is generally the primary curve, and occurs before kyphosis. Therefore, early detection and correction of primary lordosis is very important to successfully prevent the occurrence of secondary compensatory kyphosis.

Flat Back

This is a postural defect in which the pelvis is tilted backward, and the normal lumbar curve is flattened out **(Fig. 13.8)**. In fact, normal curves of lumbar and thoracic regions are both diminished, and the spine often appears nearly straight. That is why flatback is known as a reverse deformity to lordosis, which sometimes also results in lumbar kyphosis when lumbar spine slightly bulges backward.

Fig. 13.9: Sway back

Sway Back

This is a postural defect in which the pelvis is tilted anteriorly, but the patient, instead of compensating by lumbar lordosis, bends the spine sharply backward at the lumbosacral angle **(Fig. 13.9)**. As such, there is a sharp hyper-extension at the lumbosacral junction, probably due to the swaying back of the upper part of the body, but this angulation can not be called as lordosis.

There is often an existing thoracolumbar kyphosis along with it, therefore this posture is often confused with either kyphosis or lordosis, but a careful examination reveals that it has element of neither.

Forward Head

In this postural fault, the head and neck are held in a forward, downward position, the chin is pulled in toward the neck, and the face as a whole is downward at an angle of 30 to 45° from the horizontal. Thus, there is increased flexion of lower cervical and upper thoracic region. Forward head is usually associated with a rounded upper back. Sometimes, this condition is associated with the hyperextension of the upper most cervical region. Faulty sitting postures for study, improperly placed computer screen or occupational factors are the main causes for this condition.

Round Shoulders

Round shoulders is a term often connected and described with kyphosis. Round shoulders, in proper terminology is a deformity of forward deviation of shoulder girdle in which the scapulae are separated beyond normal and the acromian points of the shoulder come forward, falling anterior to the line of gravity. The sternoclavicular joint (the only joint support for the shoulders) is depressed. The vertebral borders of scapulae protrude backward and become prominent, presenting the picture of *winged scapulae*.

This condition very often accompanies kyphosis. Therefore, round shoulders should not be confused with kyphosis as both of these occur independently of the other. Round shoulders, in true sense is a deformity of shoulder girdle without any spinal deviation and entirely different muscle-groups are responsible for this deformity. However, both of these deformities often appear commonly together as an integral defect as one defect generally begets the other one.

Lateral Defects

Lateral curves of the spine are generally known as *scoliosis*. In these defects the problem may not necessarily begin in the spine itself. The origin of scoliosis may be in the feet, knees, hips or even in the arms. For example, if a person has one leg shorter than the other or habitually puts more weight on one foot, bending the knee of the leg so that it is relatively shortened, and as this position is maintained, the pelvis will be lower on one side. Let us suppose that the right leg is shortened somehow, if he has to keep his spine straight over the pelvis at right angles, the whole body would be carried over to the right. This produces a curve with a convexity to the left. To compensate this curve and to keep the head in the same relationship to the pelvis and bring the line of gravity back to the middle position, he compensates by flexing the lumbar spine to the left and thus develops a second curve with convexity to the right. This lumbar curve in turn may produce a third curve in thoracic region with its convexity to the left. This again may lead to a compensatory fourth curve in the cervical region with convexity to the right. Like anteroposterior defects, this process of compensation may take place from above downward (beginning with the cervical spine or shoulder girdle) or from below upward beginning from feet or knees.

When there are more than one lateral curves in the spine, usually the curve that develops first and has got more deviation is termed as *primary curve*, where as the compensatory curve that occurs later and has got relatively less deviation is termed as *secondary curve*.

The most common postural defects occurring in the spine, shoulder girdle and lower limbs are briefly introduced here.

Scoliosis

The term scoliosis is derived from a Greek word which means twisting or bending. A lateral or sideward bending or deviation of the spine to one or more sides from the midline is called *scoliosis* (**Fig. 13.10**). Actually, scoliosis is

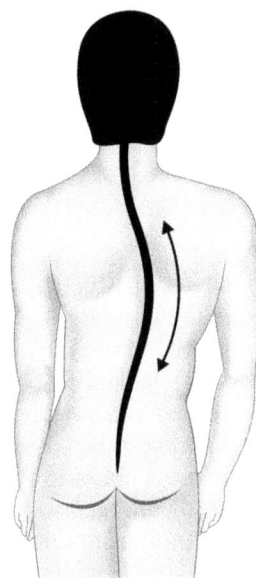

Fig. 13.10: Scoliosis: (C curve–right thoracic)

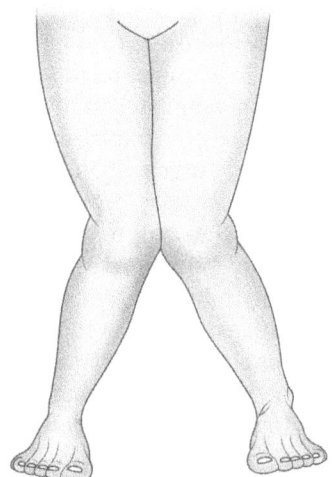

Fig. 13.11: Knock-knee (genu valgum)

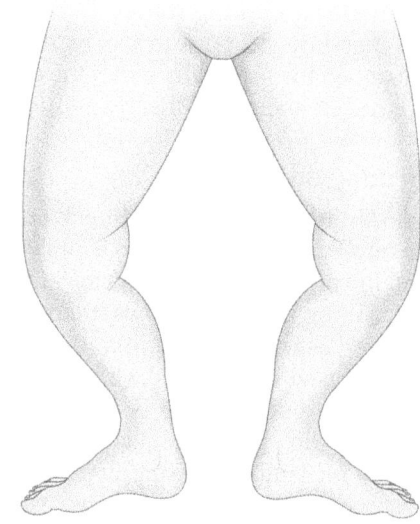

Fig. 13.12: Advanced stage of bow legs (genu varum)

a rotaro-lateral condition of the spine in which apart from bending of the spine, there is also rotation of the involved vertebral bodies. The structure of the spine is such that it is impossible for the spine to be bent even slightly to one side without being twisted or rotated.

Knock-Knee

Knock-knee or genu-valgum **(Fig. 13.11)** is a postural defect in which there is an outward deviation of the leg at the knee so that the tibia remains abducted in relation to the femur. Knock-knee is position that tends to shift the weight towards the medial border of the foot and bring about a foot position of pronation. This is usually not a congenital deformity, rather it is an acquired condition that develops a few months after the child begins to walk.

Bow Legs

Bow leg is a postural defect in which there is lateral bowing or angulation of the leg in relation to the knee. In this defect, with the subject standing with his feet together, the knees remain wide apart. In fact, bow leg is a deformity of the lower leg having maximum bowing only at the middle of the leg, whereas *genu varum* is a deformity of the whole lower limb with maximum curvature and bowing at the knee **(Fig. 13.12)**. Both the terms, bow legs and genu varum are often used indiscriminately, though they are technically different.

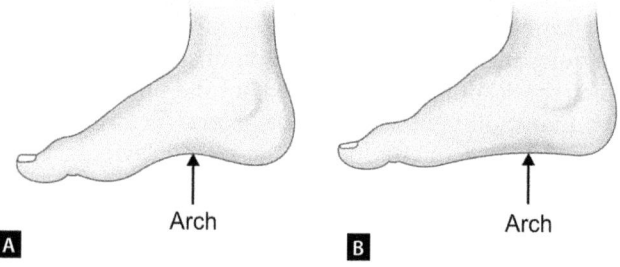

Figs. 13.13A and B: (A) Normal foot; (B) Flat foot

Bow legs are quite common during infancy and childhood. The deformity is more common than knock-knee and is generally bilateral.

Flat Foot

Flat-foot is generally defined as a foot condition in which the medial longitudinal and transverse arches of the foot are collapsed, and the entire sole of the foot (mostly the medial border) comes into complete or near-complete contact with the ground **(Fig. 13.13B)**. This condition is often associated with some degrees for eversion. It is believed that almost 60% of the people have a foot disability of one or the other type.

For the remedial measures for the specific postural defects mentioned above, the interested readers may refer to a standard textbook on 'Correctives' or the author's another book entitled 'Corrective Physical Education, Therapeutic Exercise and Rehabilitation.'

CHAPTER 14

Locomotion

INTRODUCTION

Locomotion is the progressive movement of the entire body from one place to another, by means of self-propulsion. Usually this propulsion is provided by the lower limbs, but sometimes this may involve all four limbs (as in creeping or crawling) or very rarely, with the upper limbs alone (as walking on the hands or hand-traveling on a horizontal bar). As such, the locomotion in humans includes all activities in which the entire body is moved by the action of its levers with the assistance of gravitational force.

The essential feature in all locomotion is the movement of entire body from the point of contact between body part and the supporting surface. For this, some body part must apply force to the surface, and the surface must resist that applied force. This resisting force from the surface pushes back and opposes, and thus this reaction moves the body in accordance with *Newton's third law of motion* (which states that whenever one body exerts a force on another, the other one also exerts a force on the previous one equal in magnitude and opposite in direction).

Human locomotion is an interaction of body's neurologic, musculoskeletal and cardiopulmonary systems. Locomotion may be on the ground or in the water (swimming), but the most common forms of locomotion which occur on the ground (in which body's center of gravity (COG) is transferred from one foot to another) are—*walking and running.*

WALKING

Walking is the simplest form of locomotion in which the COG of the body is carried alternately over the right foot and the left foot. At all times, one foot is in contact with the supporting surface, and for a brief period both feet contact the surface. In addition to muscular force, other forces aid in the progression. These forces are gravity and the momentum of the body once forward movement is underway.

- Walking is probably the most efficient act in which the human being engages. Walking is a process of pushing the body out of balance while it is supported over one leg and then bringing the driving leg forward in time to form successive new base of support to save the body from falling. This process is repeated with every step. The walking requires not only the smooth, co-ordinated movement of both lower extremities, but also of the head, trunk and upper extremities.
- Walking is accomplished by the alternating action of the two lower extremities. It is an example of translatory motion of the whole body brought about by rotary motion of some of its parts. It is also an example of a pendulum like movement in which the each lower extremity starts at zero, passes through its arc of motion and falls to zero again at the end of each stroke.
- The action and the movements of both lower extremities can not occur efficiently unless the pelvic-girdle co-operates with the movements of the lower extremities. The pelvis, on one hand places the acetabulum in a favorable position for the action of the corresponding femur; on the other hand, it also has a role of transmitting the weight of the body alternately first on one leg, and then over the other leg.

Components of Walking/Gait

Walking is the manner in which one moves from place to place with his feet *(Gait is the process or components of walking).* According to traditional description, *there are two phases of walking,* or *gait cycle.* In other words, in walking, each lower extremity undergoes two phases:
1. *The swing or recovery phase,* and
2. *The stance or the support phase.*

Stance Phase

Stance phase is the activity that occurs when the foot is in contact with the ground. It begins with heel strike of one foot and ends when that foot leaves the ground. *The stance phase accounts for about 60% of the gaitcycle.*

The stance or the support phase is further divided into subphases, (i) *heel strike,* (ii) *foot flat* (iii) *midstance,* (iv) *heel-off,* and (v) *toe-off.* These five components of stance phase with corresponding RLA terms are shown in **Figure 14.1**.

Many experts describe the stance phase into four components by combining heel-off and toe-off into one calling it *'push-off'.*

Swing Phase

Swing phase occurs when the foot is not in contact with the ground and refers to its non-weight bearing activities. It begins as soon as the foot leaves the ground (toe-off), and ends when the heel of the same foot strikes or touches the ground. *The swing phase makes up about 40% of the gait cycle.*

The swing phase is further divided into three components:

(i) acceleration, (ii) midswing, and (iii) deceleration.

These components of swing phase with their corresponding RLA terms are shown in **Figure 14.2.**

Description of Components of Stance (Support) Phase

Heel Strike

Heel-strike indicates the beginning of stance phase the moment the heel comes in contact with the ground. At this point, the *ankle* is in a neutral position between dorsiflexion and planterflexion, and the *knee* begins to flex. This slight kneeflexion provides some shock absorption as the foot strikes the ground.

The *hip* is in about 25° of flexion. The head and trunk are erect and remain so throughout the entire gaitcycle. The leg is in front of the body, and the trunk is rotated toward the contralateral (opposite) side, and the pelvis rotates forward to the same side. The opposite sided arm is forward, and the same sided (ipsilateral) arm is back in shoulder hyperextension. At this point, body weight begins to shift onto the stance (supporting leg). *In Rancho Los Amigos (RLA) terminology, this is the period of 'initial contact'.*

Foot Flat

Flat foot phase occurs shortly after heel strike when the entire foot is in contact with the ground.

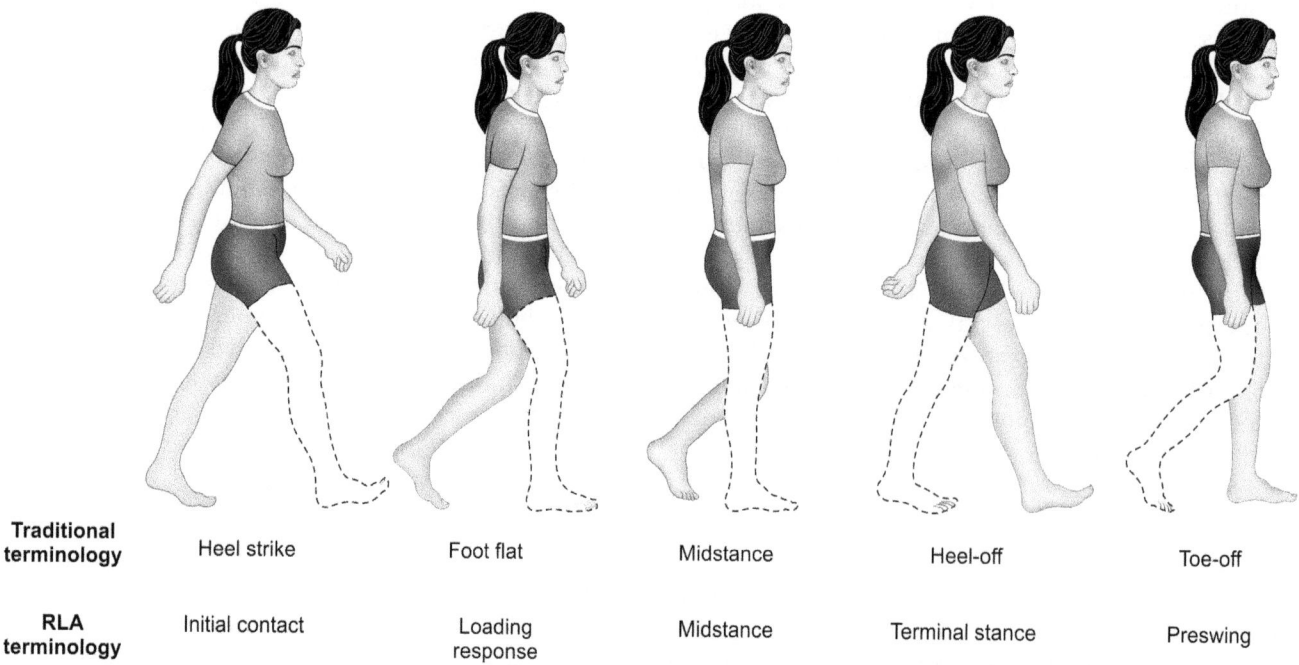

Fig. 14.1: Five components of stance phase with corresponding RLA terms

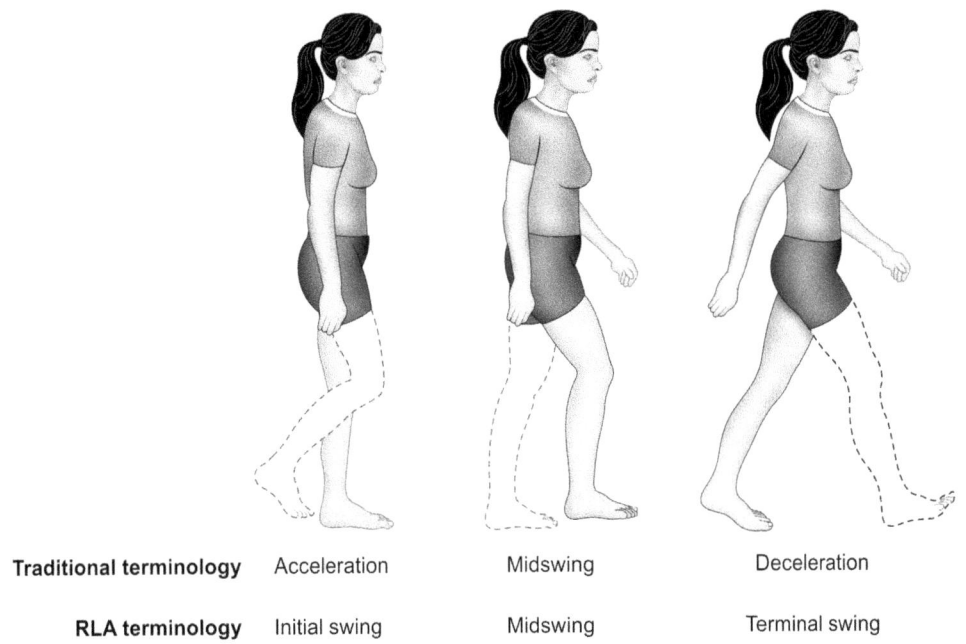

| Traditional terminology | Acceleration | Midswing | Deceleration |
| RLA terminology | Initial swing | Midswing | Terminal swing |

Fig. 14.2: Components of swing phase with corresponding RLA terms

- The *ankle* moves into about 15° of plantarflexion, with the dorsiflexors contracting eccentrically to the controlled lowering of the foot and to prevent the foot from *'slapping'* down on the ground.
- The *knee* moves into about 20° of flexion to help absorb the shock. The hip is moving into extension, allowing the rest of the body to begin catching up with the leg. The weight shift onto the stance or supporting leg continues.
- The *ipsilateral (same-sided) arm* begins to swing forward. *Foot flat is roughly comparable to the RLA terminology called 'Loading response', which is that period between the end of heel strike and the end of foot flat.*

Midstance

This is that part of stance phase at which the body passes over the weight bearing foot. In this phase, the ankle moves into slight dorsiflexion.
- The *knee and hip* continue to extend, and the pelvis is in neutral position. Both *arms* are in shoulder extension essentially parallel with the body. The trunk is in a neutral position of rotation.
- The *tibialis posterior muscle is very active during this phase,* preventing eversion (pronation) of the foot. In fact, the peroneus longus and tibialis posterior muscles apparently co-operate in stabilizing the leg and foot in weight bearing.

Heel-off

This phase begins after midstance, in which the heel rises off the ground. The ankle will slightly dorsiflex (about 15°) initially, and then begin to plantarflex. This is the beginning of the 'push-off' phase which is sometimes also called the propulsion phase, as the ankle-plantar-flexors are actively pushing the body forward.

The *knee* is in nearly full extension, and the hip has moved into hyperextension. The *leg* is now behind the body. The body is ahead of the stance leg. The pelvis on the same side has begun to rotate backward. The trunk has begun to rotate to the same side, and the same sided arm is swinging forward into shoulder flexion.

In RLA terminology, 'terminal stance' is that period between the end of midstance and the end of heel-off.

Toe-off

This is the end of the push off portion of the stance phase. The toes are in extreme hyperextension at the metatarsophalangeal MTP joints.
- The *ankle* moves into about 10° of plantarflexion and the *knee* and *hip* are flexing. The thigh is perpendicular to the ground.
- There is lateral pelvic tilt on the same side, and the same sided arm begins to move forward.

In RLA terminology, 'pre-swing' is the period just before and including when the toes leave the ground, indicating the end of stance phase and the beginning of swing phase.

Description of Components of Swing Phase

Acceleration

This is the first part of the swing phase, and after the toe-off phase when the leg was behind the body, it is now moving to catch up.

The *ankle* is dorsiflexing, and the *knee* and *hip* continue to flex, to aid forward movement of the leg. The *pelvis* is beginning to rotate forward. The *ipsilateral (same sided) arm* is beginning to swing backward.

In RLA terminology, 'initial swing' is that period between the end of toe-off and the end of acceleration.

Midswing

At this phase, the ankle-dorsiflexors have brought the *ankle* to a neutral position. The *knee* is at its maximum flexion (about 65°) which later begins to extend, and the *hip* is also in about 25° flexion. These movements act to shorten the leg, allowing the leg to pass under the body, and for foot to clear the ground as it swings through. Further hip flexion moves the leg in front of the body and places the lower leg in a vertical position. The *pelvis* is in neutral position, and the *arms* are parallel with the body and moving in the opposite directions.

In RLA terminology, 'mid-swing' is that period between the end of acceleration and the end of midswing.

Deceleration

This is the last of the swing phase in which the ankle dorsiflexors are active to keep the ankle in a neutral position in preparation for heel-strike.
- The *knee* is extending, and the hamstring muscles are contracting eccentrically to slow down the leg, keeping it from snapping into extension.
- The *leg* has swung as far forward as it is going to the *hip* remains in flexion.
- The *pelvis* on the same side is rotated forward. The *same sided arm* is backward, and the opposite arm is forward.

In RLA terminology, 'terminal swing' is that period between the end of midswing and the end of deceleration.

Rancho Los Amigos (RLA) Terminology of Gait

Other than the traditional method, many sets of terms have also been developed to describe the components of walking or gait. One such terminology developed by the gait laboratory at *Rancho Los Amigos (RLA) Medical Center* has been gaining the acceptance. Perhaps the biggest difference is that the traditional terms refer to *'points in time'* whereas Rancho Los Amigos (RLA) terms refer to *'periods of time'.* Traditional terms reflect accurately key points within the gait cycle, whereas RLA periods accurately reflect the moving or dynamic nature of gait.

According to RLA, the following terms are being used to describe the components of gait:

Initial Contact: Initial contact represents that the foot of the leading lower limb touches or contacts the ground first. The other similar term "foot strike" alternately describes the same motor function during this phase of walking.

Loading Response: In this phase, just after the initial contact the foot comes in full contact with the ground and the body weight is fully transferred onto the stance limb. This phase ends when the opposite foot leaves the ground.

Mid Stance: This phase begins when the opposite foot leaves the ground and continues as the body is directly over the weight-bearing limb.

Terminal Stance: This phase begins with heel rise of the weight bearing leg and ends when the opposite foot makes the initial contact with the ground. The body has moved in front of the weight bearing leg.

Preswing: This phase begins when the opposite foot after the initial contact has body weight shifted onto it, and ends with the toe-off of weight bearing leg (just before the toes of weight-bearing limb leave the ground).

Initial Swing: This phase begins as the toes of the weight bearing leg leave the ground and continues until its maximum knee flexion occurs. It ends when the swinging leg is directly under the body, and opposite to the stance or weight bearing leg.

Midswing: This phase begins following maximum knee flexion and the swing foot is opposite the weight-bearing foot. It ends when the swing leg has moved in front of the body and the tibia is in a vertical position.

Terminal Swing: In this phase, the tibia moves from the vertical position, and the knee is fully extended in preparation to, and just prior to heelstrike (or initial contact).

Functional Tasks of Gait

Perry (1992) has identified three functional tasks that need to be accomplished during the various phases of gait-cycle. These tasks are as follows:
1. **Weight acceptance:** This is the first functional task. Weight acceptance occurs at the very beginning of the stance phase when the foot touches the ground and the body weight begins to be shifted onto that leg. *Two phases of the stance phase of RLA, i.e. initial contact*

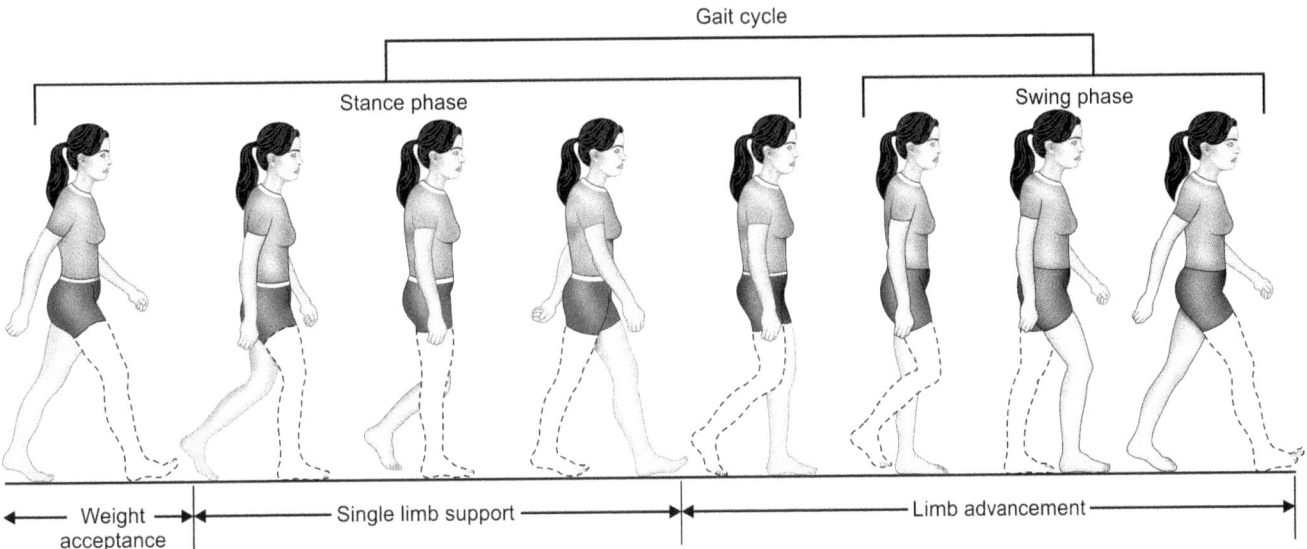

Fig. 14.3: Phases of one complete gait cycle with corresponding functional terms

and loading response are involved in the performance of weight acceptance phase.

2. **Single limb support:** This phase occurs next as the body weight shifts completely onto the stance leg so that the opposite leg can swing forward. *The two phases of RLA, i.e. midstance and terminal stance* are associated with this functional phase.
3. **Limb advancement:** During this phase, the stance limb leaves the ground and advances forward in preparation for the next initial contact or heelstrike. *The four phases of RLA, i.e. preswing, initial swing, midswing and terminal-swing all contribute to limb advancement phase.* The phases of one complete gait cycle with corresponding functional terms are shown in **Figure 14.3**.

The summary of corresponding terms of Traditional, Rancho Los Amigos, and Functional gait terminology is given in **Table 14.1**.

Determinants of Gait

Walking has been defined as the translation of body's center of gravity (COG) through space in a manner requiring the least energy expenditure. In normal walking, the COG describes a *sinusoidal curve* in both the vertical and the horizontal planes; which is drawn by marking the COG level in each phase and the line connecting all the points gives a wavy curve called *sinusoidal curve*.

The vertical displacement of body's centre of gravity during gait cycle is shown in **Figure 14.4**.

Table 14.1: Summary of corresponding terms of Traditional, Rancho Los Amigos, and Functional gait terminology

Traditional	Rancho Los Amigos (RLA)	Functional
Stance Phase:		
Heel strike	Initial contact	Weight acceptance
Footflat	Loading response	
Midstance	Midstance	Single leg support
Heel-off	Terminal stance	
Swing Phase		
Toe-off	Preswing	Limb advancement
Acceleration	Initials swing	
Mid-swing	Midswing	
Deceleration	Terminal swing	

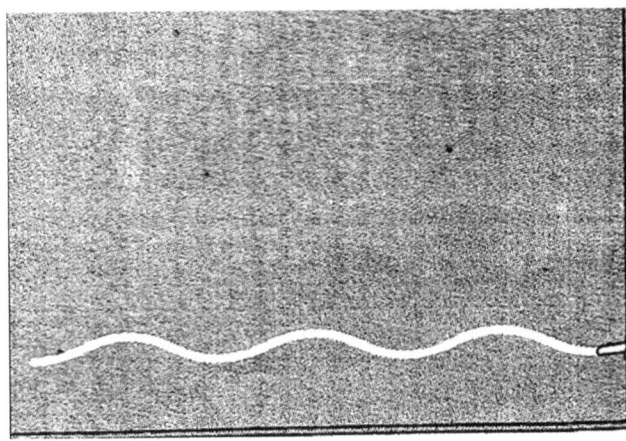

Fig. 14.4: Vertical displacement of body's centre of gravity during gait cycle

The amount of vertical displacement of COG is about 4.5 cm, and the horizontal displacement is also nearly the same. Since these two displacements are about equal, the movement of the COG in the body forms a *figure eight* occupying approximately 5 cm^2.

The longer the stride, the greater the movement of the COG with each step. The following *two principles* help us understand the energy efficiency of various determinants:
1. Any displacement that elevates, depresses or moves the COG beyond normal maximum range wastes energy, and
2. Any abrupt or irregular movement will waste energy even when that movement does not exceed the normal maximal displacement limits of the COG.

By translating the COG through a smooth sinusoidal pattern of low amplitude, the body succeeds in conserving energy. The following variables or determinants have been identified that affect energy expenditure. Variations in all these determinants, i.e. pelvic rotation, pelvic tilt, knee flexion, foot-knee motions and lateral displacement of pelvis all affect energy expenditure and the mechanical efficiency of walking.

Pelvic Rotation

In walking, the pelvis alternately rotates right and left in the horizontal plane. The forward rotation of pelvis starts during *acceleration* of swing phase and ends in its deceleration. During the midswing, the pelvis comes to the neutral position, meanwhile the opposite pelvis goes for backward rotation, and the total range of pelvic rotation is about 8°. Since after the midstance, there will be sudden dropping of body's COG level, however, the forward and backward rotations of the pelvis that use the pelvic width to extend both support points help to prevent further drop of the COG level.

During *deceleration*, forward rotation of the pelvis enables lengthening of the swinging leg. The same time, the opposite leg is also relatively lengthened (being in the midstance). As such, the relative lengthening of both the legs the sametime prevents further drop of the COG level.

Pelvic Tilt

In normal walking, the pelvis tilts downward about 5° on the side of the non-weight bearing limb at the toe-off (preswing) phase. As such, this lowers the COG level and conserves the energy.

During the *midstance*, the COG reaches the highest level, and the total body is supported on one lower extremity. To lower the COG level, the opposite sided pelvis (swinging leg) tilts laterally downward; meanwhile

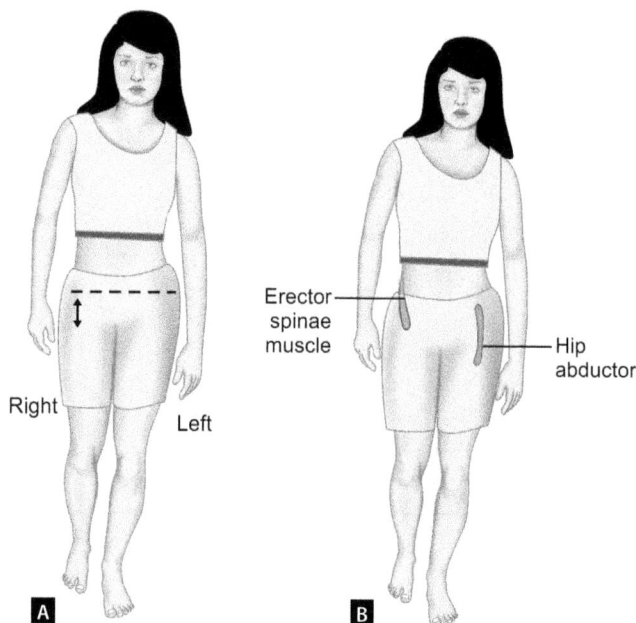

Figs. 14.5A and B: (A) Lateral pelvic tilt; (B) Showing muscles working to minimize lateral pelvic tilt

the hip abductors of the stance leg work along with the erector spinae of the swinging leg to prevent the pelvic drop of the swinging leg. **Figures 14.5A and B** show the lateral pelvic tilt, and the muscles working to minimize lateral pelvic tilt respectively.

Knee Flexion

The stance limb enters initial contact (heel strike) with knee nearly fully extended. It then flexes as the foot shifts to foot-flat position, and as the body weight passes over the supporting leg at a time when the knee is being flexed, the magnitude of kneeflexion is about 15° at about midstance which lowers the COG level and saves the energy.

It is estimated that the above mentioned three determinate, i.e. pelvic rotation, pelvic tilt and knee flexion at midstand result in a combined saving of about 2 cm vertical displacement of body's COG.

Foot and Knee Mechanisms

The foot-knee interaction prevents the abrupt hike of upward displacement of the COG when the foot passes from the 'heel strike' to 'toe-off'.

At the heelstrike, the ankle is dorsiflexed and the knee joint is fully extended so that the COG is maximally lowered. As the weight passes onto the forefoot and the heel is raised, the knee flexes so that the abrupt increase of the COG level is prevented, and the COG travels smoothly into sinusoidal waves.

Lateral or Horizontal Displacement of the Pelvis

When one walks, he does not place his feet one step in front of the other foot, but slightly apart. In other words, there is also a lateral or horizontal displacement of the COG as the body weight shifts from side to side from one lower limb to the another. This displacement is greatest during the single leg support phase at *midstance*. In other words, this represents the distance the body must shift horizontally onto one foot so that the other foot can swing forward. This side to side displacement of COG is usually about 5 cm. The displacement of walking base width for corresponding horizontal or lateral pelvic shift/body's center of gravity is shown in **Figure 14.6.**

Horizontal deviations of the COG also aid in replacing the sharp changes into smooth sinusoidal curves in the horizontal plane.

Generally, during walking, forward placing leg will have mild *knee-valgus* (called as *physiological - valgus*), but the vertical alignment of the limb (tibia and fibula) provides more base of support than the normally placed limb. Therefore, to overcome the reduced base of support by the physiological valgus (normally placed limb), the lateral shifting of the body occurs to shift the COG from one lower extremity to another.

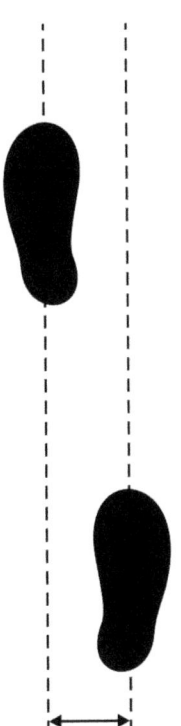

Fig. 14.6: Showing displacement of walking base width for corresponding horizontal or lateral pelvic shift/body's center of gravity

The gait-cycle has two periods of double support and two periods of single support.

Double Support

When both feet are in contact with the ground at the same time, a period of "double-support" occurs. This occurs as one leg is beginning its stance phase, and the other leg is ending its stance phase. For example, the first period of double support occurs as the right leg is beginning stance phase, and the left leg is ending stance phase. Similarly, the second period of double support occurs as the right leg is ending its stance phase, and the left is beginning its stance phase.

Each period of double support takes up about 10% of the gait cycle at an average walking speed, with increased walking speed, one spends less time in double support.

Single Support

The period of *'single support'* occurs when only one foot is in contact with the ground. Thus, two periods of single support occur in a gait cycle; once when the right foot is on the ground as the left foot is swinging forward, then again when the left foot is on the ground in the stance phase, and the right leg swings forward.

Each single support period takes up about 40% of the gait cycle.

Non-support

A period of "non-support" is a time during which neither foot is in contact with the ground, and this does not occur during walking. The period of non-support occurs during running, and along with the speed, this constitutes the biggest difference between running and walking.

The activities like hopping, jumping, skipping also have a period of non-support but there is no progression in terms of distance that walking and running have.

Description of Arm Swing

The arms swing alternately similar to the feet. However, the arms swing in opposition to the feet, i.e. the left arm swings forward with the right leg, and the right arm with the left leg. *The action of the arm swing serves to reduce the rotation of the shoulders, which indirectly aids in keeping the head facing forward.*

- The arm swing varies considerably with variations in the speed of the walk. When one walks fairly slow and an easy walk, the arm swings forward from the shoulder about 20°, with the arm hanging more or less relaxed. The swinging of the arm is almost in a sagittal plane, but it may move slightly medial ward.

- There is sometimes a little elbow flexion at the end of the swing.
- The arm then drops back and onto a position slightly posterior to its resting position. The backward movement of the arm is about 9°, usually in almost a true sagittal plane.
- With increasing speed of walking, the frequency of the arm swing also increases, and the action becomes much more vigorous. In the short and broad shouldered individuals, the arm swing tends to be across the body to a slight degree, which helps to compensate for the sideways movement of the body caused by the swinging legs.
- The movements of the shoulders are similar to those of the pelvis though somewhat little less clear.

Neuromuscular Considerations

Walking is mostly a reflex action, requiring very little conscious control. However, if too much attention is focused on any part of the gait, tension is likely to develop and the natural rhythm and co-ordination are disturbed.

- Reflexes (particularly the *stretch reflex* and the *extensor thrust reflex*) control not only the movement of the limbs but also the extension of both the supporting limb and the trunk in resisting the downward pull of the gravity. This extension serves to give stability to the body in the stance phase of locomotion, which provides for effective muscular action in producing the necessary movements.
- The *stretch reflex* may be involved at the extremes in motion, particularly at the shoulder and hip joints, assisting with the change in the direction of joint motion. The *extensor thrust reflex* may facilitate the extensor muscles of the lower extremity as the body weight rides over the foot on the support leg.

Thus, like all other motions of the body, for smooth and co-ordinated movements, walking also requires properly functioning reflexes, normal flexibility of the joints, and optimal stability of the body as a whole in the stance or weight bearing phase of walking.

Muscular Analysis of Walking

To a layman and the casual observer, the movements involved in walking appear to be relatively simple, though muscular and kinesiological analysis shows them to be extremely complex. The co-operative action of various muscles and the synchronization of joint movements beautifully show the team work present in all bodily movements. Even the most complex machinery designed by the most skilled engineers does not exceed the movements of the human machine in the perfection, nor in the smoothness of function on a wide variety of surfaces and in all conditions.

Muscular Analysis of Swing Phase

Forward swing of the leg which starts the swing phase is initiated mainly by the hip flexors, i.e. iliopsoas, assisted by the rectus femoris, sartorius, gracilis and adductor longus muscles. However, there is no appreciable action of hip flexors in normal walking on level ground in the latter part of swing phase.

- There is slight action of hip adductors, i.e. adductor longus, pectineus and gracilis muscles, probably nearing midswing phase in preventing abduction of the leg which might result from the forward swing of the pelvis. These muscles keep the leg swing in a straight line.
- Sartorius is more important in hip and knee flexion.
- Easy and quick flexion of thigh is possible only if the pelvis is fixed by the action of abdominals.
- Knee flexion in the beginning is largely passive due to the tension of the hamstrings. This tension becomes more increased with hip flexion, aided by the sartorius. Sartorius and short head of biceps femoris are mainly active following toe-off.
- The knee extension which comes in the later part of the swing is partly passive, initiated due to the gravitational force and continued by the momentum of the leg and also due to the relaxation of the hamstrings. However, knee extensor i.e. quadriceps femoris muscles do contract slightly at the end of this phase.
- At the end of the swing, the hamstrings act to halt the forward movement of the leg.
- The gluteus maximus and medius muscles contract slightly at the very end of swing phase.
- Ankle dorsiflexors (i.e. tibialis anterior, extensor digitorum longus, extensor hallucis longus, and probably the peroneus tertius) contract with slight to moderate intensity at the beginning of the swing phase and taper off during the middle part of the swing phase. Since there is also some eversion of the foot during early swing and midswing phases, these muscles as dorsiflexors and invertors (including tibialis posterior) provide adequate clearance between the foot and the ground.

Muscular Analysis of Stance Phase

During heelstrike, the guteal muscles and the hamstrings contract statically with moderate intensity. These contractions taper off gradually and disappear at midstance.

- Shortly after the heel strikes the ground, the gluteus maximus begins to contract to prevent the pelvis from

dropping forward and keeping the hip from flexing more.
- Hip extension is mainly taken care by the hamstrings and when the stride is long, the gluteus maximus contracts more strongly than when the stride is short.
- The quadriceps femoris (knee extensors) initially contract concentrically and moderately at the time of heel strike, and later switch to eccentric contraction to minimize the amount of knee flexion.
- The hip muscles, particularly the adductor magnus, longus and possibly brevis are the only hip muscles that contract appreciably during the last part of the stance phase.
- During the stance phase, tensor-fascia-latae, gluteus-mediaus and minimus are also acting to assist the inward rotation of the hip which must take place with the reversal of stride. However, the greater role of these muscles is in fixation of the pelvis to prevent the unsupported (opposite) side from dropping.
- Initially at the time of heelstrike, the ankle dorsiflexors (tibialis anterior, extensor digitorum longus, and extensor hallucis longus) contract concentrically, and later during foot-flat and midstance, they contract eccentrically for controlled lowering of the foot to the ground. Tibialis posterior is also very active during midstance to prevent eversion of foot.
- Ankle plantarflexion during 'heel-off' and 'take-off' phases comes from the action of gastrocnemius, soleus, peroneus longus, and tibialis posterior muscles. These muscles provide force to elevate the heel and weight shift onto the forefoot.
- The peroneus longus and brevis are largely responsible for the slapping action of the foot shown by the shifting of weight from lateral (outside) to the medial side (inside) of the foot.
- The final 'take off' results from the flexion of the toes which is produced by the action of both long and short toe-flexors. However, the extensor-digitorum also contracts to oppose the tendency for flexion of the toes.
- The trunk extensors (erector spinae) are also active in keeping the trunk from flexing, and also for stabilizing the pelvis to give firmer origin to the leg extensors.

Muscular Analysis of Arm Swing

In the easy walking, the arm swing is more of a reflex action and a free pendular type, with very little muscle action required for it.
- Generally, the arms tend to swing in opposition to the legs, the left arm swinging forward as the right leg swings forward and vice-versa.
- The arm swing mainly serves to balance the rotation of the pelvis. When the arm swing is prevented, the upper trunk tends to rotate in the same direction as the pelvis, causing a tense and awkward gait.
- Although, the arm swing appear to be performed without muscular effort at a normal pace on level ground, however, actually its pendular action is caused by a combination of muscular activity and gravity.
- However, the main force for the forward swing of the arm is produced by the posterior deltoid muscle which contracted eccentrically toward the end of the forward swing, probably serving as a brake to check the movement (and not by the anterior deltoid and pectoralis major muscles as earlier thought of).
- The backward swing is mainly produced by the concentric action of posterior deltoid and the teres major muscles.
- Middle deltoid, though appears primarily as an arm abductor, assists flexion and extension movements of the arm at the shoulder joint, probably to keep the arms from rubbing the sides of the trunk as they swing past it.
- Researchers have found out that the increased amplitude of arm swing accompanying the faster speeds is mainly due to the increased shoulder hyperextension in the backward swing, and increased elbow flexion in the forward swing.
- Maximum shoulder and elbow flexion occur at the moment of heel strike of the opposite foot, and their maximum extension at the moment of heel strike of the foot on the same side.
- The oscillations of the shoulders or upper trunk are more passive and dependent upon arm swing. In faster walking, the arm actually assists in oscillation of the hip by reaction on the trunk.

Anatomical Principles in Walking

There are certain anatomical factors that affect the efficiency of the walking gait. A few such important factors to be considered in any analysis of a walking pattern are the following:
- Alignment:
  - Good alignment of the lower extremities reduces friction in the joints and decreases the possibility of strain and injury.
  - The stability of the supporting limb and the balance of the trunk over this lower limb are important factors for the smoothness of the gait, and these are improved with good alignment of the weight bearing segments.

- Unnecessary lateral movements decrease gait economy. These lateral movements are usually due to excessive motion in the shoulders or pelvis:
  - Excessive trunk rotation may be caused either by exaggerated or limited arm swing. Normally the arm swing exactly counterbalances the hip swing.
  - The pelvis may drop on one side if the gluteus medius muscle is not contracted for support, and this results in an increased hip sway.
  - Pelvic rotation should be just enough to enable the leg to move straight forward. A straight forward movement of the leg is ensured by keeping the knee and foot pointing straight forward during all phases of the gait. Deviation from this will produce an inefficient gait.
- Normal flexibility of the joints (i.e. sufficiently long and flexible muscles, ligaments and fasciae) reduces internal resistance and therefore reduces the amount of force required for walking. In particular, the tendons of two-joint muscles of the lower limb contribute to the economy of muscle work during walking.
- Properly functioning reflexes contribute to a well-coordinated and efficient gait. An injury, disease or excessive alcohol or drug-abuse can interfere with the walking reflexes.

Mechanical Analysis

Walking is characterized by the translation of body's COG forward as a result of the alternating and rhythmic pattern of the lower extremity joint movements during stance and swing phases. The forces controlling walking include the external forces of weight, normal reaction, friction, air resistance, and internal muscular forces. The direction and interaction of these external and internal forces determine the nature of the gait.

- *Gravity* and *momentum* are the chief sources of motion for the swing phase; hence this phase represents a ballistic type of movement, particularly when the individual is walking at a natural pace.
- The *source of motion* for the first half of support or stance phase is the momentum of the forward moving trunk (provided by the propulsive action of the opposite leg), and the source of motion for second half of stance phase is the contraction of the extensor muscles of the supporting leg.
- The *inertia* of the stationary body is overcome by the horizontal component of the propulsive force. Since periodic movement is characterized by an alternating increase and decrease of speed, the inertia of the body must be overcome at every step. The greater the weight of the body, the greater is the inertia to be overcome.
- As the *body is inclined forward*, there is also a forward movement of the trunk and of the COG of the body. The very purpose of the forward inclination of the trunk is to place its COG more nearly in a direct line with the force exerted by the stance or moving leg.

  As the body is inclined forward and its COG also moves forward, the COG momentarily passes beyond the anterior margin of the base of support and a temporary loss of balance results. At this point, the downward pull of gravity threatens a complete loss of equilibrium and tends to cause the body to fall forward and downward. However, at this point, a timely recovery of balance occurs as the other leg is swung forward and placed on the ground. This provides a much wider base for support and saves the body from the fall.
- The *downward force of the body's weight* is counteracted by the vertical reactive force through the feet. However, if the vertical force exceeds that needed to balance the gravitational force at the time of push off, an exaggerated lift to the body results, causing a gait characterized by a bounce.
- When *forward motion* has been imparted to the trunk by means of the ground reaction to the backward thrust of the leg and foot, it tends to continue unless restrained by another force. As the body's COG passes beyond its base of support, it is essential to restrain the action of the trunk until a new base of support is established. Therefore, as the foot is brought to the ground in front of the body at the close of its recovery phase, a restraining phase is constituted, which diminishes as the leg approaches midstance.

  During the period that the foot is in front of the COG, there is a forward component of force in the thrust of the foot against the ground. This results in a backward reactive force of the ground against the foot, which is transmitted to the leg and therefore to the trunk.
- The degree to which the *pressure of the foot* actually imparts motion to the body in the propulsive phase, and restrains the body in the restraining phase is in *direct proportion to the counterpressure of normal reactive force of the supporting surface*. If the surface lacks solidity (as in case of mud, sand and soft snow), it offers too little resistance to give the needed counterpressure, and the pressure of the foot then results in slipping or sinking, and more pressure needs to be applied to achieve even a slow forward movement. Therefore, the efficiency of the gait depends on the right balance between the pressure of the foot and the counterpressure of the supporting surface.
- *Friction* is also an essential factor in the effective application of the forces needed in walking. Because

of the diagonal thrust of the leg at the beginning and end of the stance phase, friction between the foot and the ground is essential so that the counterpressure of the ground may be transmitted to the body. For efficient walking, friction must be sufficient to balance the horizontal component of force. If the friction is insufficient, the thrust of the foot results in a slipping of the foot itself rather than in the desired propulsion of the body. If the horizontal component of force is greater (as when walking with a long stride), there is greater dependence on friction for efficient locomotion.

- The forward moving trunk meets with *air resistance,* which tends to push the trunk backward. Therefore, by inclining the body forward, the pull of gravity is utilized to balance the force of air resistance. In case, the individual fails to incline the trunk forward, along with the forward moving leg, the lower part of the trunk will also be carried forward enough and since the COG of the trunk is above and behind the force exerted by the moving or stancing leg, the inertia of the body will not be overcome directly and the torque exerted about the hips in backward direction will tend to cause a person to fall over backward.

When walking against a *strong wind,* it is necessary to incline the body farther forward to maintain balance. In case the air resistance is not balanced by the force of gravity, it must be balanced by the contraction of the abdominal and other anterior muscles of the neck and trunk. However, if the body is inclined too far forward, the force of gravity acts too strongly on the body and this force of gravity must be counteracted by tension of the posterior muscles. Thus, the proper amount of forward inclination is a factor in muscular economy.

Mechanical Principles in Walking

A number of mechanical principles affect the walking gait. A mechanical analysis of walking should include all of the following important principles:

- **A body at rest will remain at rest unless acted upon by a force.** The force for walking is ground reaction force produced by extension of the lower extremity against a resistive force. Since walking is an alternating pendular motion of the lower extremities, the inertia of the body must be overcome at every step.
- **A body in motion will remain in motion until acted upon by a force.** Once motion is imparted to the trunk by the backward and downward thrust of the legs, the trunk has a tendency to continue in motion, even beyond the base of support, and a brief restraining action of the forward limb serves as a brake on the momentum of the trunk.
- **Translatory movement of a lever is achieved by the repeated alternation of two rotary movements,** the lever first turning about one end and then the other. In walking, the lower extremity alternates between rotating about the foot's point of contact (support phase) and the hip joint (swing phase).
- **Force applied diagonally has both horizontal and vertical components.** The vertical component of the ground reaction force used in walking serves to counteract the pull of gravity; whereas the horizontal component of this force serves to, (i) check forward motion of the trunk during heel strike and (ii) produce forward motion in the trunk during heel-off and toe-off.
- **The speed of walking is increased by increasing either the stride length or the stride rate or both.**
- **The speed of the gait is directly related to the magnitude of the pushing force and to the direction of its application.** This force is provided by the extensor muscles of the lower extremity, and the direction of the force application is determined by the angle of the lower extremity at the time of application of the force.
- **The efficiency of walking partially depends on the friction and ground reaction force provided by the supporting surface,** since the propulsion of the body is brought about by the backward and downward push of the foot against the supporting surface.
- **The efficiency of the gait is related to its timing with reference to the length of the limbs.** The most efficient gait is one that is so timed as to permit pendular motion of the lower extremities.
- **Walking is described as an alternating loss and recovery of balance** and therefore, a new base of support must be established at every step.
- **Stability of the body is directly related to the size of the base of support.** In walking, the lateral distance between the feet is a minor factor in maintaining balance, as momentum carries the body forward.

Individual Variations in Gait

Each person has a unique style of walking. For some people, their walking pattern or style is so unique that they can be identified from a distance well before their face can be clearly seen. However, regardless of the many different styles of walking, the components of normal walking are the same.

- Individual differences in walking appear as soon as a baby takes its first steps. The variations in gait or walking may be either *structural or functional* in origin. The structural differences include unusual body proportions, differences in the limbs themselves (such as knock-knees and bow legs) and restriction of movements in the joints. Extreme variations in the angle between the neck and shaft of the femur, and in the obliquity of the femoral shaft are also responsible for variations in the gait.

- *Walking style* may even change slightly with the *mood* of an individual. When one is happy, his steps are lighter; and when sad or depressed, his steps may be heavy. Walking is an individual matter, with each person tending to assume a type and speed which are the most efficient for his particular structure. According to certain kinesiologists, the ideal gait changes with body weight and length of the leg, and since these vary in the individuals, their gait patterns are also bound to vary.
- *Age* is also a factor in gait variations. As the human body ages, there is some decrease in both strength and flexibility as tissues change. Balance also becomes a significant concern. As a result of these changes, the gait of an elderly person is often modified from that of a healthy young adult. Though the elderly person will have the same cadence as that of a young person, however, his step length will be shorter, and he will spend more time in double support phase. These changes in gait pattern are thought to produce greater stability in the gait of the elderly person.

Energy Cost (Expenditure) in Walking

There are large individual differences in energy cost of walking as many factors such as foot length, muscular training and efficiency, weight of clothing and shoes, differences in posture, frequency of stride and rhythm of movement, etc. all affect the energy expenditure. However, according to the experts, *the energy expenditure of walking in calories per minute per kilogram of body weight may be accurately calculated by the equation:*

$$\Sigma_w(cal/min/kg) = 29 + 0.0053\, v^2$$

The energy expenditure in calories per meter walked per kilogram of body weight may be calculated by another form of the same equation:

$$\Sigma_w(cal/meter/kg) = 29/v + 0.0053\, v$$

In both of the equations:
Σ_w = *Energy expenditure in calories*
v = *Speed in meters per minute*

- There are no significant differences between men and women in the energy cost of walking and running when the expenditure is expressed per unit of body weight.
- As the age advances, particularly at about 65 years of age or so, walking appears to become more restrained, probably in an effort to obtain maximum stability. Walking speed becomes slower, stride length shortens, the time of the support phase (stance) is increased, and that of the swing phase reduced, and the amplitude of the oscillations of nearly all parts of the body becomes less.
- According to an estimation, for a person whose daily work requires him to be on his feet, he takes *about 19,000 steps daily*.
- *If he weighs 150 pounds, he will beat into his shoes about 29 lacs pounds per day.*
- *In walking at the rate of 120 three-foot steps per minute, he advances at the rate of 4.1 miles per hour.* His each foot is at rest on the ground for one half second, then moves forward 6 feet, and comes to rest again in the next one-half second. The moving foot passes the stationary foot at a maximum speed of about 12.8 miles per hour.

Definitions

Certain *terms* related to walking or gait are described here under and shown in **Figure 14.7**:

- **Stride:** A stride or gaitcycle is one full lower extremity cycle. A stride is defined as the activity occurring from heel-strike of one leg to the next heel strike of the same leg.

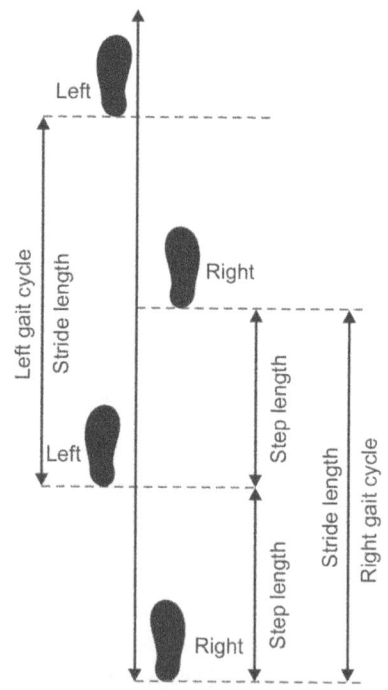

Fig. 14.7: Terms of gait cycle. One right and one left step make up one complete gait cycle (also called stride)

- **Stridelength:** A stride length is the distance traveled during a single stride (gait-cycle).
- **Step:** A step is basically one-half of a stride. Two steps (one of right leg and one of left leg) make a complete stride or gaitcycle. A step starts with the heelstrike of one leg and ends with the heelstrike of the opposite or contralateral leg.
- **Step-length:** A step-length is the distance between heelstrike of one leg and heelstrike of other (opposite) foot. With increased or decreased speed of walking, the step length will increase or decrease respectively. However, step length of each leg should remain equal, irrespective of walking speed.
- **Cadence:** Cadence or walking speed, is the number of steps taken per minute. It can vary greatly, i.e. slow walking may be as slow as 70 steps per minute, and fast walking may have upto 130 steps per minute. However, regardless of walking speed, all the parts of gaitcycle occur, at the proper time.

RUNNING

Running is an important skill that is used in everyday life as well as various sports activities. Running is a modification of walking. It differs from walking with reference to foot contact mainly in two aspects:
1. In running, there is a phase when none of the feet is in contact with the ground meaning that both the feet are in air and this is referred to as a period of *on-support*.
2. Both the feet are never in contact with the ground at the sametime, meaning that there is *no period of double support* in running.

As a result of the above, running is sometimes described as a series of jumps in which the body is alternately supported first on one foot and then on the other.

Description of Running

Like walking, running is also described in two phases (i) *Swing* and (ii) *support:*

Swing Phase

The swing phase begins with the toe-off and ends with the foot landing. The swing phase is more muscular (than pendular), and is longer than the support phase. The components of running during swing phase of right leg are shown in **Figure 14.8**.
- In fast running, the initial foot contact may be the ball of the foot, whereas in slow running, the whole foot.
- The flexed leg in the swing phase brings the mass of the leg close to the hip, reducing the moment of inertia and increasing the angular velocity of the forward swinging thigh, which in turn drives the COG of the body forward.

Support Phase

The support phase begins with the contact of the forward foot and ends at toe-off when the body is driven into the air. During this time, the knee and ankle are somewhat flexed and then extend as the body passes over the foot, and is driven into the air. The support phase decreases as the speed of running increases. The components of running during support and drive phase of right leg are shown in **Figure 14.9**.

Fig. 14.8: Components of running during swing phase of right leg

Fig. 14.9: Components of running during support and drive phase of right leg

Muscular Analysis of Running

From the anatomic view-point, the major differences in walking and running are—(i) greater range of motion of joint actions in running than walking, particularly in the actions of swinging leg and (ii) the difference in co-ordination, particularly evident in the period of non-support, and the absence of the period of double-support.

Swing Phase

During this phase, the *iliopsoas* is the main flexor of the *hip*. During the first half of the phase, there is decreased activity in *rectus femoris*, and all three muscles of hamstrings. However, in the second half of the phase, the *medial hamstrings* (i.e. *semimembranous* and *semitendinosus*) sharply increase in their activity, and slow down the flexion and then change in direction to extension.

- At the *knee*, the sharp flexion initially occurs probably due to the reflex action, and transfer of momentum from the forward moving thigh. There seems to be little muscle-work of both *medial hamstrings* in knee flexion. However, later the eccentric work of both medial hamstrings (i.e. *semimembranosus* and *semitendinosus*) appear to be controlling knee extension.
- There is initially some work of *ankle dorsiflexors*, i.e. *tibialis anterior, extensor digitorum longus* and *extensor hallucis longus* in dorsiflexing the ankle, and later in the last third of swing phase, these muscles contract eccentrically with increased activity in preparation for the foot contact and in *prevention* of plantarflexion.
- The same *muscles of spine and pelvis* that work in walking also work in the swing phase of *running* too, though somewhat in greater range.

Support Phase

Gluteus maximus works at heel strike. The *hamstrings* and *quadriceps* also contract simultaneously at the time of heel strike to stabilize the hip and pelvis.

- The *three muscles of quadriceps* namely *rectus femoris, vastus medialis* and *vastus lateralis* are active eccentrically during *knee flexion,* and reaching maximum levels of concentric contraction in knee-extension just before heel-off.
- The *plantar flexors* i.e. the *soleus* and *gastrocnemius* are initially active eccentrically in controlling the *dorsiflexion,* and later show their strongest concentric activity *after heel-off as the knee is flexing.*
- The same *muscles of spine* and *pelvis* also work in this phase of running which work in walking, though with more vigor in their reaction to leg movements.

Mechanical Analysis of Running

The *speed of running* is controlled by the *length of the stride* and the *frequency of the strides.* Better runners have a greater stride length than the poor runners.

The *stride length* is determined by the length of the leg, the range of motion in the hip, and the power of the leg extensors, which drive entire body forward.

- In order to *achieve the increased speed* (which is characteristic of running), more force must be exerted in the horizontal direction. The vertical force is also increased (but not as much as the horizontal force). And for this, the angle which the leg makes with the ground during extension should be smaller and pelvis must be necessarily kept (carried) lower.
- To keep the *lower position of the pelvis,* it is required to have greater flexion of the knee of the supporting

- leg when the COG of the trunk passes over the point of support.
- The increased flexion of the knee makes possible a more powerful extension of the driving leg, which also increases the length of the stride. The increased kneeflexion is also necessary to provide for clearance of the swinging leg due to the lower position of the pelvis.
- The *distance* the runner's body will move after it is driven into the air depends on the *angle of take-off* (the distance that COG is ahead of take-off foot), the *speed of body's projection,* and the *height of the COG* at take-off and landing.
- In running, the *swinging leg* makes first contact with the ground by the ball of the foot rather than the heel. It cushions (absorbs) the shock of impact, and also because body's COG is more nearly over the ball of the foot at the time of contact (rather than it would be when first heel touches the ground).
- In running, *more speed* (than walking) alongwith period of non-support would result in greater impact when the foot of swinging leg strikes the ground with the heel. In running, the force of push-off is greater and the body is actually suspended momentarily in the air. *Therefore, the force of impact at the time of contact with the ground is greater in running than in walking.* Therefore, to minimize the shock, the swinging foot does not strike the ground with the heel as in walking.
- *The bending of ankle, knee and hip while landing on the ball of the foot further adds to shock absorption.*
- The *stride-frequency* of the run is affected by the speed of muscle contraction and the skill (technique of the runner).
- In running, like walking, the *forces* exerted to produce and control the movement are the internal muscular forces, and the external forces of gravity, normal reaction, friction and air resistance.
- Since the COG of the body passes over the supporting leg, heel of the foot may or may not make contact with the ground. In slow running, the heel may touch the ground where as in fast running or spring, the heel may not touch the ground at any time.
- When the supporting leg reaches complete extension and the foot is lifted from the ground, the knee is flexed more (as compared to walking), and the foot is brought up quite high in the rear (back side). This serves to shorten the swinging leg, and reduces the amount of force needed to bring the leg forward. As a result, the runner while exerting his maximum effort, can swing the leg forward much faster and greatly increase his speed.
- The *inertia* of the body can be overcome by a crouching or a standing start. By a crouching start, the maximum horizontal force can be exerted at take-off. While starting a run from a standing position, the feet should be in a forward-backward stride with the toes straight ahead so that the force can be exerted straight backward and downward. The knees should be bent so that the legs can exert more force and the weight is on the forward foot.
- The *angle of body* is more forward in running (than walking) for reducing the air-pressure and increasing the forward component of the propulsive force. The forward lean should be from the ankles so as to keep the trunk in line with the driving leg. If one runs faster, then there should be more forward angle of the body.
- When *running arc/circle,* the body should lean in toward the center of the arc to overcome body's inertia for changing the direction.
- The *arm swinging* serves the same purpose in running as walking except that it must be done with far greater speed in order to co-ordinate with the faster moving legs. The arm swinging is essential to counterbalance the rotatory effect of leg swinging on the trunk.
- The *greater speed of arm action* in running is made possible by increased muscular effort, and by increasing the elbow flexion. Flexion of the arm (like flexion of knee in swing phase), shortens the lever and permits greater speed. The tendency to draw the arm across the body is increased with the speed.
- There is no optimal *speed* in running. Running itself has crossed the most economical rate of horizontal locomotion. The energy expenditure increases with the speed increase as the energy needed to run is proportional to the square of the speed. Whether the run is an easy jog or a full speed sprint, economy of effort should be highly desirable objective and that should be achieved by observing the mechanical principles applicable to efficient running.
- While *stopping the run,* the forward momentum must be absorbed gradually and body's stability should be established again. This can be done by flexing the forward leg so as to allow absorption of force and lowering of COG. The subsequent extension of forward knee moves the COG back over the center of the base.

Mechanical Principles in Running

The following mechanical principles must be observed for efficient and economic running:
- As according to the **'Law of Inertia',** a body remains at rest unless acted on by a force. Therefore, the force

required to overcome inertia should be greatest at take-off and least after acceleration has ceased. The problem of inertia decreases as the speed increases.

- In accordance with the **'Law of Acceleration'**, acceleration in the run is directly proportional to the force producing it. Therefore, the greater the power of the leg drive, the greater the acceleration of the runner.
- As according to the **'Law of Reaction'**, every action has an equal and opposite reaction. The force for the run is provided through the upward and forward ground reaction-force in response to the downward, backward drive of the foot. The smaller the vertical component of this force, the greater the horizontal or driving component. In the most efficient running, vertical movements of the COG are reduced to a minimum.

  In an *efficient run*, the foot should strike the ground as close as possible to the line of gravity. If the foot strikes ahead of the line of gravity, the reaction force to this forward and downward thrust will be a backward and upward force, which will act to retard the forward motion.

- The more completely the horizontal force is directed straight backward, the greater its contribution to the forward motion of the body. Lateral or sideward motions are inefficient for forward propulsion.

  Therefore, the knees should be lifted directly upward and forward, with the motion of the entire lower extremity occurring within the sagittal plane. The arm-swing should exactly counterbalance the twist of the pelvis and should not cause additional lateral motion.

- Since a *long lever develops greater speed* at the distal end (than does a short lever), the length of the leg in the driving phase should be as great as possible when speed is a factor. Leg drive should be maximized as early as possible in the stance phase.
- Since **efficiency in running** requires the elimination of all unnecessary forces, the internal resistance forces due to the tightness of the tendons, ligaments, and fascia should be reduced by proper and sufficient warm-up and stretching activities.
- **Resistance force** due to the moment of inertia of the free leg during the swing phase should be minimized. By flexing the free leg at the knee and carrying the heel high up under the hip, the leg should be moved more rapidly as well as more economically. The high knee lift should increase as speed increases.
- The force of **air-resistance** should be altered by shifting the COG. A forward lean of the body will work to counteract a headwind; whereas a tailwind often enhances performance.
- There should be **friction** between the feet and the supporting surface, which is more important in running than in walking, as the propulsive force is shorter in running and is exerted in more horizontal direction.

Energy Cost of Running

The shift from walking to running takes place at a speed of 8.5 km/h or 5.3 miles per hour. At this point, a running gait becomes less tiring or fatiguing than the forced rate of walking, although the energy consumption increases.

The greater part of energy expended in raising the body in running is utilized in propelling the runner forward.

The formula for calculating the number of calories used in running is given:

$C = [(RS \times 0.001) - 0.028] \times BW$

C = Calories per minute,

RS = Running speed in meters per minute

BW = Body weight in kilograms

From this formula it is very clear that the speed of running has very little effect upon caloric consumption as compared to the body weight and duration of running exercise.

Marathon Running and Sprinting

Mechanical features of running differ considerably with change of speed, in case of marathon runners and sprinters.

In a *marathon runner* or distance runner, endurance is a major factor for successful performance, the runner is greatly concerned about economy and must therefore adjust his style and pace for greatest efficiency. *Sprinters*, on the other hand, do not exhaust themselves after 100 meters (they ignore efficiency from the energy expenditure point of view and make great sacrifices in economy to gain a little added speed.

Sprinting

In sprinting, the main object is the maximum horizontal velocity. It is important to do a fast start as a yard or meter lost at the out-set because of slow start can make a difference. For a quick start, a short reaction-time and great driving power are the chief prerequisites. The spiked shoes and starting blocks are the two artificial aids that are employed for advantages.

- The *spiked shoe* provides a firm base for the driving leg so that the magnitude of the horizontal component of force will not suffer.

- In the sprint start, the runner pushes against starting blocks fixed to the track surface. The *starting blocks* provide a surface against which the foot can push horizontally while utilizing maximum hip, knee and ankle extension.
- The *crouched start* having low position of the body makes it possible to greatly increase the horizontal force component of driving force at take-off as the body is more nearly directly in line. This provides maximum acceleration against inertia. The legs can also provide a stronger drive from this position (upon extension) as gluteus maximus is brought into action.

COMMON DEVIATIONS OF NORMAL GAIT

Some of the most common deviations of gait are briefly introduced in Table 14.2.

Table 14.2: Most common gait deviations

Portion of phase	Deviation	Description	Possible causes	Analysis
Stance phase				
• Foot and ankle (as seen from a lateral view)				
– Initial contact	Foot slap	At heel strike, forefoot slaps the ground	Weak dorsiflexors or atrophy of dorsiflexors	Look for low muscle tone at ankle. Look for steppage gait (excessive hip and knee flexion) in swing phase
	Foot flat	Entire foot contacts the ground at heel strike	Excessive fixed dorsiflexion; flaccid or weak dorsiflexors	Check range of motion at ankle. Check for hyperextension at the knee and persistence of immature gait pattern
– Midstance	Excessive positional plantar flexion	Tibia does not advance to neutral from 10° plantar flexion	No eccentric contraction of plantarflexors; could be due to flaccidity/weakness in plantar flexors; surgical overrelease, rupture or contracture of Achilles tendon	Check for hyperreflexive or weak quadriceps, hyperextension at the knee; hip hyperextension; backward or forwad-leaning trunk. Check for weakness in plantar flexors or rupture of Achilles tendon
	Heel lift in midstance	Heel does not contact ground in midstance	Hyper-reflexive plantarflexors	Check reflexs of plantarflexors, quadriceps, hip flexors, and adductors
	Excessive positional dorsiflexion	Tibia advances too rapidly over the foot, creating a greater than normal amount of dorsiflexion	Inability of plantarflexors to control tibial advance; knee flexion or hip flexion contractures	Look at ankle muscles knee and hip flexors, range of motion, and pisition of trunk
– Push-off (heel-off to toe-off)	No roll-off	Insufficient transfer of weight from lateral heel to medial forefoot	Mechanical fixation of ankle and foot; flaccidity or inhibition of plantar flexors inverters and toe flexors; rigidiy/co-contraction of plantar flexors and dorsiflexors pain in forefoot	Check range of motion at ankle and foot. Check muscle function and reflexes at ankle. Look at dissociation between posterior foot and forefoot
• Knee (as seen from a lateral view)				
– Initial contact (heel strike)	Excessive knee flexion	Knee flexes or buckles rather than extends as foot contacts ground	Painful knee; hyper-reflexive knee flexors or weak or flaccid quadriceps; short leg on contralateral side	Check for pain at knee; reflexes of knee flexors; strength of knee extensors; and leg lengths; anterior pelvic tilt
– Foot flat	Knee hyprextension (genu recurvatum)	A greater than normal extension at the knee	Flaccid/weak quadriceps and soleus compensated for by pull of gluteus maximus; spasticity of quadriceps; accomodation to a fixed ankle plantar flexion deformity	Check for strength and reflexes of knee and ankle flexors and range of motion at ankle

Contd...

Contd...

Portion of phase	Deviation	Description	Possible causes	Analysis
– Midstance	Knee hyperextension (genu recurvatum)	During single limb support, tibia remains in back of ankle mortice as body weight moves over foot, ankle is plantar flexed	Same as above	Same as above
– Push off (heel-off to toe-off)	Excessive knee flexion	Knee flexes to more than 40° during push-off	Center of gravity is unusually forward of pelvis; could be due to rigid trunk, knee/hip flexion contractures; flexion-withdrawal reflex; dominance of flexion synergy in middle of recovery from CVA	Look at trunk posture, knee and hip range of motion and flexor synergy
	Limited knee flexion	The normal amoutn of knee flexion (40°) does not occur	Overactive quadriceps and/or plantar flexors	Look at reflexes of hip, knee and ankle muscles
• Hip (as seen from a lateral view)				
– Heel strike to foot flat	Excessive flexion	Flexion exceeding 30°	Hip and/or knee flexion contractures; knee flexion caused by weak soleus and quadricpes; hyper-reflexive of hip flexors	Check hip and knee range of motion and strength of soleus and quadriceps. Check reflexes of hip flexors
– Heel strike to foot flat	Limited hip flexion	Hip flexion does not attain 30°	Weakness of hip flexors; limited range of hip flexion; gluteus maximus weakness	Check strength of hip flexors and extensors. Analyze range of hip motion
– Foot flat to midstance	Limited hip extension	The hip does not attain a neutral position	Hip flexion contracture, hyper-reflexive hip flexors	Check hip range of motion and reflexes of hip muscles
	Internal rotation	An internally rotated position of the extremity	Hyper-reflexive internal rotators; weakness of external rotators; excessive forward rotation of opposite pelvis	Check reflexes of internal rotators and strength of external rotators. Measure range of motion of both hip joints
	External rotation	An externally rotated position of the extremity	Excessive backward rotation of opposite pelvis	Assess range of motion at both hip joints
	Abduction	An abducted position of the extremity	Contracture of the gluteus medius; trunk lateral lean over the ipsilateral hip	Check for abduction pattern
	Adduction	An adducted position of the lower extremity	Hyper-reflexive hip flexors and adductors such as seen in spastic diplegia; pelvic drop to contralateral side	Assess reflexes of hip flexors and adductors. Test muscle strength of hip abductors
• Trunk				
– Stance	Lateral trunk lean	A lean of the trunk over the stance extremty (gluteus medius gait/trendelenberg gait)	A weak or paralyzed gluteus medius on the stance side cannot prevent a drop of pelvis on the swing side, so a trunk lean over the stance leg helps compensate for the weak muscle; a lateral trunk lean also may be used to reuce force on hip if a patient has a painful hip	Check strength of gluteus medius and assess for pain in the hip
	Backward trunk lean	A backward leaning of the trunk, resulting in hyperextension at the hip (gluteus maximus gait)	Weakness or paralysis of the gluteus maximus on the stance leg, anteriorly rotated pelvis	Check for strength of hip extensors. Check pelvic position
	Forward trunk lean	A forward leaning of the trunk, resulting in hip flexion	Compensation for quadriceps weakness, the forward lean eliminates the flexion moment at the knee; hip and knee flexion contractures	Check for strength of quadriceps
		A forward flexion of the upper trunk	Posteriorly rotated pelvis	Check pelvic position

Contd...

Contd...

Portion of phase	Deviation	Description	Possible causes	Analysis
Swing phase				
• *Foot and ankle (as seen from a lateral view)*				
	Toe drag	Institution dorsiflexion (and toe extension) so that forefoot and toes do not clear floor	Flaccidity or weakness of dorsiflexors and toe extensors; hyper-reflexive of plantar flexors; inadequate knee or hip flexion	Check for ankle, hip, and knee range of motion. Check for strength and reflexes at hip, knee and ankle
	Varus	The foot is excessively inverted	Hyper-reflexive invertors; flaccidity or weakness of dorsiflexors and evertors; extensor pattern	Check for muscle reflexes of invertors and plantar flexors. Check strength of dorsiflexors and evertors. Check for extensor pattern of the lower extremity
• *Knee (as seen from a lateral view)*				
– Accleration to midswing	Excessive knee flexion	Knee flexes more than 65°	Diminished preswing knee flexion, flexor withdrawal reflex, dysmetria	Look at reflexes of hip knee, and ankle muscles; test for reflexes and dysmetria
	Limited knee flexion	Knee does not flex to 65°	Pain in knee, diminished range of knee motion, extensor spasticity; circumduction at the hip	Assess for pain in knee and knee range of motion; test reflexes at knee and hip
• *Hip (as seen from a lateral view)*				
	Circumduction	A lateral circular movement of the entire lower extremity consisting of abduction, external rotation, adduction and internal rotation	A compensation for weak hip flexors or a compensation for the inability to shorten the leg so that it can clear the floor	Check strength of hip flexors, knee flexors, and ankle dorsiflexors. Check range of motion in hip flexion, knee flexion, and ankle dorsiflexion. Check for extensor pattern
	Hip hiking	Shortening of the swing leg by action of the quadratus lumborum	A compensation for lack of knee flexion and/or ankle dorsiflexion; also may be a compensation for extensor spasticity of swing leg	Check strength and range of motion at knee, hip and ankle. Also check reflexes at knee and ankle
	Excessive hip flexion	Flexion greather than 20–30°	Attempt to shorten extremity in presence of footdrop; flexor pattern	Check strength and range of motion at ankle an foot. Check for flexor pattern

15
CHAPTER

Application of Kinesiology to Selected Daily Life Activities and Sports Skills

Kinesiology is not only applicable to the sports, correctives and therapeutic aspects, but it also has got a useful place in the home, garden, office, shop etc. The principles of kinesiology and body mechanics are applicable whenever and wherever man moves and performs various activities whether they relate to a sport, daily life skills, or any specific occupations. Whenever the human body is used in accordance with the principles of body mechanics, the result is a strong, efficiently upright, and good looking body, whereas their violation may lead to postural strains, injuries, abnormal tensions and other pathological conditions of muscles, bones and joints.

Some of the activities that relate to the daily life and motor skills of sports having practical application of kinesiology are briefly described hereunder:

- *Daily Life Activities:*
  - Stair climbing (ascending and descending stairs)
  - Stooping
  - Being seated (sitting) and rising
  - Lifting
  - Handling an object overhead
  - Carrying, and
  - Moving weights (pushing and pulling).
- *Sports Skills*:
  - Landing
  - Falling
  - Catching
  - Throwing, and
  - Striking.

DAILY LIFE ACTIVITIES

Stair Climbing (Ascending and Descending Stairs)

This is a special type of locomotion and an activity of daily life in which the major concern is the alignment of weight bearing joints and other anatomical aspects along with the viewpoint of mechanics. Poor body alignment causes strain and fatigue while ascending and descending stairs, particularly when it forms a relatively large proportion of a person's daily occupation such as house keeping or office-work which involves frequent use of staircases in multi-storeyed building (without a lift), and in all cases, the good alignment of feet, knees and hips is of great importance.

The major problem in going up the stairs (*ascending*) is economy of effort, whereas in coming down (*descending*) is the safety. The correct manner of going up or ascending the stairs, and descending the stairs is shown in **Figure 15.1A and B** respectively. While *ascending*, the economy of the effort can be achieved by slightly inclining the body forward from the ankles which keeps body's

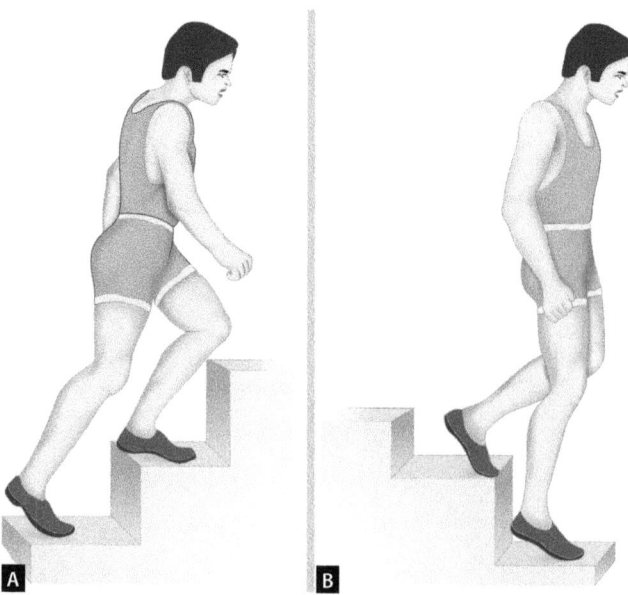

Figs. 15.1A and B: Correct manner of going up or ascending the stairs (A), and descending the stairs (B)

center of gravity forward, thus minimizes the weight or resistance arm of the thigh lever which ultimately reduces unnecessary muscle and joint action, particularly in the plantar flexion of the ankle. Depending more on knee-extension for lifting the body while climbing stairs up can greatly reduce the plantar flexion action at the ankles.

- Raising the body onto the next higher tread is accomplished by contraction of the soleus, quadriceps femoris, hamstrings, and gluteus maximus muscles. At the sametime the gluteus medius contracts in order to prevent the body from falling to the unsupported side. The tibialis anterior muscle dorsiflexes the foot and helps the limb clear the stair on which the supporting limb rests. The hamstrings flex the knee and control the terminal part of knee extension. The erector spinae contract to control the forward bending of the trunk. All these muscles also eccentrically contract while descending stairs.
- Placing the ball of the foot on the higher stair tread and then lifting the body by contraction of the gastrocnemius (plantar flexion) is considerably more fatiguing than placing the entire sole of the foot on the tread and raising the body by extension of the knee joint.
- Elderly people often grasp the hand rail for support and lean well forward. The increased hip flexion thus obtained enables gluteus maximus to contract better for extending the hip joint.

While descending the stairs with safety, one should keep his body more erect so that the body's center of gravity is approximately over the center of the supporting base. Therefore, leaning too far forward (particularly while hurring) places the body's line of gravity to fall over the anterior margin of the supporting base and thus body's equilibrium is disturbed which should be avoided. This may also cause fall particularly when there is slippery surfaces and high heels are worn.

While descending stairs, the hip of the swinging leg is slightly flexed, and the knee and ankle are extended. The extensors of the hip, knee, and ankle of the supporting leg gradually reduce the work which they are exerting, and permit a slow flexion. This is done by eccentric contraction of these extensor muscles to control the gravitational force. The body is lowered until the ball of the foot of the swinging leg makes contact with the step. The weight is then transferred to this leg, the knee of the other leg is flexed, the foot is lifted and swung forward, and when the knee is extended, the cycle is repeated.

As described above, the safety is endangered in home while using the staircase or otherwise on the plane surfaces when there is insufficient friction between the feet and supporting/walking surface or when the floor is slippery. The falls therefore can be avoided by paying attention to this aspect and accordingly suitable material for the floor or their coverings can be appropriately used as per the economic status of the family.

According to an estimation, the energy cost of ascending stairs has been reported much higher than the energy cost of descending the stairs (about 60% of the energy cost of ascending).

The experts have found out the ratio of oxygen costs for ascending and descending stairs at 160 steps per minute as following:

	Ascending	:	Descending
Males:	4.7	:	1
Females:	5.7	:	1

This shows that the oxygen cost for ascending the stairs is approximately 4.5 to 5.5 times higher than that of descending the stairs. It has also been reported that even descending stairs requires a greater energy expenditure than walking on a horizontal surface, probably due to maintenance of body posture while moving from one step to another.

Stooping

So far as stooping or bending forward is concerned, there are three main interrelated objectives: (i) *economy of effort*, (ii) *maintenance of equilibrium* and (iii) *avoidance of strain*. In stooping to pick up something from the floor, stability will be increased if the feet are slightly separated, both sideward, and front-back directions to enlarge the supporting base.

For the maximum efficiency and minimizing danger of strain while bending forward, one should involve bending of knees (rather than extended knees), and also inclining the trunk slightly forward. The incorrect and correct manner of stooping is shown in **Figures 15.2A and B** respectively.

By effecting the two as described above, the thigh becomes the major lever of which fulcrum is at the knee joint, and the force is supplied by the knee-extensors (at the proximal attachment due to the reverse contraction). The resistance is the weight of the body, and its point or application depends upon the position of the body. When the trunk is inclined slightly forward, the line of gravity falls through approximately the center of supporting base. The lever effect is advantageous and similar as explained in case of ascending and descending of stairs. Thus, stooping with the trunk held vertically requires more muscular effort than when stooping with trunk slightly inclined forward.

While bending forward from the waist with straight knees to reach the floor, center of gravity of the body is

Figs. 15.2A and B: (A) Incorrect; (B) Correct manner of stooping

relatively higher than when one stoops (with knees bent), and thus is less stable.

Bending forward from the waist with straight knees is a position of mechanical disadvantage as in this position, the force of gravity now acts on a horizontal spine. This calls upon the spinal extensors and hip extensors to work strongly in stretched position to stabilize the flexible spine and control the movement of the trunk, respectively. Since the spinal extensors are relatively not very strong muscles, and also that in this position the powerful knee-extensors do not help at all in this movement. Stoowping forward with knees bent and inclining forward has an advantage as the strong knee-extensors perform the major work, assisted by hip extensors. With minimal work of spinal extensor muscles in this position does not subject them to strain than when they work in a stretched position.

Being Seated (Sitting) and Rising

Since body position of an individual reflects his anatomical, psychological, cultural, kinesiological, and environmental factors, being seated in the chair is one of the most common activities of daily life of the western and most of the modern world.

In sitting down and rising, one is confronted with the problem of supporting the body while the center of gravity moves backward and downward (in being seated), and forward and upward (in rising).

In sitting down on a chair, the individual stands with his back toward the chair, and by placing one foot slightly

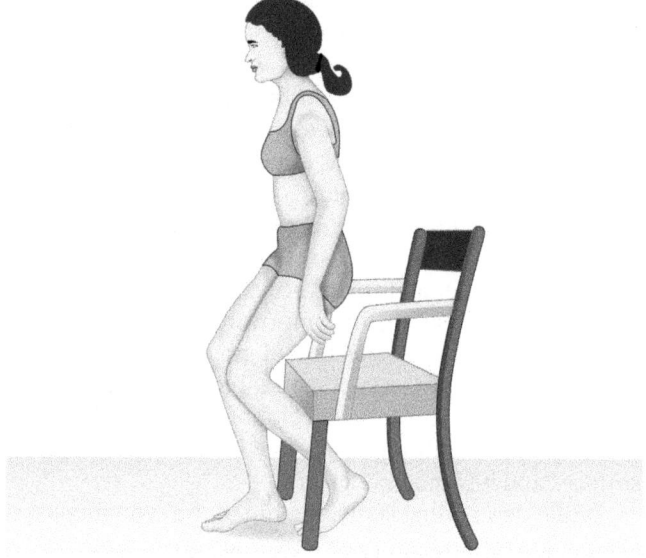

Fig. 15.3: Attemping to being seated gracefully with body well balanced and controlled

back under the chair, and inclining the body forward from the hips, the individual enables to keep his center of gravity over the base of support. He then lowers his body by relaxing the knee extensors and permitting his knee joints to bend. This way, the individual is seated with ease into the chair rather than just dropped into it **(Fig. 15.3)**.

In rising from the chair the problem is to provide a base of support with the feet immediately, so that the transfer of weight can be made gradually rather than throwing the body forward or using the hands for a push off from the seat of the chair. This is done by placing one foot well back under the chair, inclining the body slightly forward, and pushing the feet down against the floor and extending the knees. With the knees extending, the body weight is shifted from the back foot to the forward foot.

Lifting

Lifting is a form of pulling. It involves pulling of a movable object vertically or obliquely upward. A number of lifting activities occur in everyday tasks. Weight lifting is the prime example of a sport activity that involves lifting.

Since the levers of the human body are adapted for range, speed and precision of movement, rather than for weight lifting or handling, it is not surprising that the incidence of back injuries attributed to lifting is extremely high.

When the object to be lifted is on the floor and the hands are to make the contact, they must be definitely lowered, usually by flexing the joints of lower limbs, and later by their extension, the needed force can be provided. Thus, the work is done mostly by the large muscles crossing the hips, knees and ankles. In all heavy lifting, the back is kept straight and acts as a single lever, with the fulcrum at the hips.

In activities like lifting, the direction and point of application of force are important and interrelated. These have an important bearing on:
- Effectiveness of the force applied,
- The economy of effort, and
- Avoidance of strain.

While lifting a heavy object from the floor:
- Stand close to the object
- Object should be in front of the individual
- Lower the body by bending the legs
- Position should be such that the body parts exerting force (hands, arms, thighs) are directly under the object to oppose the force of gravity
- Body balance must be maintained while lifting without loss of basic body alignment
- For lifting, apply vertically upward force rather a diagonal force
- Use stronger leg muscles than weaker back muscles
- See if the momentum can support the lifting force.

For lifting an object, the trunk is to be lowered to reach the object to be lifted, the stooping (crouching) posture (that involves bending of knees rather than bending from the waist) utilizes the stronger leg muscles instead of weaker back muscles. The incorrect and correct manner

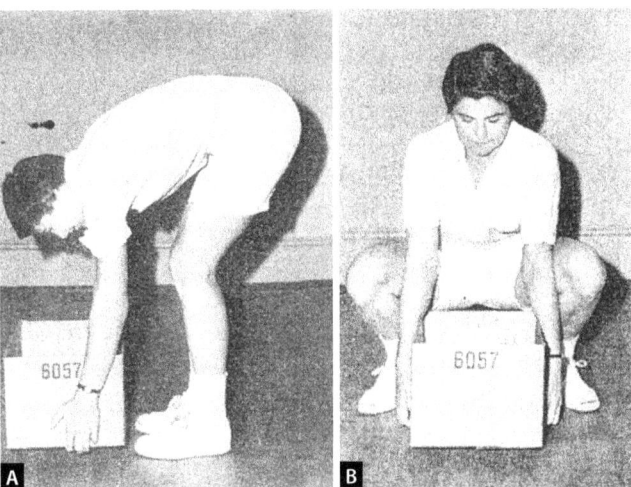

Figs. 15.4A and B: (A) Incorrect manner of lifting with both hands strains the back; (B) Correct manner of lifting with both hands using the strong leg muscles (weight in front)

of lifting with both hands is shown in **Figures 15.4A and B** respectively. The back muscles are not very powerful muscles and get easily strained when they are required to work in a stretched position. This way less effort is applied to lift the object and there is less strain on the muscles and joints of the lower back.

In bending from the knees, the weight arm of the lever is reduced than when bending from the waist. The object to be lifted greatly adds to the weight arm at the extreme end of the lever, i.e. the hands (in bending from knees, the thigh serves as the lever, knee as the fulcrum, and the weight arm of lever is the horizontal distance from the center of knee joint to the body's line of gravity, whereas in bending from waist, the weight arm of the lever is the horizontal distance from the center of hip joint to the body's line of gravity).

Research has shown that while bending from the waist to touch the floor without bending knees creates a tensile force of 450 pounds in erector-spinae muscles, and a compression force of nearly 500 lbs on L5 vertebra. If a 50 lb weight is held in the hands, the tensile force is increased to 750 lbs, and the compression force on L5 to 850 lbs.
- While lifting the object it also takes less effort to lift with the trunk slightly inclined forward than when the trunk is held vertically upright. The trunk held inclined forward is mechanically more efficient and a more comfortable position. when the trunk is held slightly inclined forward, the weight arm of the lever is slightly shorter than when trunk is held vertically upright, and therefore, a mechanical advantage results.

- With knees bent and trunk held slightly inclined forward, it places the body in an easy normal position rather than the stiff position and when the knees are bent, it is impossible to arch the lower back as the trunk inclines forward and therefore, there is less possibility to strain the lower back. The correct and incorrect manner of lifting a suitcase with one hand is shown in **Figures 15.5A and B** respectively.
- If the object to be lifted does not have a handle (like in a suitcase), then the hands can be placed under the object to be lifted.
- The lifting is more efficient when the pull applied is more nearly vertical, and more in line with the center of gravity of the object to be lifted. The main principle involved here is to shorten the weight or the resistance arm of a lever so as to reduce the amount of effort needed to lift a given weight. For example, it takes less effort to lift and hold a heavy object or weight close to the body than to lift and hold it at arm's length, and also in the desired direction of motion.
- When lifting, the feet should be separated some what into a forward-backward stride position to widen the base in the direction that the body segments will be moving and therefore increase the stability of the body.

Handling an Object Overhead

If a heavy object such as a box from a high shelf is to be lowered, the main problems are: (i) *To avoid strain*, and (ii) *to keep the movement of the object or the box under the control.*

When the individual pulls the object off the shelf, he is imparting horizontal movement to the object, and the overhead position of the object also raises the center of gravity of the mover, thus decreasing his stability and rendering him more likely to be tipped in the direction in which the object is moving. If the individual pulls quickly, or the object/box comes suddenly, as soon as the box is free of the shelf, its momentum will be transferred to the individual. And, if he is not prepared for this, it may swing his trunk backward and strain his lower back. To avoid such a possibility, the individual should observe the principles of widening the base of support in the direction of oncoming force. In other words, he should stand with one foot forward than the other foot. The box or the object, this way will be first moved with the body's center of gravity supported by the forward foot. As the object comes forward, his body weight would be enabled to be shifted from the forward foot to the back (rear) foot and also to take a step backward if necessarily required. This provides additional space in which the box can lose its horizontal momentum. The incorrect and correct manner of lowering a box down from a high shelf is shown in **Figures 15.6A and B** respectively.

Whether the object or the box is lowered off the shelf slowly or quickly, there will always be a problem of maintaining equilibrium when reaching forward for the box, and when the weight of the box is first transferred to the hands. In this situation, the center of gravity, at least for a brief moment is near the front margin of the base of support. Therefore, by placing one foot well forward, the

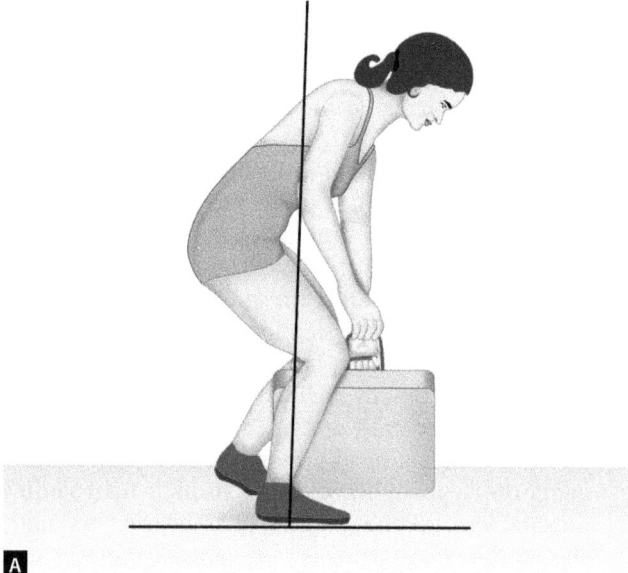
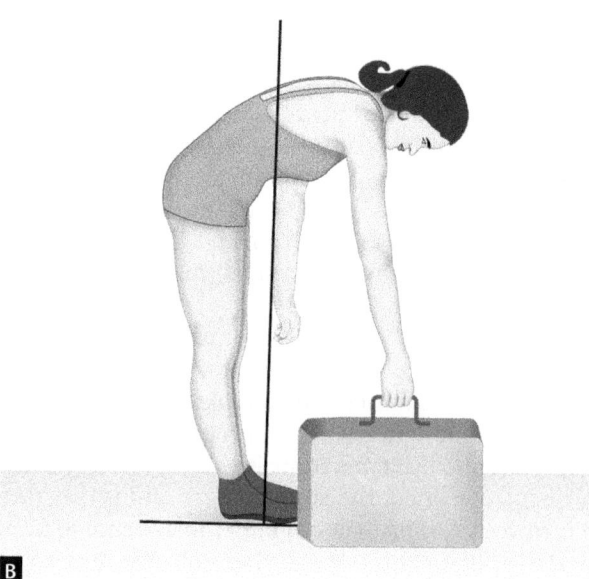

Figs. 15.5A and B: (A) Correct; (B) Incorrect manner of lifting a suitcase with one hand

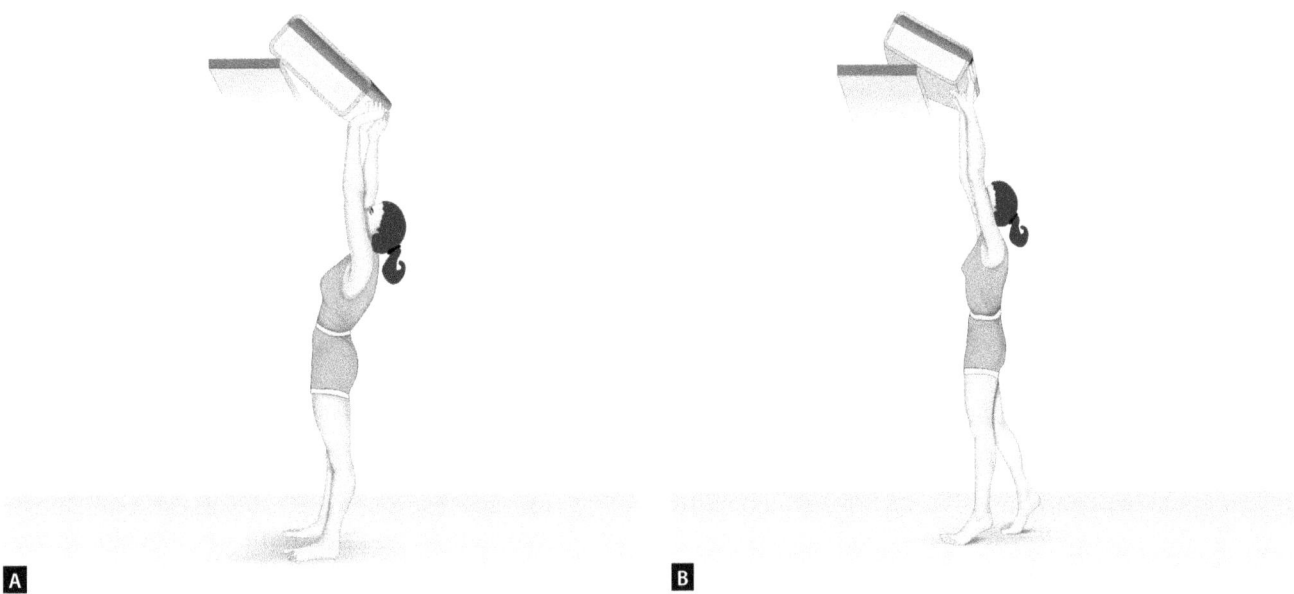

Figs. 15.6A and B: (A) Incorrect; (B) Correct manner of lowering a box down from a high shelf

individual provides an adequate base of support for the anteriorly displaced center of gravity.

To lift or lower an object overhead while standing on a stool or similar support is potentially a hazardous activity, as there is no space to allow reduction of momentum in the backward direction, and as such there would always be a danger of loss of stability.

In all overhead lifting, an effort should be made to avoid thrusting the hips forward or increasing the lordotic curve of the lower back.

Carrying

Carrying the objects or loads involves addition of weight to the body, and therefore the equilibrium gets upset and the body weight must be adjusted. If some weight is to be carried, the normal erect posture is modified to adjust the load in the line of center of gravity of the body. The most common cause of accidents in carrying the load is loss of balance due to muscular strains. Therefore, economy of the effort, and avoidance of strain should always be the major concerns while carrying the objects or loads. The correct manner of carrying the weight close in front and back of the body is shown in **Figures 15.7A and B** respectively.

Holding an object involves exerting or application of an upward force to balance the gravity's vertically downward force, so as to prevent vertical downward movement of the object being held. Carrying differs in that the object acquires the momentum of the body as it moves. Actually, the object is held while it is carried. The force required to hold and carry an object depends upon its weight.

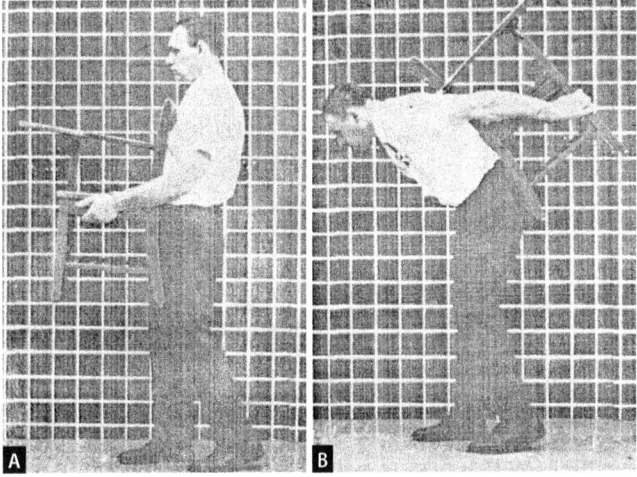

Figs. 15.7A and B: Correct manner of carrying the weight close in front (A) and back of the body (B)

There are certain general rules that should be applied while carrying the object for the ease and economy of effort:

- The object or the load should be carried nearer to body's center of gravity so as to decrease the rotational force exerted on the body otherwise the difficulty in carrying the load becomes greater in direct proportion to its distance away from the body. For example, even a relatively lightweight is difficult to hold at arm's length from the body. Therefore, the load or the object should be held and carried as close to the body as possible to keep the weight arm shorter. If the object is held farther

from the body, it will lead to a longer weight arm and more force will be required to balance the weight.
- Since gravity acts in a vertically downward direction, the body force must be applied in the upward direction.
- The hands and arms must be placed under the object (as much as possible) in order to resist the downward pull of the gravity.
- The feet should be used in such a way that there is a firm base.
- The object that is carried should not have any motion or momentum independent of the person carrying it. Any swinging or swaying motions of the object may cause disturbance of the equilibrium and may require gross adjustments by the performer.
- If possible, the load to be carried must be divided into two equal bags and carried in each hand so that the spine is kept straight.
- When carrying packages in the hands, the elbows should be kept slightly bent to take some of the strain off the elbow joints.
- The load or the object should be carried as a part of the body weight or the mass. The load or the object that is carried becomes a part of the total body weight, and the center of gravity of the body-plus-weight shifts in the direction of the weight. To maintain stability, this new center of gravity must be shifted back over the center of supporting base by an adjustment of the body. When this adjustment is made by shifting the body as a unit from the ankles, the relative alignment of various body segments is not disturbed. Thus, the body as a whole should be used to counterbalance the weight of the object, and there will be a minimum strain on various muscles and joints.
- Some loads, because of their size and shape must be carried in front of the body. Sometimes a heavy load may be supported by the thighs also (if the knees are somewhat bent). In this case, the whole body leans backward as a unit from the knees to balance the weight and the COG of the body weight is kept over the supporting feet.
- Carrying a single heavy object at one side of the body shifts the center of gravity to that side, resulting into falling of center of gravity of the body at the edge or outside the base of support. By raising the free opposite arm sideways counterbalances the weight of the object, and the line of gravity again falls through the center of the supporting base, and the spine is kept as erect as possible.
  - Students often habitually carry books in one arm up against the chest, which results into the shoulder of the carrying side being lifted (high shoulder) and ultimately to the development of a compensatory curve in the upper dorsal spine. It is therefore advised to use the arms alternately rather than using only one and the same arm constantly to carry the books.
  - Even, sometimes carrying heavy books in both arms in front of the chest pulls the shoulders forward and rounds the upper back. This further leads to a tendency for the individual to thrust the pelvis forward.
  - Carrying heavy objects on the head has some advantages and disadvantages. The main advantage is that it keeps the weight directly over the COG of the body. Since the COG of body plus weight is raised high, it causes instability and the slightest rotatory effect will cause the line of gravity to fall outside the base. Similarly, a very flexible neck below the weight requires strong neck muscles to prevent any movement which can otherwise shift the COG of the load away from the line of gravity of the body and thus disturb the stability of the body.
  - The most common cause of accidents in carrying loads is loss of balance resulting in insupportable muscular strains that are placed on the body.
  - Although the amount of weight that can be safely carried or handled depends upon the individual's constitution type, strength, age, experience, compactness of load, etc. however as according to experts, the optimum load to be carried by both men and women should be about 35% of their body weight. Similarly, it is also opined that since taller men are normally heavier, they may carry somewhat greater amounts of load/weights. Carrying weights at faster rates also increases the greater energy consumption.

Moving Weights

When possible, heavy objects should be pushed rather than pulled, and slid rather than carried. Theoretically, it is estimated that the force required to lift a weight is nearly 34 times that required to slide it. Though in actual practice this difference is reduced by the amount of energy necessary to overcome the friction involved. The two main activities of moving weights, i.e. pushing and pulling are described hereunder.

Pushing and Pulling

In general, pushing and pulling tasks are more common in normal everyday activities than in sports. People push heavy furniture, carts full of groceries, and many other objects. They pull rakes, drawers, doors and windows. However, the examples of pushing movements in sports are found in aquatic sports, basketball, track (shot put), etc. Archery is an example of pulling activity.

Pushing and pulling are both the skills in which force plays quite an important part but the hands generally keep contact with the object to be pushed or pulled.

Many pushing and pulling tasks are light but many of them involve application of considerable amount of force, and as such economy of effort and avoidance of strain are the main considerations. In heavy tasks, the initial force overcomes the inertia and additional force must be applied as long as movement is desired. To achieve successfully the particular purpose involved in pushing or pulling an object, the desired direction of movement, type of movement (linear or rotatory), distance to be moved, speed of movement, as well as the friction and other resistive forces must be considered. The friction between the feet and the supporting surface is also important. The speed of movement and distance moved are actually determined by the magnitude of the force applied in relation to the magnitude of the resistive forces. The direction of movement is dependent upon the point of application of the force and the direction in which it is applied.

- An anatomic principle that applies to both pushing and pulling when great amount of force is required is to use the stronger muscles and mass of the body to supplement the weaker muscles. In other words, the leg muscles should be used to supplement the action of the arms. As such, the object should be pushed by contracting the extensors of the hips, knees and ankles and straightening the legs, and not by extending the arms. Likewise, by inclining the body from the ankles toward the object being pushed, one can add to the force of push by using the leg extensor muscles. The correct manner of pushing heavy object using legs is shown in **Figure 15.8**. Similarly, the force of pulling can be increased by inclining the body away from the object and extending the legs in as nearly horizontal plane as possible.
- Generally speaking, the hands of the pusher or mover should be placed at the level of the object's center of weight. If the hands are placed above the center of weight, the object will tend to tip forward rather than to slide forward.

Pushing may take place in any plane and apply to any type of object. Most usually, the force is applied as nearly as possible in line with the desired movement and near the center of gravity. If the object is light and

Fig. 15.8: Correct manner of pushing heavy object using legs

only requires slight pressure, the hand is placed on it and extension of the elbow is sufficient. The elbow extension is accompanied by some movement in the shoulder. The point of application of the force and the direction in which it is applied depend upon the object and its possibilities for the movement. For example, a window must always move in the same groove, and it makes practically no difference whether the force is applied at the center or to one side of center or at top or bottom, if it slides easily. Similarly, if an object which is on a slippery surface, or which has a very low center of gravity may be pushed almost at any point, to move it parallel to its supporting surface. However, an object which offers considerable friction between its base and supporting surface or the one in which its center of gravity is high, probably over a narrow base, is much more likely to rotate and fall.

- The weight of the performer may be put entirely over the object when the push is required in downward direction. If the pushing is to be done upward, the body is placed as nearly under the object as possible, to give a vertical direction of the force. The knees should be bent so that leg extension can be added to the force of the arms. For a push in forward direction, the hands are placed about in front of the chest with elbows flexed. The body leans forward with chest or shoulder against the object, and the main force is created by leg extension.
- Usually, the force should be applied squarely in the direction in which the object is expected to move. When this is not possible, the undesirable component of force should be as small as possible. For example, if a low trunk is to be pushed across the floor, it would be difficult to stoop low enough to push with the arms

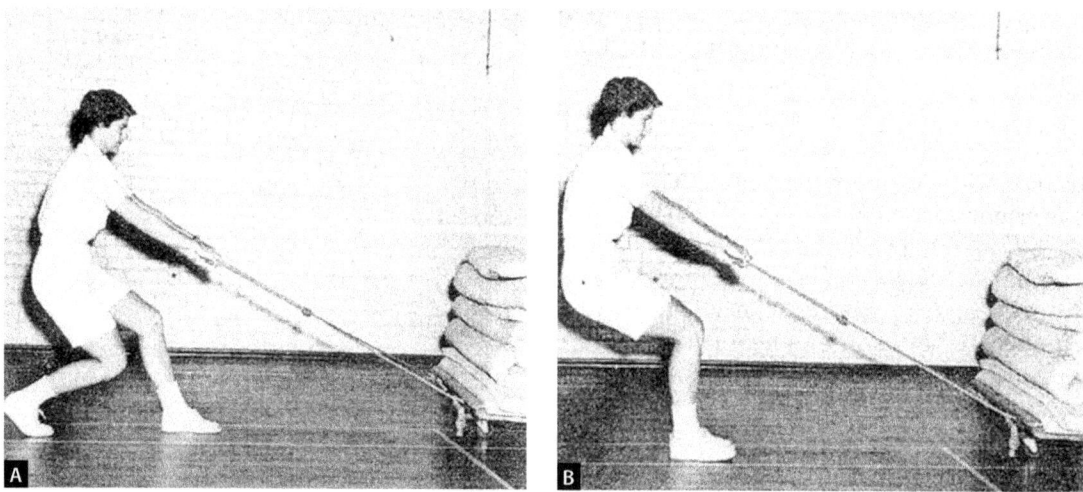

Figs. 15.9A and B: (A) Correct; (B) Incorrect manner of pulling heavy load held in front of the body

in a horizontal position. However, one should stoop as low as conveniently possible to reduce the downward component of force which would tend to increase friction. Sometimes, by using a long rope (tied to the trunk or some object to be moved or pulled), more of the horizontal component of force would be necessary, and a very little of vertical or lifting component of force would be desirable.
- While pushing an object, the spine should be kept as straight as possible, and the hips should be kept low. The feet should be placed a comfortable distance apart, with one foot near to the object to be pushed and the other foot extended backward.

Pulling is similar to pushing in many ways. If the resistance is light, the arm action alone may be sufficient. If the pulling is downward from shoulder level or above, there is depression of the arm at the shoulder and usually with elbow flexion. Pulling in all planes practically requires elbow flexion, however shoulder action may vary considerably, being either elevation, depression or so.

When the resistance is greater, the force can be increased by leaning away from it. Leg extension may also be brought into use. If someone is facing the object and moving in backward direction, the back extensors fix the trunk. If one is facing the other direction, the abdominal muscles are active. When someone is pulling a very heavy object downward, the body weight may be practically suspended on the object by removing his feet from the floor. The correct and incorrect manner of pulling heavy load held in front of the body is shown in **Figures 15.9A and B** respectively; and the correct manner of pulling heavy load held in back side of the body is shown in **Figure 15.10**.

Fig. 15.10: Correct manner of pulling heavy load held in back side of the body, using the legs (trunk inclined forward from the ankles places the COG ahead of the pushing foot)

SPORTS SKILLS

Landing

Landing is a type of fall that enables a performer to strike the surface without injury to the body parts. In the act of landing in a long jump, the jumper wants to dissipate gradually the force of the ground reacting against him, and also to ensure that the feet strike the pit at the fartherest distance possible. Therefore, the jumper flexes the hips to add to the distance upon landing. The correct manner of landing in the long jump is shown in **Figure 15.11**.
- Just before landing, the knees must flex to bring the feet (to a contact position) ahead of body's center of gravity.

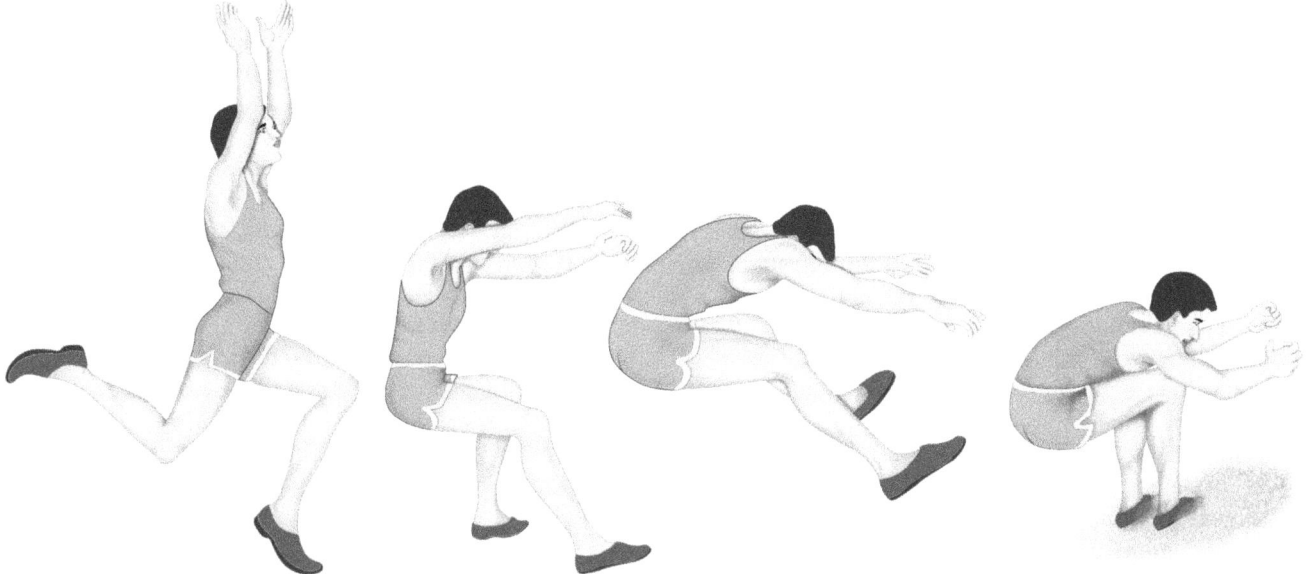

Fig. 15.11: Correct manner of landing in the long jump

- The desired forward movement will be enabled by hip flexion and continued knee flexion on landing. These actions lower the body's center of gravity (and thus shorten the distance between COG of body and ankle joint which will be the axis as the body moves forward by rotating around the base).
- The hip flexion, by moving the trunk forward also assists at landing by moving body's center of gravity forward.
- In any vertical landing from a jump or fall, the center of gravity should be over the base. At the moment of impact, when the line of gravity intersects the base of support, it provides more opportunities for adjustment of body's line of gravity in case forces apply upon landing.
- While performing a high jump, the forceful extension of the legs leaves them in a position directly below the center of gravity of the body so that the equilibrium on landing is not a problem. When the legs are extended, they are in a position to flex at all of the joints on contact with the floor so as to reduce the downward momentum gradually, and therefore absorb the force of landing without an impact to the body.
- When falling downward with little or no forward momentum, the feet should be kept under the body if at all possible so that they will contact the ground first.
- Since balance on landing is a problem, if possible, the feet should be adjusted to form a wide base (no more than width of the pelvis-if in sideways) before landing is made. (If the feet are more wider on landing, the force of landing causes pressure on the inside of knees and ankles). If possible the larger supporting base should be made in the direction of the force, and it is preferred that there is an equal distribution of body weight over both the feet. However, when one lands with so much of force that it is difficult to establish an adequate base of support with the feet alone, one or both hands should be used to establish a temporary base large enough to assure a quick recovery of equilibrium.
- By landing with the ankles plantarflexed (extended), the weight can be taken first on the balls of the feet followed by in rapid succession, on the ankles, knees, and hips by bending them so that the downward motion can be slowed gradually.
- While falling from a great height, more of the momentum must be absorbed by placing the hands on the ground (this adds the area of contact) and, bending of wrists and elbows further increases the time and distance over which the force is absorbed. Sometimes it becomes necessary to tuck the body and go into a roll so as to spend the force even more gradually.

  When forward momentum is involved, i.e. when one trips, or becomes overbalanced while running, or falls from some moving object, injury can be avoided or minimized by relaxing to allow the joints to give (bend) and muscles to be soft while curling into a ball and rolling so that the momentum is gradually slowed down.
- Whether it is a forward roll, shoulder roll or a hip roll, the area absorbing the force is gradually enlarged and the time for reduction of momentum is increased.

- When body does fall forward onto the hands and arms, the wrists and elbows must bend (give) as they take weight.
- Every effort must be made to avoid landing on head, elbows or knees as these areas are solid (as no give is possible). It is suggested that an injury to the head must be avoided even at the cost of injury to the arm.
- While one is falling in such a way that knee would be the first part of the body to hit the ground, a slight twist of the body will allow the body weight distributed over well padded buttocks.
- Likewise, while falling onto the elbow, a twist will throw the weight onto the back of the shoulders which also allows the body to roll.
- While falling backward, the body should be relaxed. The body weight should be taken on the hands and arms, allowing the weight to come down onto the buttocks so that the momentum can be absorbed gradually.

 In general, when falling, the body should be relaxed, and flexed so that it will roll rather than stop suddenly. When arms or legs contact the ground, all of the joints give, each absorbing some of the force so that the momentum is gradually slowed down.
- While landing from a jump, the controlled flexion at the joints, i.e. ankles, knees and hips of the landing extremities through eccentric contraction of the extensor muscles helps in more gradual loss of body's momentum and kinetic energy, and thus avoid injury.
- *While landing on hard or unyielding surface, body's kinetic energy and momentum can be more gradually lost by increasing the distance and time such as rolling.* This way the downward motion of the body is transferred to the horizontal motion and the injury is avoided.
- In case of landing when there is limited opportunity for increasing the distance over which the kinetic energy is lost, it is important to increase the size of the body area which receives the force of impact. More serious injury is often caused when the same amount of force is borne on relatively smaller body surface than when the same force is spread over a larger body area. Similarly, one should attempt to land on the more heavily padded parts of the body while landing from a fall.
- *For landing from jumps, one should wear rubber soled shoes, and also use smooth landing pit (sand) or gymnasium mat.*
- The landing surface should not be uneven or such to cause the performer to turn or throw the body weight sideward. The knee has to be flexed for the landing, and any twisting of the flexed knee can cause severe internal knee injuries. Ankle sprains may also be caused in the same way. The body weight sideward. The knee has to be flexed for the landing, and any twisting of the flexed knee can cause serve internal knee injuries. Ankle sprains may also be caused in the same way.

Falling

The main difference between landing from a jump and falling is that in falling the equilibrium of the body gets out of control. During landing from a jump, if the equilibrium becomes out of control, a fall results. The dangers of falling result largely from impact or the reception of forces.

The main problems that arise with falling are: *(i) regaining of equilibrium and (ii) avoidance of injury.*

While analyzing the application of various kinesiological principles involved in falling, the following factors must be considered:
- The velocity at the moment of impact
- The mass of the falling object
- The distance through which the deceleration takes place
- The surface area over which the impact is absorbed
- The part of body subjected to the impact
- The type or properties of landing surface.

Since the problem is to do work so as to absorb the kinetic energy of fall to do it safely, and the formula of kinetic energy is given as:

Work = Force × Distance = $\frac{1}{2} mv^2$ = Kinetic energy

Therefore, anything that can reduce the velocity will be beneficial. It is to remember, that most of the time the body is falling under the influence of gravity, the velocity of fall (or gravitational acceleration) constantly increases rapidly during the fall.
- Since the impact of landing is proportional to the mass of falling body, obese persons because of their greater mass are injured more seriously by falls if they do not possess equivalent agility.
- When there is longer distance through which deceleration takes place, there would be less danger of trauma (as per the formula of kinetic energy, $\frac{1}{2} mv^2$ = fd, the distance through which force is applied has to be increased to reduce the deceleration force, or ultimately the kinetic energy of fall is reduced).
- The arms or legs are sometimes used as shock-absorbers since these are controlled by the muscles. Therefore, the limbs should be nearly half way extended at the moment of contact.
- Landing or falling on both feet is safer than landing on one foot as the impact of fall spread over a large surface area reduces the magnitude of force per unit surface area of the body. In some cases, landing flat on the

back is preferable than to land on small surface area of hands and/or feet (i.e. circus acrobats).
- Similarly the *Judoka* is taught to land on well padded side of the thigh and take as much of shock as possible on his hands and feet.
- In many falls, a forward, sideward or backward roll is advisable when there is lateral as well as vertical momentum.

Therefore, it can be summarized that during falling, the avoidance of injury and regaining of equilibrium can be mostly handled by reduction of shock of impact through— (a) increasing the distance, and time over which the momentum of kinetic energy is reduced and (b) increasing the area of body receiving the force, that too mostly on the softer or more padded areas of body.

Catching

Catching is an activity that involves getting the ball or an object to strike the hands properly and then to absorb the energy of the moving object by means of a short recoil. Catching is an act of reducing the momentum of an object which is in flight to a zero or near zero velocity, and retaining possession of the object, at least momentarily. This is accomplished by use of the hands, body, feet, and any auxiliary equipment (such as gloves). The momentum of the ball or object is transferred to that of the receiving mechanism. If the weight of the object is light and it is traveling at a slow speed, a human being can easily stop it. However, if the object is heavy and/or traveling at great speed, the momentum must be transferred gradually in order to prevent injury to the hands or parts of the body, and also to enable control of the object once it is in possession of the catcher. The correct manner of catching a large object, i.e. football with both hands is shown in **Figure 15.12**.

In catching, the shock of impact should be reduced by:
- Effecting a more gradual loss of kinetic energy of the ball through using as much distance as possible to reduce its velocity
- Increasing the time over which the force is absorbed
- Increasing the surface area which receives the force of impact
- Using all the 'shock absorbers' of the body such as wrists, elbows, shoulders, hips, knees and ankles.
  - The first part involves running in order to get the hand on the object and drawing the head backward as the object is caught. In this way, the momentum of the object is reduced over a longer distance and therefore shock to the catcher is much less.
  - The other important issue involved in catching is the maintenance of equilibrium and catching the object with accuracy and control.

Fig. 15.12: Correct manner of catching a large object, i.e. football with both hands

- *Avoidance of injury in catching is achieved by increasing the distance over which the object's kinetic energy is lost.* It is not desirable for a player to hold his hands rigidly in front while catching a fast moving object, rather he should give with it. When he moves his hands toward his body, he is making the object's kinetic energy lost gradually. When catching a ball with both hands, pull the arms toward the body by bending or flexing the joints at the time of impact. This will control the momentum or the kinetic energy of the fast moving ball gradually by the eccentric contraction of the antagonist muscles.
- The skill of catching also involves accuracy and control and can be achieved partly by keeping an eye on the oncoming object or ball to judge its speed and direction, and accordingly adjusting the body position and the body parts responsible for catching. The catcher should place himself in the most favorable position in relation to the direction of the approaching ball or object.
- Maintenance of equilibrium is also an important factor which must be considered while catching a fast moving object. The stance of the catcher at the time of impact of the object is important and can be moderately increased by widening the base in the direction of the oncoming object so as to enable the catcher to shift his body weight from the forward foot to the back foot at the time of impact. It will help in ensuring stability of the catcher and

also allow the gradual loss of the object's motion. If possible or necessary, one or two backward steps may also be taken to shift the body weight backward.
- Every attempt should be made to place the body in line with the approaching ball so that the force can be taken close to the center of the gravity of the body.
- Lowering the center of gravity by taking the crouch position and by flexing the knees also improves the stability of the catcher.
- The line of gravity of the catcher should be near the front edge of supporting base.
- Similarly, *while catching a high moving object* with one hand particularly by the extended arm (that acts as a lever), the force of impact of the object is being applied on the palm. Therefore, it is likely to cause strain to the shoulder joint and also to the hand. This can be avoided by 'giving' by reaching somewhat forward and drawing the arm backward at the time of impact.
- In catching, the *position of hands* is very important for avoiding injury. Fingers should never be pointed toward the approaching ball as the area of fingertips is very small to absorb the force of impact. The hands should be placed in such a position that the palms (that present relatively larger area) face the direction of the approaching ball, and depending upon the height of the ball or object, the palms/hands should be either pointed downward or upward, if the object is below or above the waist level respectively.
- *When catching fast balls*, wear thickly padded gloves to reduce the shock of impact for the hands. The gloves distribute and reduce the force of ball over a larger area, before the force is transferred to the hands. As the mass of the balls is increased, the thicker the padding is needed.

Throwing

Throwing is a skill common to many activities, however it is primarily a sport activity in which an object i.e. ball, javelin or discus held in the hand acquires the speed and direction of the hand, and when released, continues to move at this velocity and in the same direction until acted upon by other forces. In different sports, usually balls of different sizes and weights are rolled or thrown by one hand or both. However, regardless of size, shape or weight, the essential mechanics of throwing remain the same. Momentum is transferred from the body to the object

Fig. 15.13: Correct manner of a javelin throw

thrown. The object is held in one or both the hands, and the momentum of the hand/hands is developed, and the object is released. The correct manner of a javelin throw is shown in **Figure 15.13**.

- Faster a hand is moving when a ball is released, the greater is the speed of the thrown ball.
- Maximum *speed* results from the use of all of the body parts that can contribute to the movement when timed in sequence, fast muscular contraction, the longest back swing possible, full body rotation, follow through, and transference of weight in the direction of the throw.
- *Friction* between the feet and supporting surface is essential if the force exerted by the body against the surface is to be transferred back through the body to the ball.
- A step, run or hop before the throw adds speed because the momentum of the body is added to that of the throwing movement.
- Follow-through is essential as it enables a release which coincides with the instant of maximum momentum of the hand.
- The extent to which these various factors are applied depends upon the amount of force called for by the purpose of a particular throw.
- Flattening the arc through which the hand moves increases the possibilities for accuracy of the throw. In the underhand throw, the arc is flattened by moving the shoulder forward, by stepping forward onto a bent leg, and following-through as far as possible in the forward direction. In the over hand throw, the sequential extension of the joints of the arm as the

shoulder moves forward affects some flattening of the arc, whereas in the side arm throw, it is the rotation of the trunk which moves the shoulder forward and out to the side of the throwing hand is more effective.
- The distance which an object thrown travels depends upon the speed and angle of release, and the action of outside forces. When maximum distance is desired, all factors which contribute to the speed should be used, and the ball should be thrown nearly at 45° angle.
- If the purpose of throw involves having the ball reach a certain destination in the shortest possible time, the angle should be flattened as much as possible. The degree to which this angle can be flattened depends upon the force that can be produced by the thrower. As such, both force and angle applied depend upon the purpose of the throw.
- Since the head wind resists the forward moment of an object, the angle of release should be flattened to increase the forward component of force. A tail wind adds its force to the momentum of the ball, and therefore makes throw advantageous. If distance is the objective, to keep the object in the air longer and, therefore, the angle of release is increased so that a greater proportion of the force is effective in resisting the pull of gravity.

Mechanical Principles

In throwing, many applications of the laws of leverage, motion and balance may be found. Motion is basic to throwing when the angular motion of the levers (bones) of the body (trunk, shoulder, elbow, and wrist) is used to give linear motion to the ball when it is released.
- *Newton's law of motion* apply in throwing as the inertia of the thrower and that of the ball must be overcome by the application of the force (which is provided by the muscles of the body to move the body parts and the ball held in the hand).
- *The law of acceleration (Newton's second law)* comes into application with the muscular force necessary to accelerate the arm, wrist and hand. The greater the force that a person can produce, the faster the arm will move, and thus a greater speed will be imparted to the ball.
- *The law of reaction (Newton's third law)* is applied during the reaction of feet against the surface on which the person stands.
- The *leverage* factor in throwing a ball or an object is also very important. Longer the lever, the greater the speed that can be imparted to the ball. (For all practical purposes, the body from the feet to the fingers of the hand can be considered as one long lever). The longer lever can be either from the natural body length or from the movements of extension of shoulder and elbow joints (as in the extended backward position of the body in throwing).

In certain cases when a ball is only to be thrown for a short distance, the shorter lever is advantageous as it takes total less time to release the ball.

Balance or *equilibrium* is a factor in throwing when the body is rotated to the backside in the beginning of the throw. This motion moves the body nearly out of balance to the rear, and then balance changes again in the body with the forward movement. Balance is again re-established with the follow through when the feet are spread, and the knees and trunk are bent to lower body's center of gravity.

Throwing Patterns: There are three basic throwing patterns.
1. The *overhand pattern* utilizes all of the factors of maximum force to advantage and therefore is the most effective for high speed and long distance.
2. The *under hand pattern* is the most effective for throwing tasks involving a high degree of accuracy (as the throwing hand follows a straight path, however, in order to follow the straight, path many factors leading to maximum speed are omitted and speed is sacrificed to accuracy).
3. The *side-arm pattern* uses many of the factors that develop speed and also enables to throw large objects (as these can not be grasped by the fingers and hence can not be thrown with overhand pattern). In this pattern, the hand moves through a horizontal arc, and therefore left-right accuracy is more difficult to control. This pattern is most useful in throwing large objects for distance.

Striking

While there are some household tasks that involve striking such as beating a rug, and swatting a fly, this is a much more important skill in the field of sports. The striking activities in sports represent a wide variety of skills. The tool may be the hand only (as in volleyball or handball); or the foot (as in football), or a racket (as in tennis or badminton) or a more solid stick (as in hockey, baseball and golf).

In some of these activities, a stationary ball is hit by a moving object, and in others a moving object (ball or shuttle) is struck by a moving object (bat, hand, racket etc.). In some, the striking surface is flat, in others it is rounded. The correct manner of striking of tennis ball in forehand batting is shown in **Figure 15.14**. Regardless of

Fig. 15.14: Correct manner of striking of tennis ball in forehand batting

these differences, the problem for the player is to produce the required force and apply it to the object either directly or through the use of an implement, at such an angle that the particular purpose will be served. An example of overhead striking pattern is shown in **Figures 15.15A and B** how the entire body acts as a lever to impart maximum force to the ball in tennis serve.

- A stationary object upon striking will move only if the force applied to it is of sufficient magnitude to overcome its inertia (Newton's first law of motion). The force must be great enough to overcome not only the inertia (due to the object's weight and speed), but also the restraining forces such as friction, air resistance, etc.
- The direction in which an approaching object or ball travels is determined by the direction of the striking surface at the instant of contact only if a sufficient force is applied to overcome the force of the object. As per the angle of the striking surface, the ball or object may leave at a right angle to the surface, forward, up or down, right or left.
- The other factor in determining direction is the rotation of the ball, usually caused by the way the striking surface moves over the ball at the moment of contact.
- When a bat meets an oncoming ball, the direction of the resulting motion will be in the direction of the greater momentum possessed either by bat or ball. If the momentum of the bat (the weight of the bat × its speed) is greater than that of the ball (the weight of the ball × its speed), the forward momentum of the bat is decreased, and the ball is given a reverse momentum (which is equal to the sum of the two original momentums minus the momentum which the bat still has). This is in accordance to the *law of conservation of momentum* which states that when two or more objects

Figs. 15.15A and B: An example of overhead striking pattern showing how the entire body acts as a lever to impart maximum force to the ball in tennis serve

collide with each other, momentum is conserved. The total momentum after the impact is equal to the total momentum before the impact.

The effectiveness of striking is judged in terms of the speed, distance and direction of the struck ball. *The six major factors are being described which apply to the speed of a struck ball.* These are: (i) The speed of oncoming ball, (ii) the mass of the ball, (iii) the speed of the striking implement at the moment of contact, (iv) the mass of the striking implement, (v) the elasticity of the ball and (vi) the elasticity of the striking implement.

Applications

- If the velocity of the approaching ball is greater, there will be greater velocity of the ball in the opposite direction after being struck, other things being equal.
- If the velocity of the striking implement is greater at the moment of contact, there will be greater velocity of the struck ball, other things being equal.
- If the mass of the ball is greater (upto a point), there will be its greater velocity after being struck, other things being equal.
- If the mass of the striking implement is greater (upto a point), there will be greater striking force and therefore a greater speed of the struck ball, other things being equal.
- If the elasticity of the ball and the striking implement is greater, there will be greater speed of the struck ball, other things being equal.
    - As with regard to the distance traveled by the struck ball, if the ball's speed of departure is greater, greater will be the distance traveled, when other things are equal.
    - Assuming that maximum distance or speed of the ball is desired in each case, the player attempts to carry the striking tool through as long an arc as can be properly controlled and in the least amount of time. This ensures maximum momentum of the tool, which in turn is imparted to the ball. A longer tool will probably give a greater speed.
    - *Newton's third law of motion* (the law of action and reaction) also applies to striking, meaning that during impact there are two equal and opposite forces set up between the two objects. One force is exerted by the striking implement on the object struck, and the other force by the struck object on the striking implement. The reaction of the object against the striking surface is just as great as the force which projects the object.

16

Prevention of Sports Injuries: The Mechanical and Kinesiological Viewpoint

Almost in all physical activities and sports there is some likelihood of injury. Athletes are subjected to injuries during training as well as during competition. Many activities of the gymnasium, different sports and play ground have possibilities of injuries. Stress injuries account for maximum percentage of injuries which an athlete encounters. Wrong technique, substandard equipment, mishap or excessive loading are some of the factors responsible for these injuries. Quite often it is observed that the mechanical and physiological principles are disregarded. For example, the fatigue in a wrestler forces him to violate sound mechanical and kinesiological principles causing serious injury to his pectoralis major muscle.

Whether such injuries are due to lack of flexibility, accidents, structural limitations or otherwise, these can be avoided to some extent if proper precautions and preventive measures are taken. A serious injury to a budding sportsman can end his sports career or sometimes it can be life-threatening. As such, an understanding of these possibilities, and precautions that need to be taken by the physical education students, teachers and coaches, and sportsmen will certainly minimize the incidence of sports injuries, if not completely omitting them.

The incidences of musculoskeletal injuries as reported from various studies and surveys indicate that the injuries to the upper and lower extremities comprise the great majority of the total injuries. In the lower extremity, the knee ligaments or the menisci suffer tears most when the knee is bent or struck in such a way as to cause twisting when flexed. Ankle sprains and fractures in the feet are common from landing on uneven surfaces. In the upper extremity, irritations such as bursitis, tennis elbow or tendinitis are quite common. Shoulder dislocations may occur from forced hyperextension of the arm, fractures of the wrist from falls, and finger fractures may occur most frequently from extending fingers firmly towards a ball in an attempt to catch it.

Injuries can be studied from many angles as from training point of view, medical viewpoint, physical and physiological fitness, or from mechanical and kinesiological viewpoints. In this chapter only the mechanical and kinesiologcal aspects of prevention of sports injuries will be dealt with.

From the mechanical and kinesiological viewpoint, stress injury to a body part occurs when excessive pressure is exerted on it. This may be due to the fact that implied force is too large or the implied force is not absorbed properly, i.e. it is not distributed to other parts of the body and remains localized over a small area. To minimize injuries, there are two possibilities, (a) Force may be reduced so that effective implied force is less and (b) Force is distributed to other body parts and its effects are not localized. **Thus, the knowledge of mechanical factors which are responsible for dissipation of force is helpful to minimize the occurrence of injuries in sports.**

DISSIPATION OF FORCE

There are three following aspects in the effective dissipation of force:
1. Area of absorption
2. Distance of absorption
3. Time of absorption.

Area of Absorption

According to the definition of pressure, it is the force per unit area, i.e.

$$P = \frac{F}{A}$$

Where *P* is the pressure
F is the force
A is the area over which force acts.

In conditions where the force is constant, pressure is inversely proportional to the area. In other words, more the area of absorption, lesser the pressure exerted and vice versa. If the same amount of force is applied with the fist and a finger on a part of the body, pressure exerted is more with the finger than with the fist, as the area of absorption of force on the body part is less when force is applied with the finger. Pressure (pain) is still more if same force is applied with a needle without puncturing the skin. *Thus, to reduce pressure on a body part to avoid injury, area of absorption of force should be increased.*

Distance of Absorption

According to the definition of energy, it is the capacity to do work or in other words, when energy is absorbed, it does equivalent amount of work on the absorbing part. For example, a moving ball has kinetic energy because of its mass and velocity. When this ball is caught, it is able to do equivalent work on hands equivalent to its kinetic energy possessed. Now

$W = F \times D$

Where *W* is work done
F is the implied force
d is the distance of absorption
or
$F = \dfrac{W}{D}$

Under given condition, kinetic energy is constant as mass of the ball is constant and its velocity is constant which is determined at the time of release. Now force is inversely proportional to the distance. More the distance of absorption of energy, lesser the implied force and vice versa. For example, if the moving ball is able to do work equivalent to 250 newton-meters by virtue of its kinetic energy and is stopped over a distance of 10 cm, then the force imparted on the hands is 2500 newtons, but if it is stopped over a distance of 50 cm, then the force imparted is only 500 newtons. *Thus to minimize the effective implied force, distance of absorption should be increased.*

Time of Absorption

According to Newton's second law of motion, impulse is equal to the change of momentum.

Impulse = Change of momentum
or
$F \times t = m (V_f - V_i)$

Where *F* is the force
t is the time of absorption
V_f is the final velocity
V_i is the initial velocity.

A moving ball or an athlete when in motion, has momentum because of its mass and velocity. When the ball is stopped or the athlete lands, momentum is changed to zero or near zero, mass cannot be changed and the initial velocity also cannot be changed in a given situation. This change of momentum is equal to impulse ($F \times t$). If the time of catching or landing is increased, force implied is reduced. *Thus, to decrease the force, time for which force is absorbed should be increased.*

Landing and *catching* are the two important actions in sports which, if not properly executed, lead to injury. These constitute the major portion of injuries which the athlete encounters.

LANDING

Landing when not properly executed becomes a fall and as a result, injury occurs. Landing or falling is of two categories; first, landing on a firmer surface like ground, floor or mat, and secondly, on the non-resistive surface like entry into the water. In the first category, landing can be from a high jump, long jump, pole vault, jump in basketball or volleyball, fall in judo, wrestling or gymnastics.

When *falling downward* with little or no forward momentum, as landing after a jump in volleyball or basketball, feet should be kept under the body so that they contact the floor first. Since balance is a problem on landing; a wide base (not more than shoulder width) should be formed before landing is made. When landing is done with ankles in planterflexed position, weight force can be taken first with the balls of the feet and then in rapid succession by the ankles, knees, and hips by flexing these. Thus, the *downward momentum* is reduced slowly by increasing the time of absorption so that impact is minimized. When *forward momentum* is involved as in many exercises in gymnastics, injury can be avoided or minimized by relaxing the joints to give in and muscles not remaining stiff while curling into a ball and rolling, thus increasing the time of absorption and area of absorbtion of force. Whether a forward roll, shoulder roll or hip roll, as in wrestling or judo, area of the force absorption is greatly increased to minimize the pressure on a body part, alongwith the time of absorption.

Every effort should be made to avoid landing on head, knee or elbow, as these are solid areas and possible dissipation of force is negligible.

In *jumping*, shoes with soles which give good cushioning effect should be used. Cushions in the shoes transmit the force, thus, increasing the area of absorption of force and reducing the pressure. During training where repeated jumps are involved, mats or other soft material should be used especially when state of conditioning is not good, otherwise there are greater chances of stress injuries. In jumps, as high jump or pole vault where great amount of force is to be absorbed, soft jumping pits must be used to have the cushioning effect. The landing surface should also not be uneven or in any way cause the performer to turn or throw the weight sideward.

Falling into water differs from landing on a firmer surface as ground or mat. In the case of firmer surface, all the dissipation is within the body while in the case of non-resistive surface, force is reduced by gradual giving of the water. The larger the area of the body with the water contact, the greater is the water resistance and vice versa. With small area of contact, shock of impact is less because of less water resistance. This is just the reverse when fall is on the resistive surface. When falling into water, if arms are held rigid, they cut a hole for the body to move in. In opposite to fall on the resistive surface, body should be kept rigid and straight as far as possible while falling on non-resistive surface.

Falling is most dangerous when the performer is in an inverted position, however, if this does happen, the performer should attempt enough support with the hands to protect the head, or to give time to get it out of the way, or to tip the body over the side. Similarly, landing may be made safely relaxed on all fours, or the more skilled performer may even use a roll for recovery. Mats should be used at times, to provide a less resistant surface for falling, as in certain gymnastic events.

In *swimming* also sprains and strains occur due to water resistance, particularly during diving, the force with which the swimmer enters the water depends upon the height of the dive and of the board or platform. The entry of the hands first is designed to cut the water, and to allow the body to enter with the minimum of resistance. The water resists passage of any broad flat surface trying to progress through. Such resistance may cause strains.

If the head is pulled back at the moment of entry to the water, the neck may be strained. Strains in the lower back are also the most frequent, caused by hyperextended legs and flexed knees which drag in the water, causing further hyperextension of the lumbar spine.

The lower back strains are usually in the region of the lumbosacral or sacroiliac joints. Since the spine is fairly flexible in the lumbar region, far more hyperextension is possible in this region than the standards of dive would permit. Such strains can be avoided by the building up of control on simpler dives and by having stronger abdominal muscles to hold the pelvis well.

CATCHING

Catching is involved in many sport activities, i.e. basketball, cricket, handball, football, etc. A small ball coming at high speed has considerable *momentum* (M = mv). Therefore, small children or inexperienced players should not be expected to catch it. Even an experienced player should be provided with a catching glove to prevent bruises of the hand. A larger, heavier ball at a very much lower velocity may have the same or greater *kinetic energy* (KE = ½ mv^2), and its momentum may be such as to give the athlete a severe jolt or to cause him to fall. However, the larger ball is less dangerous to the catcher and less likely to cause severe injury as the larger ball spreads its impact over a wider area resulting in lower amount of force per unit of surface area, and therefore, is less apt to bruise tissues or break bones. It is largely due to this distribution of force that the football pads are effective. The momentum of the oncoming tackler or blocker then reacts on the opponent's body mass more nearly as a whole instead of being localized at the point of contact.

In catching, the shock of impact can be minimized by gradual loss of kinetic energy which can be achieved by using as much distance as possible, increasing the time of absorption and the area of contact. While catching, sometimes maintenance of stability is important. Thus, the area of the base should be enlarged in the direction of oncoming ball by having forward-backward stance.

The distance and time of absorption can be increased by pulling the hands towards the body (flexion of elbow joints and extension of shoulder joints); position of the hands in catching is the most important factor in avoiding injuries. Since the area of the tips of fingers is small, the impact of the oncoming ball will lead to injuries of the fingers. Thus, fingers should not be pointed towards the oncoming ball. Palms of the hands provide greater area and more absorbent surface. The padded gloves reduce the force of impact by increasing the area or absorption of force and providing more absorbing surface, when used.

Other situations in which impact is involved are *striking* or *punching* as in boxing, hitting force of ball as batting in cricket and goalkeeping in hockey due to rebound. In these situations, protective devices such as face mask, head gears, pads, padded gloves etc. should be used. These devices not only provide greater area for force absorption at sensitive area of the body, but also provide more absorbing surface.

- *Speed and momentum become very great at the end of a long lever.* Therefore, practice with tennis or badminton rackets, baseball bats or any long striking implement should always be conducted with ample space between all players. Every player should understand the length of his reach and the potential danger, as the players frequently fail to decrease the momentum of the bat/racket on follow through before dropping them.
- *Organization of class, and carefully spaced practice areas are also very important in the prevention of injuries* in the activities such as softball, and striking and throwing events in the track. The striking activities in soft ball, and throwing events in track involve more or less rotatory motion. Therefore, there is a large possible range in which the ball, discus or javelin may be headed when the delivery is poorly executed. Various events of the track should be so spaced and divided that the minimum spacing of participants should be that of their maximum possible range.
- Other types of injuries occur when an athlete slips and falls on the surface on which he is running. Usually, the slippage occurs when the body is suddenly stopped or when the direction is suddenly changed. One of the reasons for this is insufficient friction of shoes with the ground. These injuries can be reduced by selection of proper shoes according to the playing surface (which give firm grip) and by bringing down the center of gravity of the body by widening the base at the time of stopping and change of direction.

Strains or injuries occur when sportspersons attempt to *lift loads* that are too heavy or to lift loads in such a way that the back muscles must do all the work, or continue heavy work when fatigued. In lifting stress injuries occur when excessive pressure is exerted on the weaker body parts (e.g. back muscles). These injuries can be minimized by proper execution of technique. As a rule, at the time of initiation, major force should be applied by strong hip and knee muscles so that weaker muscles are not involved to break the inertia of the weights. Furthermore, load should be increased progressively, depending upon the conditioning state of the athlete.

- *Rifle shooting* is another sport where shoulder joint injury occurs because of backward impact of the rifle. This injury can be prevented by reducing the backward velocity of the rifle by increasing the effective mass (addition of mass). This is achieved by tightly holding the rifle against the shoulder.

Other type of injuries occur when a body limb (arm or leg) is stopped suddenly after maximum acceleration because of the jerk, as in *throwing* or *kicking*. This can be minimized by the follow through action.
- *Many gymnastic equipments are quite prone to injuries because of the angular momentum developed in the moving apparatus.* Therefore, these should be used carefully by the beginners, and their safety habits should also be enforced.

Any type of spinning equipment tends to throw the sportsperson off at a tangent to its arc. An athlete on the equipment runs to get started, and considerable speed is created. As speed develops, the rider swings out above the ground. The pull increases with the speed and the finger flexors may not be sufficient to overcome it.

The to-and-fro swing of the equipment such as flying rings also presents the similar danger, throughout the arc there is more or less outward momentum, and the gravity also tends to pull the person off. While being suspended by the hands, security is dependent upon the finger flexors. On the horizontal bar, centrifugal force tends to pull the performer off on very fast swings. The grip must counteract this effect.

- *Rotating equipment* frequently causes dizziness or disturbance of the sense of equilibrium of the rider. This may be a source of danger as the rider may fall from the equipment. The beginners should be taught to focus the eyes on a point above the ground, to avoid such difficulty.
- Prevention of injuries in *stunts and tumbling* is also dependent upon an understanding of momentum and kinetic energy, of factors in stability, and of human anatomy.

Stability in any position is dependent upon an adequate base and maintenance of the center of gravity over the base. Therefore, the base should be as large as possible, steady and sufficient to resist the superimposed weight or any additional force exerted by jumping or turning.

- *Failure to understand anatomical structure and its limitations also make sportspersons susceptible to sustain injuries.* For example, excessive hyperextension of the spine or of the hip joint, maintenance of weight in a hyperextended position, maintenance of heavy weight entirely by trunk muscles or movements requiring flexibility beyond the normal range in any joint may cause severe strains and injuries. Therefore, any sporting activity which involves such movements repeatedly must be carefully planned and exercised according to the age, muscle-strength and all these factors.

- The injuries in stunts and tumbling are mainly due to falling, poorly controlled momentum or due to strains from excessive weights poorly supported.

SUMMARY OF MECHANICAL AND KINESIOLOGICAL PRINCIPLES RELATED TO PREVENTION OF INJURIES

Certain major mechanical and kinesiological principles pertaining to prevention of sports injuries which if followed will minimize the incidence of injuries. Though some of them are described in the preceding chapter, they are briefly and collectively summarized hereunder also for the convenience of the students and other readers.

Principles relating to the maintenance of equilibrium and prevention of falls:

- Maintain an adequate base of support.
- Lower the center of gravity when this is possible.
- Keep the center of gravity well centerd over the base of support.
- Increase the size of supporting base in the direction of force or motion.

Principles relating to the range of movement:

- Keep the range of motion within normal for each joint in order to avoid the injury to the ligaments and fascia. (A brief knowledge of anatomic structures and their limitations is required for this aspect).
- Strengthen the muscles in order to lessen the chances of exceeding the normal range of movement. For example, strengthen the abdominal muscles in order to help prevent an overthrow in diving (swimming).
- If the range of movement in a joint or muscle is to be increased, it should be done gradually and only after an adequate warm up.

Principles relating to the intensity and quantity of muscular exercise:

- Never exercise past the point of excessive fatigue, as fatigue dulls the senses and hence chances of injuries are increased.
- Adequate training should be provided before permitting participation in certain sports, as an untrained individual is likely to use his body poorly under the stress of the moment.
- Sudden, violent movements should be avoided whenever possible. And there should be proper and adequate warm up before all strenuous activity, as the sudden exertion particularly when the muscles are cold may cause muscle tears.

Principles relating to the transmission of weight through the body segments and weight-bearing joints:

- Adequate protection and support should be provided to the weak or previously injured joints. For example, a previously injured or poorly aligned individual with knock-knees is likely to have his knees injured in vigorous activities involving running and jumping, and therefore he may be excluded from such sports or activities.
- Rotatory components of force should be reduced to a minimum. For this, a knowledge is required of positions which put certain segments under rotatory stress. For example, while on the hands and knees supporting others, as in a pyramid, the thighs should be perpendicular to the floor.

  Similarly, while being supported by another person, the weights should be placed in line with the other individual's supporting structures. For example, a person who is standing on the back of another person who is on his hands and knees, should place his one foot over the shoulders and the one over the pelvis of the another person.
- Only the individuals with adequate strength and maturity should support the weight of the others. For example, in stunts and pyramids, too great a discrepancy should be avoided between the strength of the lower man and the weight of the upper man in the pyramids. Similarly, physically immature and teenaged boys and girls whose bony epiphyses are still not united should be avoided for supporting heavy weights, as in the base of the pyramids.

Principles relating to the reception of one's own weight

- *When landing from a downward jump*:
  - To be ensured a gradual loss of kinetic energy by landing on the ball of the feet and immediately letting the ankles, knees and hips bend, controlling the action by means of eccentric contraction of the muscles.
  - To reduce the force of impact by wearing rubber soled shoes and by having a soft landing surface.
  - The stability to be regained by keeping the center of gravity over the center of the supporting base on landing. This can be done by keeping the body weight evenly distributed over both feet or over the hands and feet, and by providing a sufficiently large base of support on landing.

- *When landing from a forward jump or when participating in any activity that involves forward momentum:*
  - A gradual loss of the forward kinetic energy to be ensured by continuing a run, roll or slide unless landing in a soft pit.
  - The stability should be regained by providing support in the direction of motion. This is done by landing with the weight forward, using the hands if necessary, in which case the elbows and the leg joints should '*give*'.
  - The force of impact should be reduced by wearing rubber soled shoes and by having a soft landing surface.
- *When falling*:
  - Ensure a gradual loss of the forward kinetic energy as described above.
  - The shock of impact should be absorbed or gradual loss of kinetic energy to be ensured by letting the arms '*give*' as they receive the weight when falling forward on the knees and the hands, arching the back and turning the face sideward make it possible to avoid the injury. The elbows should bend as soon as the hands touch the floor/ground.
  - One should attempt to land on the padded as well as large an area of the body as possible to minimize the force per square inch. One should also land on various parts of the body in a progressive sequence. For example, after falling on the knee, attempt to fall sideways on the thigh, then onto the side of the arm or the hand, and promptly sliding out with the head falling on the extended arm.

Principles relating to the lifting and carrying external weights:

- The rotatory components of the weight to be lifted from the floor should be reduced to a minimum by getting close to the heavy object. Similarly, heavy weights should be carried close to the body. To reach an object on the floor, one should stoop by bending the knees and inclining the trunk slightly forward.
- One should use the stronger muscles best suited to the task. For example, stronger thigh and buttock muscles should be used while lifting a heavy weight/object from the floor, rather than using the weaker spinal extensor muscles.
- Similarly, lifting excessive weights should be avoided. Generally, the boys should not lift weights of more than one half of their body weight, and the girls not more than one-fourth or one-third of their body weight. These amounts of weight can be increased gradually through training.

Principles relating to the receiving the impact of external forces:

- *When catching* balls and other objects, and receiving individuals in tumbling and gymnastic apparatus work:
  - The principles of reducing the kinetic energy gradually should be observed by '*giving*' of the arms, by shifting the weight back from one foot to the other or taking a step backward.
  - Point the fingers upward, outward or downward, but not downward.
  - Adequate protection such as shin guards to be worn as according to the sport.
- *Upon being hit* or struck by a moving body; i.e. another sports person or a ball:
  - Try to receive the force and ease the shock the same way as catching a ball, as described above.
  - Guard against being caught of balance.
  - Wear adequate protection, i.e. shin guards for hockey and padded clothing for football.
  - Move along with the striking object.

Principles relating to circular motion:

- Since the centrifugal force develops in all circular movements and activities such as swinging on the flying rings, only those individuals or sports persons known to have sufficient strength should be engaged to keep them from being turning away.
- The individuals or sports persons having a tendency for recurrent shoulder dislocation should not be allowed to do vigorous arm circling or swinging movements because of the centrifugal force developing in such activities.

Bibliography

1. Anderson TM. Human Kinetics and Analyzing Body Movements. William Heinemann Medical Books Ltd; 1951.
2. Barham JN, Edna PW. Structural Kinesiology. The McMillan Company, New York; 1973.
3. Bindal VD. Corrective Physical Education, Therapeutic Exercise and Rehabilitation. Associated Publishing House, Agra; 2009.
4. Bowen WP, Stone HA. Applied Anatomy and Kinesiology. Lea and Febiger, Philadelphia; 1953.
5. Broer MR. Efficiency of Human Movement. WB Saunders Company, London; 1973.
6. Brunnstrom S. Clinical Kinesiology. FA Davis Company, Philadelphia; 1966.
7. Chatterjee CC. Human Physiology. Medical Allied Agency, Calcutta; 1985.
8. Chaurasia BD. Human Anatomy. CBS Publishers and Distributors, New Delhi; 1985.
9. Cooper JM, Glassow RB. Kinesiology. The CV Mosby Company, Saint Louis; 1963.
10. Duvall EN. Kinesiology: The Anatomy of Motion. Prentice Hall Inc., Englewood Cliffs, NJ; 1959.
11. Gladys SM. Kinesiology (Analysis of Human Motion). Sports Publication - Khel Sahitya Kendra, New Delhi; 1998.
12. Hamilton Nancy, Weimar Wendi, Luttgens Kathryn. Kinesiology: Scientific Basis of Human Motion. McGraw Hill, Boston; 1997.
13. Hawley G. Kinesiology of Corrective Exercise. Lea and Febiger, Philadelphia; 1978.
14. Hoffman SJ, Harris JC. Introduction to Kinesiology –Studying Physical Activity. Human Kinetics, Champaign, IL; 2000.
15. Jensen CR, Schultz GW. Applied Kinesiology. The Scientific Study of Human Performance. McGraw Hill Book Company, New York; 1977.
16. Kelly DL. Kinesiology. Prentice Hall Inc., Englewood Cliffs, NJ; 1971.
17. Lippert Lynn S. Clinical Kinesiology and Anatomy. FA Davis Company, Philadelphia; 2006.
18. Luttgens K, Wells KF. Kinesiology: Scientific Basis of Human Motion. Saunders College Publishing, Chicago; 1982.
19. Narayanan S Lakshmi. Textbook of Therapeutic Exercises. Jaypee Brothers, New Delhi; 2005.
20. Rai Ramesh. Injuries in Sports: A Mechanical viewpoint. Published in 'Society for the National Institutes of Physical Education and Sports (SNIPES)' Journal. 1982; 5(4)96-99.
21. Rasch PJ, Burke RK. Kinesiology and Applied Anatomy. Lea and Febiger, Philadelphia; 1978.
22. Rothstein JM, Roy SH, Walf SL. et al. The Rehabilitation Specialist's Handbook. FA Davis Company, Philadelphia; 1991.
23. Smith LK, Weiss EL. Brunnstrom's Clinical Kinesiology. FA Davis. Company, Philadelphia; 1996.
24. Thompson, Clem W, Floyd RT. Manual of Structural Kinesiology. McGraw Hill, New York; 2004.
25. Turner Molly. Faulty Posture: Its Effects and Treatment. William Heinemann Medical Books Ltd, London; 1965.
26. Uppal AK, et al. Kinesiology in Physical Education and Exercise Science. Friends Publications (India), New Delhi; 2004.
27. Wells Katharine F. Kinesiology. WB Saunders Company, London; 1960.
28. Williams PL, Dyson M. Gray's Anatomy. Churchill Livingstone, Edinburgh; 1993.

Glossary

Abduction: Movement of a body segment in the frontal plane away from the midline of the body.

Achilles tendon: The tendon at the back of the leg directly above the heel.

Acceleration: The rate of change in velocity.

Adduction: Movement of a body segment in the frontal plane toward the midline of the body.

Adipose tissue: Tissue largely composed of fat cells.

Afferent nervous system: The bodily system that directs impulses from sensory receptors toward the spinal cord and brain. Also called sensory nervous system.

Agonist: A muscle primarily responsible to produce the desired movement; prime mover.

Anatomical position: A stance in which the body is at 'military attention' with the palms facing forward.

Antagonist: A muscle, which acts in opposition to the desired movement.

Angle of pull: The angle between mechanical axis of bone and the line of pull of a muscle.

Anterior: Situated before or toward the front.

Anteroposterior (AP) plane: That plane which divides the body into right and left halves.

Appendicular skeleton: The portion of the skeleton comprising the upper and lower extremities.

Aponeurosis: A fibrous sheet of connective tissue: a flat tendon.

Articulation: A joining of bones that usually allows motion between them; joint.

Archimedes' principle: Law governing buoyancy, which states that a body immersed in water is buoyed up by a force equal to the weight of the fluid displaced by the body.

Atlas: First cervical vertebra which articulates with occipital bone.

Autonomous nervous system: Part of nervous system that regulates involuntary body functions through sympathetic and parasympathetic systems that work together to maintain homeostasis.

Axis: The second cervical vertebra also a fixed point around which a moving object revolves.

Axial skeleton: The portion of the skeleton comprising head, neck and trunk.

Backward tilt: Pelvic motion in which the posterior surface moves somewhat backward and downward.

Balance: Ability to keep the center of gravity over the base of support and to maintain equilibrium.

Ball and socket joint: Articulation of a rounded portion of one bone with a cup like fossa of another, allowing multi-axial motion.

Ballistic movement: Sustained force movement usually resulting in rapid acceleration followed by a coasting motion, then deceleration near the end of the range of motion.

Base of support: Any part of the body in contact with the supporting surface and the intervening area.

Biceps brachii: The muscle at front of the upper arm that flexes or bends the elbow.

Bilateral movement: Movement on two sides.

Buoyancy: An upward force that acts to support a body immersed in water.

Cardinal plane: A primary plane, one that passes through the center of gravity.

Caudal: Toward the tail (coccyx).

Center of buoyancy: That point in the body at which the upward force of buoyancy acts.

Center of gravity: The 'balance point' of the body; the center of mass; the intersection of the three cardinal planes.

Central nervous system (CNS): The bodily system that includes the brain and spinal cord.

Centrifugal force: The 'center fleeing' force acting on an object undergoing circular motion; a reaction force to centripetal force.

Centripetal force: The 'center seeking' force constraining an object to a circular path.

Cephalad: Toward the head.

Circumduction: Movement in a circular, cone-shaped pattern.

Compression force: Force that acts to press or compact.

Concentric muscle contraction: Contraction in which the muscle fibers shorten.

Contractile force: The amount of tension applied by a muscle or group of muscles during contraction.

Concurrent forces: Forces acting at the sametime and point of application but at different angles.

Connective tissue: Supporting tissues of the body formed by ground substance and fibrous tissues. The term generally refers to tendons, ligaments and fascia though technically bone, blood and lymph are also the connective tissues.

Contralateral: Opposite side.

Coordination: The act of various muscles working together in a smooth concerted way. Correct and precise timing of muscle contractions.

Curvilinear motion: Motion that follows curved path.

Deceleration : The rate at which velocity decreases.

Dermatome: Cutaneous (skin) distribution of spinal nerve sensation.

Density: The amount of mass per unit of volume.

Displacement: A vector quantity that reflects any change in position.

Distal: Farthest from the body midline or point of reference. The hand is at the distal end of the arm.

Dorsal: Relating to the back; posterior.

Dorsiflexion: Act of moving the foot up toward the leg.

Dynamic posture: The posture or position of the body while in motion.

Eccentric contraction: Controlled lengthening of a muscle. The muscle becomes longer as it contracts because the resistance is greater than the contractile force.

Efferent nervous system: The motor nervous system, which directs impulses from the brain and spinal cord to muscles, glands and organs.

Effort arm: The perpendicular distance between the applied effort force and the axis of a lever.

Elasticity: The ability to resist deformation and to return to the original shape.

Electromyograph: A device used to measure and record the changes in electrical potential of muscles during contraction.

Endomysium: Connective tissue found surrounding the muscle cell.

Epicondyle (Bone): A projection on or above a condyle on a bone, which is used for muscle attachment.

Epimysium: Connective tissue, which surrounds the whole muscle.

Equilibrium: A state of balance.

Eversion: The act of turning the sole of the foot outward.

Extension: The return from flexion.

Fascia: A sheet of fibrous tissue that envelops the body beneath the skin, individual muscles or parts of muscles.

Fascicle: A small group of muscles or nerve fibers.

First class lever: A lever arrangement with the axis between the force and the resistance (F-A-R).

Flexion: A movement that brings the two parts of a joint into a bend position.

Force: A push or a pull.

Force Couple: Parallel forces on either side of the axis acting in opposite directions.

Friction: A resistance to rolling or sliding, based on the nature of the two interacting surfaces.

Frontal plane: Vertical plane passing through the body from side to side, dividing it into anterior and posterior halves.

Fulcrum: An axis; the point about which rotation occurs.

Ganglion: A collection or bundle of nerve-cell bodies outside the central nervous system.

Gluteus maximus: One of the main muscles of buttock, commonly called the gluteal muscle that moves the thigh backward at the hip joint.

Golgi Tendon Receptors: Receptors in the tendons that sense tension.

Gravity: An acceleration toward the center of the earth.

Hamstrings: The collective term for three muscles located at the back of thigh that bend the leg at the knee.

Horizontal abduction: Movement of a segment in the transverse plane away from the midline of the body.

Horizontal adduction: Movement of a segment in the transverse plane toward the midline of the body.

Horizontal extension: Extension of a body segment through the transverse (horizontal) plane.

Hypertrophy: An increase in the overall size of a tissue.

Hypothenar eminence: Slightly elevated area of the palm near medial or ulnar border of hand.

Inertia: A property of matter by which it remains at rest or in uniform motion in the same straight line unless acted upon by some external force.

Inferior: The portion of a body that is below or deeper than another structure or reference point.

Ipsilateral: Same side.

Kinetic energy: Energy based on motion; a product of mass and velocity.

Kinesthetic sense: A sense of awareness, without the use of the other senses of muscle and joint positions and actions.

Latissimus dorsi: An important muscle on either side of the spine in the lower and middle back, that extends and adducts the arm at the shoulder.

Law of reaction: The law of motion that states that for every action there will be an equal and opposite reaction.

Lever: A rigid bar that is fixed at a single point, about which it may be made to rotate.

Ligament: Tough connective tissue, which binds bones together forming joints.

Line of gravity: A line that extends from the center of gravity of an object toward the center of the earth.

Linear motion: Motion in a line in which all parts of the object move in the same direction and at the same speed.

Linear forces: Forces applied in the same direction and along the same line of action.

Mandible: The lower jaw bone, the largest and strongest bone of the face, and the only moveable bone in the skull.

Mass: The quantity of matter an object contains.

Mechanical axis of bone: A line drawn from the proximal joint center to distal joint center.

Mechanical advantage: The ratio of effort arm to resistance arm or the ratio of resistance to effort in a lever.

Moment arm: The perpendicular distance from any force to an axis in a lever system.

Moment of inertia: The quantity of a rotating mass and its distribution around the axis of rotation.

Momentum: The amount of motion an object possesses, the product of mass and velocity.

Motor unit: A single motor neuron and all of the muscle fibers it innervates.

Multijoint muscle: A muscle that extends over two or more joints.

Muscle-tone : A state of slight tension present in the muscles even at rest, a state of preparedness of a muscle, controlled by nervous system.

Nerve: A cable-like bundle composed of many nerve fibers.

Neuromuscular coordination: Coordination, which results from nerve impulses reaching the proper muscles, with sufficient intensity, at the correct time.

Neuron: A complete nerve cell, including the cell body and all its appendages.

Neutralizer: A muscle that acts to equalize the action of another muscle.

Origin (of muscle): The muscle attachment closer to the midline of the body; the less movable attachment.

Palmar: Relating to the palm, the anterior surface of the hand in anatomical position.

Parallel forces: Forces that act parallel to each other.

Parasympathetic: Pertaining to a major subdivision of the autonomic nervous system.

Passive flexibility: Flexibility exhibited as a result of the pull of gravity or an outside force such as pressure by another person, as opposed to flexibility exhibited as a result of muscle contractions.

Periosteum: The thick fibrous membrane covering the entire outer surface of the bone except at the articular ends.

Perimysium: Connective tissue, which surrounds fascicles of muscle fibers.

Peripheral nervous system: The body system, which includes all those parts of the nervous system not included in the brain and spinal cord.

Plantar flexion: Act of moving the foot downward away from the leg.

Posterior: Back portion of the body, dorsal.

Postural muscles: Also called antigravity muscles. The muscles used to maintain posture.

Potential energy: Energy based on the position of an object, usually to height from the surface.

Power: The rate at which work is done.

Prime mover: A muscle, which makes a primary contribution to the desired movement.

Principle of levers: For a lever to be in equilibrium, the clockwise torques must equal the counterclockwise torques:

$E \times EA = R \times RA$

Projectile: An object that is given some initial velocity and then released.

Pronation: Act of turning the palm downward, performed by the internal rotation of the forearm.

Prone: Lying face down.

Proximal: Part closer to a reference point, for example, with reference to elbow, shoulder is proximal.

Proprioceptors: Sensory receptors that detect joint and muscle activity.

Quadriceps: The collective term for a group of four muscles that lies at the front of the thigh.

Reciprocal inhibition: The automatic blocking of nerve impulses to those muscles in opposition (antagonistic) to a desired movement. This automatically occurs with all well-coordinated movement patterns. Lack of reciprocal inhibitions is a contributor to poor coordination.

Reflex: An involuntary response to a stimulus in which the thought process is by-passed.

Relative motion: Motion with respect to some reference object.

Resistance arm: Perpendicular distance between the resistance force vector and the axis of a lever.

Resultant: The combined effect of two or more vector quantities.

Rotary motion (Angular): The rotation of an object about an axis with each point of the object describing an arc or a circle around the axis.

Sacrum: The triangular bone at the base of the spine. Made up of five fused vertebrae it forms part of the pelvis.

Sagittal plane: Vertical plane passing through the body from front to back, dividing it into right and left halves.

Scapula: A thin, triangular bone located on either side of the thorax posteriorly also known as shoulder blade.

Skeletal muscle: Also called striated, motor and voluntary muscle.

Smooth muscle: Also called visceral and involuntary muscle. A muscle located in the internal organs, with the exception of the heart.

Specific gravity: Ratio between the density of an object and that of water.

Speed: Scalar quantity quantifying the rate of motion.

Stability: The ability to remain in or return to a state of equilibrium.

Stabilizer: Any muscle that acts to stabilize or fix a body segment in order for another segment to move on it.

Static contraction: Same as tonic or isometric contraction.

Static posture: The posture or position of the body while at rest.

Static strength: Strength exhibited without motion.

Static work: The condition in which no movement occurs when muscular force is applied.

Stabilizing component: That vector component of force which acts parallel to a lever.

Strain: The amount of deformation experienced by an object when stressed.

Strength: The ability to apply force with a segment of the body.

Stretch reflex: Involuntary contraction of muscle when it is suddenly stretched, due to stimulation of proprioceptors controlled by central nervous system.

Stress: The force applied per unit of area.

Soft tissue: The skin, fascia, muscles, tendons, joint-capsules, and ligaments of the body.

Superior portion: A body portion that is above or over another body portion or reference point.

Superficial: Nearer to the surface of the body, the opposite of deep.

Superficial fascia: Connective tissue layer just under the skin.

Supination: Act of turning the palm upwards, performed by outward rotation of the forearm.

Tendon: A tough, fibrous tissue that connects muscles to bones.

Tension: A form of stress that involves pulling forces.

Thenar eminence: Small elevated area of the palm at the base of thumb.

Third-class lever: A lever arrangement with the force between the axis and the resistance (A-F-R).

Torque: The tendency for rotation; turning effect.

Torsion: A form of stress that involves twisting.

Translatory motion: Motion in which an object moves from one point to another point, as opposed to rotary motion. It can occur in either a linear or a curvilinear path.

Transverse plane: An imaginary plane, which bisects the body into upper and lower halves.

Trapezius: The muscle in the upper back that supports the neck and head.

Triceps brachii: A muscle at the back of the upper arm that extends the forearm at the elbow.

Trochanter: A large projection or prominence upon a bone.

Tuberosity: A large rounded prominence upon a bone.

Velocity: A vector quantity describing the rate of displacement.

Viscosity (blood): The 'thickness' of the fluid; the resistance to flow.

Weight: The product of the mass of an object and the acceleration due to gravity.

Work: The product of a force and the distance over which that force produces motion.

Index

A

Abdominal muscles 149-150
   external oblique 150
   internal oblique 150
   rectus abdominis 150
Abduction 23
Abductor digity minimi 107
Abductor pollicis brevis 107
Abductor pollicis longus 107
Acetabular labrum 116
Acetabulum 116
Acromioclavicular joint 79
Adduction 23-24
Adductor brevis 122
Adductor longus 122
Adductor magnus 122
Adductor pollicis 107
Agonists 43
All or None Law 27
Amphiarthrodial joints 20-21
Anconeus 97-98
Angle of pull 39-40
Ankle 131-142
   joints of ankle & foot 131-132
   ligaments of ankle & foot 134-135
   movements of ankle & foot 133-134
   muscles of ankle & foot 137-142
Antagonists 43
Anterior cruciate ligament 127
Anterior longitudinal ligament 144
Appendicular skeleton 14-15
Arches of foot 136-137
Axes 52-53
Axial skeleton 14-15

B

Ball-and-socket joints 20, 86, 116
Basal ganglia 45
Base of support 51
Biarticular muscles 40
Biceps brachii 90, 96
Biceps femoris 121, 129

Biomechanical concepts 50-76
Bones
   of ankle and foot 131-132
   of elbow and radioulnar joints 94
   flat bones 16
   gross structure 15-16
   of hip 116
   irregular bones 16
   of knee 126
   long bones 16
   short bones 16
   of shoulder 86
   of spinal column 143
   types 16-17
   of wrist and hand 100-101
Bow Legs 170
Brachialis 96
Brachioradialis 96
Brain 45
   basal ganglia 46
   brain stem 46
   cerebellum 46
   cerebral cortex 46

C

Calcaneous 131
Calcaneocuboid joint 131
Capitate 101
Capitis 148
Cardinal planes 54-55
Carpal joints 100
Carpal tunnel 109
Carpometacarpal joints 101
Carrying 195-196
Cartilaginous joints 18
Catching 201-202, 208
Center of gravity 50-51
Central nervous system 45-47
   brain 45
   components of 45-46
   spinal cord 46-47
Cerebellum 46
Cerebral cortex 46

Cervical nerves 47
Cervicis 149
Cervical vertebrae 143
Circumduction 24
Coccyx 143
Collateral ligaments 127, 134
Concentric contraction 34-35
Concurrent forces 60-61
Condyloid joints 20
Coracoacromial ligament 86
Coracobrachialis 89
Coracoclavicular ligament 80
Coracohumeral ligament 86
Cranial nerves 45
Cruciate ligaments 127
Cuboid 131
Cuneiforms 132
Curvilinear motion 63

D

Daily life activities 192-207
Deceleration 65
Deep posterior spinal muscles 152
Deltoid 89
Deltoid ligament 134
Depression 25, 80
Diagonal plane 55
Diarthrodial joints 20-21
Direction of force 60
Dorsiflexion 24, 133

E

Eccentric contraction 34
Elbow joint 94-99
   movements of, 94-95
Elbow muscles 95-99
   anconeus 97-98
   biceps brachii 90, 96
   brachialis 96
   brachioradialis 96
   pronator teres 98
   supinator 99
   triceps brachii 80, 97

Elevation 25, 80
Equilibrium 55-56
    factors affecting stability 56-58
    neutral equilibrium 56
    stable equilibrium 55-56
    unstable equilibrium 56
Erector spinae 150-152
Eversion 25, 133
Extension 23
Extensor carpi radialis brevis 105
Extensor carpi radialis longus 105
Extensor carpi ulnaris 105
Extensor digiti minimi 107
Extensor digitorum 107
Extensor digitorum longus 138
Extensor hallucis lougus 138
Extensor indicis 107
External oblique 150
Extensor pollicis brevis 107
Extensor pollicis longus 107
External rotators of hip 123

F

Falling 202-203
Facet joints 144
Fast twitch fibers 30
Femur 126
Fibrous joints 18
Fibula 131
Flat back 168
Flat bones 16
Flat foot 170
Flexion 23
Flexor carpi radialis 105
Flexor carpi ulnaris 105
Flexor digitorum profundus 105
Flexor digitorum superficialis 105
Flexor digiti minimi brevis 107
Flexor digitorum longus 140-141
Flexor hallucis longus 141
Flexor pollicis longus 105
Foot 131-142
    bones of ankle and foot 131-132
    joints of ankle and foot 131-132
    ligaments of ankle and foot 134-135
    movements of ankle and foot 133-134
    muscles of 137-142
Force 58-62
    definition of 58
    direction of forces 60
    internal and external forces 59
    magnitude of 59

    parallel forces 61-62
    point of application of 59-60
    resolution of forces 60-62
Forward head 169
Friction 57
Frontal axis 53
Frontal plane 52, 54
Fusiform (spindle-shaped) muscle 32-33
Fundamentals of human motion
    anatomical and physiological 14-49
Fundamental joint movements 23-27

G

Gait 173-185
    common deviations 189-191
    determinants of 177-178
    individual variations in 183-184
Gastrocnemius 139
Gemmellus superior and inferior 123
Glenoid labrum 86
Glenohumeral joint 86
Glenohumeral ligament 86
Gluteus maximus 120
Gluteus medius 121-122
Gluteus minimus 122
Golgi tendon organ 49
Gracilis 123
Gravity 50

H

Hamstrings 120-121
Hamate 101
Hand 100-111
    bones of 100-101
    joints of 100-103
    ligaments of 101
    movements of hand 103
    muscles of 105-109
Handling an object overhead 196-197
Hinge joints 19, 94, 126
Hip 116-123
    ligaments of 116
    movements at 116-117
Hip abductors 121-122
Hip adductors 122
Hip extensors 120
Hip flexors 119
Hip joint muscles 117-124
    adductor brevis 122
    adductor longus 122
    adductor magnus 122

    external ratators 123
    gluteus maximus 120
    gluteus medius 121-122
    gluteus minimus 122
    gracilis 123
    hamstrings 120
    iliopsoas 119
    pectineus 119
    rectus femoris 119
    sartorius 119
    tensor fasciae latae 120
Humeroulnar joint 94
Humerus 86, 94
Hyperabduction 23
Hyperextension 23
Hyperflexion 23

I

Iliacus 119
Iliofemoral ligament 116
Iliopsoas 119
Iliotibial band 120, 128
Ilium 112, 116
Inertia 65
Infraspinatus 92
Injury prevention 208-213
Intercarpal joint 100
Intermetacarpal joints 101
Intermetatarsal joints 132
Internal oblique 150
Interphalangeal joints 103, 132
Interossei 108, 141
Intervertebral discs 144
Intrinsic muscles
    of foot 141
    of hand 107-109
Inversion 25, 133
Irregular bones 16
Irregular joints 19
Ischiofemoral ligament 116
Ischium 112, 116
Isokinetic contraction 35

J

Joints 17-22
    acromioclavicular joint 79-80
    ankle and foot joints 131-133
    ball-and-socket joints 20, 86, 116
    carpal joints 100
    carpometacarpal joints 101
    cartilaginous joints 18

classification of 18-22
condyloid joints 20
diarthrodial joints 20-21
elbow joint 94-99
fibrous joints 18
of fingers 101
hinge joints 19, 94, 126
hip 116
intercarpal joint 100
intermetacarpal joints 101
interphalangeal joints 103, 132
irregular joints 19
knee 126
ligamentous joints 21
metacarpophalangeal joints 103
metatarsophalangeal joints 132
midcarpal joint 100-101
midtarsal joints 132
mobility of 22
pivot joints 19
radiocarpal joint 100
radiohumeral joint 94
radioulnar joints 94-99
range of motion 26-27
saddle joints 20
shoulder joint 86
of spinal column 144
stability of 22
sternoclavicular joint 79-80
subtalar joint 131
synarthrodial joints 20-21
synovial joints 18
    talocalcaneal joint 131
    of thumb 103
    wrist joint 100-101
Joint proprioceptors 48-49
Jumping 200

K

Kinesiology
    aims and objectives 9-11
    application in selected daily
    life activities and sports skills
        190-205
    definition 3
    introduction 3-13
    major contributions 5-9
    practical application 11-12
    role and importance 12-13
Kinesthetic sensations, role 48-49
Knee 126-130
    ligaments of 127-128
    menisci 126
    movements at 128

Knee joint muscles 128-130
    gastrocnemius 129,139
    gracilis 123
    hamstring group 120,129
    popliteus 129
    quadriceps femoris 128-129
    sartorius 119
Knock-knee 170
Kyphosis 167
Kypholordosis 169

L

Landing 200, 209
Lateral collateral ligament 127
Lateral flexion 26,145
Latissimus dorsi 90
Law of acceleration 66-67
Law of inertia 65-67
Law of reaction 67
Levator scapulae 84
Levers 68-76
    anatomical levers 68-69
    classification of 69-72
    definition of 68
    first-class levers 69-70
    functions of 68
    mechanical advantage of 76
    principle of levers 74-76
    second-class levers 70-71
    in sports 76
    third-class levers 71-72
Lifting 193-194
Ligamentous joints 21
Ligaments,
    ankle and foot 134-135
    hip joint 116
    knee 127-128
    shoulder joint 86
    spinal column 144-145
    wrist and hand 101
Linear forces 60
Linear motion 63
Line of gravity 51
Line of pull of muscle 38-39
Locomotion 171-189
    running 183-187
    walking 171-183
Long bones 16
Longitudinal muscle 32
Lordosis 169
Lower extremities 112-142
    ankle and foot 131-142
    hip 116-123
    knee 126-130

Lumbar vertebrae 143
Lumbricals 108, 141
Lumbosacral joint 113
Lumbosacral angle 113
Lunate 100

M

Magnitude of force 59
Marathon running 186
Mechanical advantage of levers 76
Medial collateral ligament 127
Menisci, knee 126
Metacarpophalangeal joints 102-103
Metatarsals 131
Metatarsophalangeal joints 132
Midcarpal joints 100
Midtarsal joint 132
Momentum 66
    in law of acceleration 66-67
Motion 62-67
    angular or rotatory motion 63-64
    curvilinear motion 63
    definition of 62-63
    factors modifying motion 65
    rectilinear motion 63
    translatory motion 63
    types of motion 63-64
Motor units 36-37
Movement
    abduction 23
    adduction 23-24
    circumduction 24
    extension 23
    flexion 23
    pronation 24
    rotation 24
    supination 24
Moving weights 196-198
Multifidus 151
Multipenniform muscle, skeletal 33
Muscle belly 28
Muscle contraction,
    gradation of 37
    physiology of 36
    types 34-35
Muscle fiber 28-30
    fast twitch fibers 30
    slow twitch fibers 30
Muscular system 27-36
Muscle proprioceptors 48-49
Muscular system 27-36
    functions 27
    nomenclature 32
    roles 43

skeletal 27-28
structural classification 32-33
types 27
Muscle spindles 49
Muscle tissue, properties of skeletal 28

N

Navicular 131
Nervous control of voluntary movements 45-49
    central nervous system 45-47
    motor units 36-37, 48
    nerves 47
    peripheral nervous system 47-48
Neutral equilibrium 56
Neuromuscular concepts 36-45
Neutralizers, skeletal muscle 44-45
Newton's laws of motion 65-67
    law of acceleration 66-67
    law of gravity's attraction 50
    law of inertia 65-66
    law of reaction 67

O

Obturator externus and internus 123
Opponens digiti minimi 107
Opponens policis 107
Opposition 26, 102
Overhead striking 204

P

Palmar flextion 25
Palmaris longus 105
Parallel forces 61-62
Patella 126-127
Patellar ligament 128
Patello femoral joint 126
Pectineus 119
Pectoralis major 89
Pectoralis minor 82-83
Pelvic girdle 112-115
    movements of 113-114
    muscles of 115
    relationship to trunk/lower extremities 114
Piriformis 123
Peripheral nervous system 47-48
Peroneus brevis 140
Peroneus longus 140
Pennate muscles 33
Peroneus tertius 138

Phalanges 101, 131
Pisiform 101
Pivot joint 19
Planes 52-55
Plantaris 129, 139
Planterflexion 24, 133
Popliteus 129
Posterior cruciate ligament 127
Posterior longitudinal ligament 144
Posture 155-170
    definitions 155
    defects 163-170
    evolution of erect posture 155-156
    good posture 159-160
    maintenance of 158-159
    poor posture 160
    types of 156
Prevertebral muscles 148
Principle of levers 74-76
Pronation 24, 133
Pronator quadratus 98-99
Pronator teres 98
Proprioceptors 48-49
Protraction 25, 81
Psoas 119
Pubis 112
Pubofemoral ligament 116
Pushing and pulling 198

Q

Quadrate (quadrilateral) muscle 32
Quadratus lumborum 152
Quadriceps femoris 128-129

R

Radial deviation 26
Radial collateral ligament 101
Radial flexion 102
Radiocarpal joints 100
Radiohumeral joint 94
Radioulnar joint 94-99
Radius 94, 100
Range of motion
    of major joints 26-27
Reciprocal inhibition 38
Rectilinear motion 63
Rectus abdominis 150
Rectus femoris 119, 129
Reposition 26
Retraction 25, 81
Rhomboids 84
Rotatory motion 63-64

Rotation (lateral/medial) 24
Rotator cuff 86
Rotatores 151
Round shoulders 169
Running 183-187
    mechanical analysis 184-185
    mechanical principles 185-186
    muscular analysis 186
    sprinting 188
    support phase 183
    swing phase 183-184

S

Sacroiliac joint 112
Sacrum 112, 143
Saddle joints 20
Sagittal axis 52
Sagittal plane 52-54
Sartorius 119
Scalenes 149
Scapula 79
Scaphoid 100, 131
Scapulohumeral rhythm 88
Scapulothoracic joint 80
Scoliosis 171
Semimembranosus 121, 129
Semitendinosus 121, 129
Serratus anterior 81-82
Short bones 16
Shoulder girdle 79-85
    acromioclavicular joint 79
    movements of 80-81
    muscles of 81-85
    sternoclavicular joint 80
Shoulder girdle muscles 81-85
    levator scapulae 84
    pectoralis minor 82-83
    rhomboids 84
    serratus anterior 81-82
    subclavius 83
    trapezius 83-84
Shoulder joint 86
    ligaments of 86
    movements of 87-88
Shoulder joint muscles 88-93
    biceps brachii 90
    coracobrachialis 89
    deltoid 89
    infraspinatus 92
    latissimus dorsi 90
    pectoralis major 89
    subscapularis 92
    supraspinatus 91
    teres major 90

teres minor 92
triceps brachii 80
Sitting and rising 194
Skeletal muscle roles 43-45
    agonists/movers 43
    antagonists 43
    fixators 44
    neutralizers 44-45
    synergists 44
Skeletal muscle structure 28-32
    fusiform (spindle-shaped) muscle 32-33
    longitudinal muscle 32
    multipenniform muscle 33
    muscle fiber 28-30
    muscular attachments 31-32
    muscular tissue, properties of 28
    quadrate (quadrilateral) muscle 32
    triangular (fan-shaped) muscle 33
    unipenniform muscle 33
Skeleton
    appendicular skeleton 14-15
    axial skeleton 14-15
Skeletal muscles 27-32
    properties 28
    gross and microscopic structure 28-30
Skeletal system 14-27
functions of skeleton 14
    types, 14-15
Slow twitch fibers 30
Soleus 139
Spine 143-153
    joints of 144
    ligaments of, 144-145
    movements of 145-147
Spinal muscles 147-153
    abdominal muscles 147-149
    capitis 147-147
    cervicis 147
    deep posterior spinal muscles 150
    erector spinae 150-152
    prevertebral muscles 148
    quadratus lumborum 152
    scalenes 149
    semispinalis 151
    splenius capitis 149
    sternocleidomastoid 148
    suboccipital muscles 149
    transversospinalis 151-152
Spinal nerves 47
Splenius capitis 149
Sports skills 200-207
Sports injuries prevention 208-213

Sprinting 188
Stability 55-58
    factors affecting 56-58
Stabilizers, skeletal muscle 44
Stable equilibrium 55-56
Stair climbing 192
Standing position 160-162
Sternoclavicular joint 80
Sternocleidomastoid 148
Sternum 79
Striking 204
Stooping 192
Subclavius 83
Subscapularis 92
Subtalar joint 131
Supination 24, 133
Supinator 99
Supraspinatus 91
Sway back 168
Synarthrodial joints 20-21
Synergists, skeletal muscle 44

T

Talocalcaneal joint 131
Talonavicular joint 131
Talotibial joint 131
Talus 131
Tarsals 131
Tarsal joints 131-132
Tendon 28
Tensor fasciae latae 120
Teres femoris 116
Teres major 90
Teres minor 92
Thoracic vertebrae 143
Throwing 204
Thumb 101-103
Tibia 126, 131
Tibialis anterior 137-138
Tibialis posterior 139
Tibiofemoral joints 126
Tilt of scapula 81
Toes 132-134
Translatory motion 63
Transverse acetabular ligament 116
Transverse ligament 116, 128
Transverse plane 54
Transverse tarsal joint 132
Transversospinalis 151-152
Trapezius 83-84
Trapezium 101
Trapezoid 101
Triangular (fan-shaped) muscle 33

Triceps brachii 80, 97
Triquetral 100
Two joint muscles 40-43

U

Ulna 94, 100
Ulnar deviation 26
Ulnar collateral ligament 101
Ulnar flexion 102
Underhand throwing 203
Unipennate muscle 33
Unstable equilibrium 56
Upper extremities 86-111
    elbow and radioulnar joints 94-99
    shoulder 86-93
    wrist and hand 100-111

V

Vastus intermedius 129
Vastus lateralis 129
Vastus medialis 129
Vertebrae 143
Vertical axis 53
Voluntary movements, nervous control 45-49

W

Walking 173-185
    anatomical principles of 181
    components of 173
    energy cost of 184
    gait, individual variations 181-182
    mechanical analysis 180-181
    mechanical principles in 181
    neuromuscular considerations 178
    stance phase 172-173
    swing phase 172-174
    arm swing 177, 179
Wrist 100-111
    carpal tunnel 109
    ligaments of 101
    midcarpal and intercarpal joints 100-101
    movements of 102
    muscles of 105

Y

Y ligament 116

EU GSPR Authorised Reprsentative
Logos Europe, 9 rue Nicolas Poussin
1700, La Rochelle, France
Phone: +33 (0) 6 67 93 73 78
E-mail: contact@logoseurope.eu

www.ingramcontent.com/pod-product-compliance
Ingram Content Group UK Ltd.
Pitfield, Milton Keynes, MK11 3LW, UK
UKHW051302180426
11947UKWH00020B/1856